Springer Optimization and Its Applications

Volume 138

Aims and Scope
Optimization has been expanding in all directions at an astonishing rate during the last few decades. New algorithmic and theoretical techniques have been developed, the diffusion into other disciplines has proceeded at a rapid pace, and our knowledge of all aspects of the field has grown even more profound. At the same time, one of the most striking trends in optimization is the constantly increasing emphasis on the interdisciplinary nature of the field. Optimization has been a basic tool in all areas of applied mathematics, engineering, medicine, economics and other sciences.

The series *Springer Optimization and Its Applications* publishes undergraduate and graduate textbooks, monographs and state-of-the-art expository works that focus on algorithms for solving optimization problems and also study applications involving such problems. Some of the topics covered include nonlinear optimization (convex and nonconvex), network flow problems, stochastic optimization, optimal control, discrete optimization, multi-objective programming, description of software packages, approximation techniques and heuristic approaches.

More information about this series at http://www.springer.com/series/7393

Vijay Gupta • Themistocles M. Rassias
P. N. Agrawal • Ana Maria Acu

Recent Advances in Constructive Approximation Theory

Vijay Gupta
Department of Mathematics
Netaji Subhas Institute of Technology
New Delhi, India

Themistocles M. Rassias
Department of Mathematics
National Technical University of Athens
Athens, Greece

P. N. Agrawal
Department of Mathematics
Indian Institute of Technology
Roorkee, India

Ana Maria Acu
Department of Mathematics and Informatics
Lucian Blaga University of Sibiu
Sibiu, Romania

ISSN 1931-6828 ISSN 1931-6836 (electronic)
Springer Optimization and Its Applications
ISBN 978-3-030-06374-0 ISBN 978-3-319-92165-5 (eBook)
https://doi.org/10.1007/978-3-319-92165-5

Mathematics Subject Classification: 47-XX, 30-XX, 32-XX, 34-XX, 41-XX, 46-XX, 49-XX, 65-XX

Printed on acid-free paper

This Springer imprint is published by the registered company Springer International Publishing AG part of Springer Nature.
The registered company address is: Gewerbestrasse 11, 6330 Cham, Switzerland

Preface

Recent Advances in Constructive Approximation Theory deals with various recent problems on linear positive operators. We survey upon recent research work in this domain and also present extensions of known approximation results on linear positive operators in post quantum and in bivariate setting. This book is designed for graduate students, researchers, and engineers working on approximation theory and related areas of mathematical analysis. The book in hand is a self-contained research monograph which presents theory, methods, and various applications in mathematical analysis and approximation theory.

We would like to acknowledge the superb assistance that the staff of Springer has provided for the publication of this book.

New Delhi, India — Vijay Gupta
Athens, Greece — Themistocles M. Rassias
Roorkee, India — P. N. Agrawal
Sibiu, Romania — Ana Maria Acu

Contents

Chapter 1
Moment Generating Functions and Central Moments

1.1 Some Operators

This section deals with the moment generating functions (m.g.f.) and moments up to sixth order of some discretely defined operators. We mention the m.g.f. and express them in expanded form to obtain moments, which are important in the theory of approximation relevant to problems of convergence.[1]

1.1.1 Bernstein Operators

For $f \in C[0, 1]$ the n-th degree Bernstein polynomials are defined as

$$B_n(f, x) := \sum_{k=0}^{n} \binom{n}{k} x^k (1 - x)^{n-k} f\left(\frac{k}{n}\right). \tag{1.1.1}$$

By simple computation, the moment generating function of the Bernstein polynomial is given by

[1]The interested reader is referred to the works [162–166] relevant to other techniques for the approximation of moments of trigonometric functions with applications to problems of analytic number theory, such as Riemann's Hypothesis.

V. Gupta et al., *Recent Advances in Constructive Approximation Theory*, Springer Optimization and Its Applications 138,
https://doi.org/10.1007/978-3-319-92165-5_1

$$B_n(e^{\theta t}, x) = \left(1 - x + xe^{\frac{\theta}{n}}\right)^n. \tag{1.1.2}$$

It was observed in [135] that $B_n(e^{\theta t}, x)$ provides the moments

$$\mu_r^{B_n}(x) = B_n(e_r, x), \quad \text{where } e_r(t) = t^r,\ r \in \mathbb{N} \cup \{0\}$$

of (6.1.2) by the relation:

$$\begin{aligned}\mu_r^{B_n}(x) &= \left[\frac{\partial^r}{\partial \theta^r} B_n(e^{\theta t}, x)\right]_{\theta=0} \\ &= \left[\frac{\partial^r}{\partial \theta^r}\left\{\left(1 - x + xe^{\frac{\theta}{n}}\right)^n\right\}\right]_{\theta=0}.\end{aligned}$$

The property of change of origin of the moment generating function

$$e^{-\theta x}\left(1 - x + xe^{\frac{\theta}{n}}\right)^n$$

can be used to find the central moments

$$U_{n,m}^{B_n}(x) = B_n((t-x)^m, x).$$

Expanding $e^{-\theta x}\left(1 - x + xe^{\frac{\theta}{n}}\right)^n$ in powers of θ, we have

$$\begin{aligned}&e^{-\theta x}\left(1 - x + xe^{\frac{\theta}{n}}\right)^n \\ &= 1 + \frac{(x - x^2)\,\theta^2}{2n} + \frac{(x - 3x^2 + 2x^3)\,\theta^3}{6n^2} \\ &+ \frac{(x - 7x^2 + 3nx^2 + 12x^3 - 6nx^3 - 6x^4 + 3nx^4)\,\theta^4}{24n^3} \\ &+ \frac{(x - 15x^2 + 10nx^2 + 50x^3 - 40nx^3 - 60x^4 + 50nx^4 + 24x^5 - 20nx^5)\,\theta^5}{120n^4} \\ &+ \frac{1}{720n^5}\Big(x - 31x^2 + 25nx^2 + 180x^3 - 180nx^3 + 15n^2x^3 - 390x^4 + 415nx^4 \\ &-45n^2x^4 + 360x^5 - 390nx^5 + 45n^2x^5 - 120x^6 + 130nx^6 - 15n^2x^6\Big)\theta^6 + O[\theta]^7.\end{aligned}$$

Thus the central moments of order m can be obtained by collecting the coefficient of $\theta^m/m!$. Some of the central moments are given below:

$$U_0^{B_n}(x) = 1,$$
$$U_1^{B_n}(x) = 0,$$
$$U_2^{B_n}(x) = \frac{(x - x^2)}{n},$$
$$U_3^{B_n}(x) = \frac{(x - 3x^2 + 2x^3)}{n^2},$$
$$U_4^{B_n}(x) = \frac{(x - 7x^2 + 3nx^2 + 12x^3 - 6nx^3 - 6x^4 + 3nx^4)}{n^3}, \tag{1.1.3}$$
$$U_5^{B_n}(x) = \frac{(x - 15x^2 + 10nx^2 + 50x^3 - 40nx^3 - 60x^4 + 50nx^4 + 24x^5 - 20nx^5)}{n^4},$$
$$U_6^{B_n}(x) = \frac{1}{n^5}\Big(x - 31x^2 + 25nx^2 + 180x^3 - 180nx^3 + 15n^2x^3 - 390x^4 + 415nx^4$$
$$- 45n^2x^4 + 360x^5 - 390nx^5 + 45n^2x^5 - 120x^6 + 130nx^6 - 15n^2x^6\Big)$$

Obviously for the Bernstein polynomials, the central moments are given by

$$U_{n,m}^{B_n}(x) = \left[\frac{\partial^m}{\partial\theta^m} e^{-\theta x}\left(1 - x + xe^{\frac{\theta}{n}}\right)^n\right]_{\theta=0}. \tag{1.1.4}$$

1.1.2 Baskakov Operators

For $f \in C[0, \infty)$ the Baskakov operators are defined as

$$V_n(f, x) := \sum_{k=0}^{\infty} \binom{n+k-1}{k} \frac{x^k}{(1+x)^{n+k}} f\left(\frac{k}{n}\right). \tag{1.1.5}$$

By simple computation, we have

$$V_n(e^{\theta t}, x) = \left(1 + x - xe^{\frac{\theta}{n}}\right)^{-n}. \tag{1.1.6}$$

We observe that $V_n(e^{\theta t}, x)$ may be treated as m.g.f. of the operators V_n, which may be utilized to obtain the moments of (1.1.5). Let

$$\mu_r^{V_n}(x) = V_n(e_r, x), \quad \text{where } e_r(t) = t^r, \; r \in \mathbb{N} \cup \{0\}.$$

The moments are given by

$$\begin{aligned}\mu_r^{V_n}(x) &= \left[\frac{\partial^r}{\partial\theta^r} V_n(e^{\theta t}, x)\right]_{\theta=0} \\ &= \left[\frac{\partial^r}{\partial\theta^r}\left\{\left(1 + x - xe^{\frac{\theta}{n}}\right)^{-n}\right\}\right]_{\theta=0}.\end{aligned}$$

The property of change of origin of the moment generating function

$$e^{-\theta x}\left(1 + x - xe^{\frac{\theta}{n}}\right)^{-n}$$

can be used to find the central moments

$$U_{n,m}^{V_n}(x) = V_n((t-x)^m, x).$$

Expanding $e^{-\theta x}\left(1 + x - xe^{\frac{\theta}{n}}\right)^{-n}$ in powers of θ, we have

$$\begin{aligned}&e^{-\theta x}\left(1 + x - xe^{\frac{\theta}{n}}\right)^{-n} \\ &= 1 + \frac{(x+x^2)\,\theta^2}{2n} + \frac{(x+3x^2+2x^3)\,\theta^3}{6n^2} \\ &+ \frac{(x+7x^2+3nx^2+12x^3+6nx^3+6x^4+3nx^4)\,\theta^4}{24n^3} \\ &+ \frac{(x+15x^2+10nx^2+50x^3+40nx^3+60x^4+50nx^4+24x^5+20nx^5)\,\theta^5}{120n^4} \\ &+ \frac{1}{720n^5}\Big(x+31x^2+25nx^2+180x^3+180nx^3+15n^2x^3+390x^4+415nx^4 \\ &+ 45n^2x^4+360x^5+390nx^5+45n^2x^5+120x^6+130nx^6+15n^2x^6\Big)\theta^6 + O[\theta]^7.\end{aligned}$$

Thus the central moments of order m of the Baskakov operators can be obtained by collecting the coefficient of $\theta^m/m!$ in the above expansion. Some of the central moments are given below:

$$
\begin{aligned}
U_0^{V_n}(x) &= 1,\\
U_1^{V_n}(x) &= 0,\\
U_2^{V_n}(x) &= \frac{\left(x+x^2\right)}{n},\\
U_3^{V_n}(x) &= \frac{\left(x+3x^2+2x^3\right)}{n^2},\\
U_4^{V_n}(x) &= \frac{\left(x+7x^2+3nx^2+12x^3+6nx^3+6x^4+3nx^4\right)}{n^3},\\
U_5^{V_n}(x) &= \frac{\left(x+15x^2+10nx^2+50x^3+40nx^3+60x^4+50nx^4+24x^5+20nx^5\right)}{n^4},\\
U_6^{V_n}(x) &= \frac{1}{n^5}\Big(x-31x^2+25nx^2+180x^3+180nx^3+15n^2x^3+390x^4+415nx^4\\
&\quad +45n^2x^4+360x^5+390nx^5+45n^2x^5+120x^6+130nx^6+15n^2x^6\Big)
\end{aligned}
\tag{1.1.7}
$$

For the Baskakov operators, the central moments are given by

$$
U_{n,m}^{V_n}(x) = \left[\frac{\partial^m}{\partial\theta^m} e^{-\theta x}\left(1+x-xe^{\frac{\theta}{n}}\right)^{-n}\right]_{\theta=0}. \tag{1.1.8}
$$

1.1.3 Szász–Mirakyan Operators

For $f \in C[0,\infty)$ the Szász operators are defined as

$$
S_n(f,x) := \sum_{k=0}^{\infty} e^{-nx}\frac{(nx)^k}{k!} f\left(\frac{k}{n}\right). \tag{1.1.9}
$$

Let $f(t) = e^{\theta t}$, $\theta \in \mathbb{R}$, then

$$
S_n(e^{\theta t},x) = e^{nx\left(e^{\frac{\theta}{n}}-1\right)}. \tag{1.1.10}
$$

Here $S_n(e^{\theta t},x)$ may be treated as m.g.f. of the operators S_n, which could provide the moments of (1.1.9). Let

$$
\mu_r^{S_n}(x) = S_n(e_r,x), \quad \text{where } e_r(t) = t^r,\ r \in \mathbb{N}\cup\{0\}.
$$

The moments are given by

$$\mu_r^{S_n}(x) = \left[\frac{\partial^r}{\partial\theta^r} S_n(e^{\theta t}, x)\right]_{\theta=0}$$
$$= \left[\frac{\partial^r}{\partial\theta^r}\left\{e^{nx\left(e^{\frac{\theta}{n}}-1\right)}\right\}\right]_{\theta=0}.$$

The property of change of origin of the moment generating function

$$e^{-\theta x} e^{nx\left(e^{\frac{\theta}{n}}-1\right)}$$

can be used to find the central moments

$$U_{n,m}^{S_n}(x) = S_n((t-x)^m, x).$$

Expanding $e^{-\theta x} e^{nx\left(e^{\frac{\theta}{n}}-1\right)}$ in powers of θ, we have

$$e^{-\theta x} e^{nx\left(e^{\frac{\theta}{n}}-1\right)}$$
$$= 1 + \frac{x\theta^2}{2n} + \frac{x\theta^3}{6n^2} + \frac{(x+3nx^2)\theta^4}{24n^3} + \frac{(x+10nx^2)\theta^5}{120n^4}$$
$$+ \frac{(x+25nx^2+15n^2x^3)\theta^6}{720n^5} + O[\theta]^7.$$

Thus the central moments of order m of the Szász operators can be obtained by collecting the coefficient of $\theta^m/m!$ in the above expansion. Some of the central moments are given below:

$$\begin{aligned}
U_0^{S_n}(x) &= 1,\\
U_1^{S_n}(x) &= 0,\\
U_2^{S_n}(x) &= \frac{x}{n},\\
U_3^{S_n}(x) &= \frac{x}{n^2},\\
U_4^{S_n}(x) &= \frac{x+3nx^2}{n^3},\\
U_5^{S_n}(x) &= \frac{x+10nx^2}{n^4},\\
U_6^{S_n}(x) &= \frac{x+25nx^2+15n^2x^3}{n^5}
\end{aligned} \tag{1.1.11}$$

For the Szász operators, the central moments are given by

$$U_{n,m}^{S_n}(x) = \left[\frac{\partial^m}{\partial\theta^m} e^{-\theta x} e^{nx\left(e^{\frac{\theta}{n}}-1\right)}\right]_{\theta=0}. \tag{1.1.12}$$

1.1.4 Lupaş Operators

For $f \in C[0,\infty)$ the Lupaş operators are defined as

$$L_n(f,x) := \sum_{k=0}^{\infty} 2^{-nx} \frac{(nx)_k}{k!\,2^k} f\left(\frac{k}{n}\right). \tag{1.1.13}$$

The moment generating function of the Lupaş operators is defined as

$$L_n(e^{\theta t},x) = \left(2 - e^{\frac{\theta}{n}}\right)^{-nx}. \tag{1.1.14}$$

Let

$$\mu_r^{L_n}(x) = L_n(e_r,x), \quad \text{where } e_r(t) = t^r,\ r \in \mathbb{N} \cup \{0\}.$$

The moments are given by

$$\begin{aligned}\mu_r^{L_n}(x) &= \left[\frac{\partial^r}{\partial\theta^r} L_n(e^{\theta t},x)\right]_{\theta=0}\\ &= \left[\frac{\partial^r}{\partial\theta^r}\left\{\left(2 - e^{\frac{\theta}{n}}\right)^{-nx}\right\}\right]_{\theta=0}.\end{aligned}$$

The property of change of origin of the moment generating function

$$e^{-\theta x}\left(2 - e^{\frac{\theta}{n}}\right)^{-nx}$$

can be used to find the central moments

$$U_{n,m}^{L_n}(x) = L_n((t-x)^m,x).$$

Expanding $e^{-\theta x}\left(2 - e^{\frac{\theta}{n}}\right)^{-nx}$ in powers of θ, we have

$$\begin{aligned}
&e^{-\theta x}(2-e^{\frac{\theta}{n}})^{-nx}\\
&=1+\frac{x\theta^2}{n}+\frac{x\theta^3}{n^2}+\frac{(13x+6nx^2)\theta^4}{12n^3}+\frac{(5x+4nx^2)\theta^5}{4n^4}\\
&+\frac{(541x+570nx^2+60n^2x^3)\theta^6}{360n^5}+O[\theta]^7.
\end{aligned}$$

Thus the central moments of order m of the Lupaş operators can be obtained by collecting the coefficient of $\theta^m/m!$ in the above expansion. Some of the central moments are given below:

$$\begin{aligned}
U_0^{L_n}(x)&=1,\\
U_1^{L_n}(x)&=0,\\
U_2^{L_n}(x)&=\frac{2x}{n},\\
U_3^{L_n}(x)&=\frac{6x}{n^2},\\
U_4^{L_n}(x)&=\frac{26x+12nx^2}{n^3},\\
U_5^{L_n}(x)&=\frac{150x+120nx^2}{n^4},\\
U_6^{L_n}(x)&=\frac{1082x+1140nx^2+120n^2x^3}{n^5}
\end{aligned}\tag{1.1.15}$$

For the Lupaş operators, the central moments are given by

$$U_{n,m}^{L_n}(x)=\left[\frac{\partial^m}{\partial\theta^m}e^{-\theta x}(2-e^{\frac{\theta}{n}})^{-nx}\right]_{\theta=0}.\tag{1.1.16}$$

1.1.5 Post Widder Operators

In the year 1941, Widder [227] defined the following operators for $f\in C[0,\infty)$:

$$P_n(f,x):=\frac{1}{n!}\left(\frac{n}{x}\right)^{n+1}\int_0^\infty t^n e^{-\frac{nt}{x}} f(t)\,dt.\tag{1.1.17}$$

Let us consider $f(t) = e^{\theta t}$, $\theta \in \mathbb{R}$, then we have

$$P_n(e^{\theta t}, x) = \left(\frac{n}{x}\right)^{n+1} \left(\frac{n}{x} - \theta\right)^{-(n+1)}. \tag{1.1.18}$$

We observe that $P_n(e^{\theta t}, x)$ may be treated as m.g.f. of the operators P_n, which may be utilized to obtain the moments of (1.1.17). Let

$$\mu_r^{P_n}(x) = P_n(e_r, x), \quad \text{where } e_r(t) = t^r, \ r \in \mathbb{N} \cup \{0\}.$$

The moments are given by

$$\begin{aligned}\mu_r^{P_n}(x) &= \left[\frac{\partial^r}{\partial \theta^r} P_n(e^{\theta t}, x)\right]_{\theta=0} \\ &= \left[\frac{d^r}{d\theta^r}\left\{\left(\frac{n}{x}\right)^{n+1}\left(\frac{n}{x} - \theta\right)^{-(n+1)}\right\}\right]_{\theta=0}.\end{aligned}$$

The property of change of origin of the moment generating function

$$e^{-\theta x}\left(\frac{n}{x}\right)^{n+1}\left(\frac{n}{x} - \theta\right)^{-(n+1)}$$

can be used to find the central moments

$$U_{n,m}^{P_n}(x) = P_n((t-x)^m, x).$$

Expanding $e^{-\theta x}\left(\frac{n}{x}\right)^{n+1}\left(\frac{n}{x} - \theta\right)^{-(n+1)}$ in powers of θ, we have

$$\begin{aligned}&e^{-\theta x}\left(\frac{n}{x}\right)^{n+1}\left(\frac{n}{x} - \theta\right)^{-(n+1)} \\ &= 1 + \frac{x\theta}{n} + \frac{(2+n)x^2\theta^2}{2n^2} \\ &\quad + \frac{(6+5n)x^3\theta^3}{6n^3} + \frac{\left(24 + 26n + 3n^2\right)x^4\theta^4}{24n^4} \\ &\quad + \frac{\left(120 + 154n + 35n^2\right)x^5\theta^5}{120n^5} \\ &\quad + \frac{\left(720 + 1044n + 340n^2 + 15n^3\right)x^6\theta^6}{720n^6} + O[\theta]^7.\end{aligned}$$

Thus the central moments of order m of the Lupaş operators can be obtained by collecting the coefficient of $\theta^m/m!$ in the above expansion. Some of the central moments are given below:

$$\begin{aligned}
U_0^{P_n}(x) &= 1,\\
U_1^{P_n}(x) &= \frac{x}{n},\\
U_2^{P_n}(x) &= \frac{(2+n)x^2}{n^2},\\
U_3^{P_n}(x) &= \frac{(6+5n)x^3}{n^3},\\
U_4^{P_n}(x) &= \frac{\left(24+26n+3n^2\right)x^4}{n^4},\\
U_5^{P_n}(x) &= \frac{\left(120+154n+35n^2\right)x^5}{n^5},\\
U_6^{P_n}(x) &= \frac{\left(720+1044n+340n^2+15n^3\right)x^6}{n^6}.
\end{aligned} \tag{1.1.19}$$

For the Post Widder operators, the central moments are given by

$$U_{n,m}^{P_n}(x) = \left[\frac{\partial^m}{\partial\theta^m} e^{-\theta x}\left(\frac{n}{x}\right)^{n+1}\left(\frac{n}{x}-\theta\right)^{-(n+1)}\right]_{\theta=0}. \tag{1.1.20}$$

1.2 Kantorovich Type Operators

1.2.1 Bernstein–Kantorovich Operators

For $f \in C[0,1]$ the Bernstein–Kantorovich operators are defined as

$$\overline{B}_n(f,x) := (n+1)\sum_{k=0}^{n}\binom{n}{k}x^k(1-x)^{n-k}\int_{\frac{k}{n+1}}^{\frac{k+1}{n+1}} f(t)\,dt. \tag{1.2.1}$$

Here $\overline{B}_n(e^{\theta t},x)$ may be treated as moment generating function of the operators $\overline{B}_n$. Let

$$\mu_r^{\overline{B}_n}(x) = \overline{B}_n(e_r,x), \quad \text{where } e_r(t) = t^r,\ r \in \mathbb{N}\cup\{0\}.$$

The moments are given by

$$\mu_r^{\overline{B}_n}(x) = \left[\frac{\partial^r}{\partial\theta^r}\overline{B}_n(e^{\theta t}, x)\right]_{\theta=0}$$
$$= \left[\frac{\partial^r}{\partial\theta^r}\left\{\frac{1}{\theta}(n+1)\left(e^{\frac{\theta}{n+1}}-1\right)\left(1-x+xe^{\frac{\theta}{n+1}}\right)^n\right\}\right]_{\theta=0}.$$

The property of change of origin of the moment generating function

$$e^{-\theta x}\frac{1}{\theta}(n+1)\left(e^{\frac{\theta}{n+1}}-1\right)\left(1-x+xe^{\frac{\theta}{n+1}}\right)^n$$

can be used to find the central moments

$$U_{n,m}^{\overline{B}_n}(x) = \overline{B}_n((t-x)^m, x).$$

Expanding

$$e^{-\theta x}\frac{1}{\theta}(n+1)\left(e^{\frac{\theta}{n+1}}-1\right)\left(1-x+xe^{\frac{\theta}{n+1}}\right)^n$$

in powers of θ, we have

$$e^{-\theta x}\frac{(n+1)}{\theta}\left(e^{\frac{\theta}{n+1}}-1\right)\left(1-x+xe^{\frac{\theta}{n+1}}\right)^n$$
$$= 1 + \frac{(1-2x)\theta}{2(1+n)} + \frac{\left(1-3x+3nx+3x^2-3nx^2\right)\theta^2}{6(1+n)^2}$$
$$+\frac{\left(1-4x+10nx+6x^2-30nx^2-4x^3+20nx^3\right)\theta^3}{24(1+n)^3}$$
$$+\frac{1}{120(1+n)^4}\Big(1-5x+25nx+10x^2-125nx^2+15n^2x^2$$
$$-10x^3+200nx^3-30n^2x^3+5x^4-100nx^4+15n^2x^4\Big)\theta^4$$
$$+\frac{1}{720(1+n)^5}\Big(1-6x+56nx+15x^2-420nx^2+105n^2x^2$$
$$-20x^3+1120nx^3-420n^2x^3+15x^4-1260nx^4+525n^2x^4$$
$$-6x^5+504nx^5-210n^2x^5\Big)\theta^5$$
$$+\frac{1}{5040(1+n)^6}\Big(1-7x+119nx+21x^2-1295nx^2+490n^2x^2$$

$$-35x^3+5215nx^3-3045n^2x^3+105n^3x^3+35x^4-9765nx^4$$
$$+6685n^2x^4-315n^3x^4-21x^5+8589nx^5-6195n^2x^5+315n^3x^5$$
$$\left.+7x^6-2863nx^6+2065n^2x^6-105n^3x^6\right)\theta^6+O[\theta]^7$$

Thus the central moments of order m of the Bernstein–Kantorovich operators can be obtained by collecting the coefficient of $\theta^m/m!$ in the above expansion. Some of the central moments are given below:

$$\begin{aligned}
U_0^{\overline{B}_n}(x) &= 1,\\
U_1^{\overline{B}_n}(x) &= \frac{(1-2x)}{2(1+n)},\\
U_2^{\overline{B}_n}(x) &= \frac{\left(1-3x+3nx+3x^2-3nx^2\right)}{3(1+n)^2},\\
U_3^{\overline{B}_n}(x) &= \frac{\left(1-4x+10nx+6x^2-30nx^2-4x^3+20nx^3\right)}{4(1+n)^3},\\
U_4^{\overline{B}_n}(x) &= \frac{1}{5(1+n)^4}\left(1-5x+25nx+10x^2-125nx^2+15n^2x^2\right.\\
&\quad \left.-10x^3+200nx^3-30n^2x^3+5x^4-100nx^4+15n^2x^4\right),\\
U_5^{\overline{B}_n}(x) &= \frac{1}{6(1+n)^5}\left(1-6x+56nx+15x^2-420nx^2+105n^2x^2\right.\\
&\quad -20x^3+1120nx^3-420n^2x^3+15x^4-1260nx^4+525n^2x^4\\
&\quad \left.-6x^5+504nx^5-210n^2x^5\right),\\
U_6^{\overline{B}_n}(x) &= \frac{1}{7(1+n)^6}\left(1-7x+119nx+21x^2-1295nx^2+490n^2x^2\right.\\
&\quad -35x^3+5215nx^3-3045n^2x^3+105n^3x^3+35x^4-9765nx^4\\
&\quad +6685n^2x^4-315n^3x^4-21x^5+8589nx^5-6195n^2x^5+315n^3x^5\\
&\quad \left.+7x^6-2863nx^6+2065n^2x^6-105n^3x^6\right).
\end{aligned} \tag{1.2.2}$$

For the Bernstein–Kantorovich operators, the central moments are given by

$$U_{n,m}^{\overline{B}_n}(x)=\left[\frac{\partial^m}{\partial\theta^m}e^{-\theta x}.\frac{(n+1)}{\theta}\left(e^{\frac{\theta}{n+1}}-1\right)\left(1-x+xe^{\frac{\theta}{n+1}}\right)^n\right]_{\theta=0} \tag{1.2.3}$$

1.2.2 Baskakov–Kantorovich Operators

For $f \in C[0,\infty)$ the Baskakov–Kantorovich operators are defined as

$$\overline{V}_n(f,x) := (n-1)\sum_{k=0}^{\infty}\binom{n+k-1}{k}\frac{x^k}{(1+x)^{n+k}}\int_{\frac{k}{n-1}}^{\frac{k+1}{n-1}} f(t)\,dt. \qquad (1.2.4)$$

The moment generating function is given by

$$\begin{aligned}\overline{V}_n(e^{\theta t},x) &= \frac{(n-1)\left(e^{\frac{\theta}{n-1}}-1\right)}{\theta\,(1+x)^n}\left(1-\frac{x\,e^{\frac{\theta}{n-1}}}{1+x}\right)^{-n}\\ &= \frac{1}{\theta}\,(n-1)\left(e^{\frac{\theta}{n-1}}-1\right)\left(1+x\left(1-e^{\frac{\theta}{n-1}}\right)\right)^{-n}. \qquad (1.2.5)\end{aligned}$$

This may be utilized to obtain the moments of (1.2.4). Let

$$\mu_r^{\overline{V}_n}(x) = \overline{V}_n(e_r,x), \quad \text{where } e_r(t)=t^r,\ r\in\mathbb{N}\cup\{0\}\,.$$

The moments are given by

$$\begin{aligned}\mu_r^{\overline{V}_n}(x) &= \left[\frac{\partial^r}{\partial\theta^r}\overline{V}_n(e^{\theta t},x)\right]_{\theta=0}\\ &= \left[\frac{\partial^r}{\partial\theta^r}\left\{\frac{1}{\theta}\,(n-1)\left(e^{\frac{\theta}{n-1}}-1\right)\left(1+x\left(1-e^{\frac{\theta}{n-1}}\right)\right)^{-n}\right\}\right]_{\theta=0}.\end{aligned}$$

The property of change of origin of the moment generating function

$$e^{-\theta x}\frac{1}{\theta}\,(n-1)\left(e^{\frac{\theta}{n-1}}-1\right)\left(1+x\left(1-e^{\frac{\theta}{n-1}}\right)\right)^{-n}$$

can be used to find the central moments

$$U_{n,m}^{\overline{V}_n}(x) = \overline{V}_n((t-x)^m,x)\,.$$

Expanding

$$e^{-\theta x}\frac{1}{\theta}\,(n-1)\left(e^{\frac{\theta}{n-1}}-1\right)\left(1+x\left(1-e^{\frac{\theta}{n-1}}\right)\right)^{-n}$$

in powers of θ, we have

$$e^{-\theta x}.\frac{1}{\theta}(n-1)\left(e^{\frac{\theta}{n-1}}-1\right)\left(1+x\left(1-e^{\frac{\theta}{n-1}}\right)\right)^{-n}$$

$$=1+\frac{(1+2x)\theta}{2(n-1)}+\frac{\left(1+3x+3nx+3x^2+3nx^2\right)\theta^2}{6(n-1)^2}$$

$$+\frac{\left(1+4x+10nx+6x^2+30nx^2+4x^3+20nx^3\right)\theta^3}{24(n-1)^3}$$

$$+\frac{1}{120(n-1)^4}\Big(1+5x+25nx+10x^2+125nx^2+15n^2x^2$$

$$+10x^3+200nx^3+30n^2x^3+5x^4+100nx^4+15n^2x^4\Big)\theta^4$$

$$+\frac{1}{720(n-1)^5}\Big(1+6x+56nx+15x^2+420nx^2+105n^2x^2$$

$$+20x^3+1120nx^3+420n^2x^3+15x^4+1260nx^4+525n^2x^4$$

$$+6x^5+504nx^5+210n^2x^5\Big)\theta^5$$

$$+\frac{1}{5040(n-1)^6}\Big(1+7x+119nx+21x^2+1295nx^2+490n^2x^2$$

$$+35x^3+5215nx^3+3045n^2x^3+105n^3x^3+35x^4+9765nx^4$$

$$+6685n^2x^4+315n^3x^4+21x^5+8589nx^5+6195n^2x^5+315n^3x^5$$

$$+7x^6+2863nx^6+2065n^2x^6+105n^3x^6\Big)\theta^6+O[\theta]^7$$

Thus the central moments of order m of the Baskakov–Kantorovich operators can be obtained by collecting the coefficient of $\theta^m/m!$ in the above expansion. Some of the central moments are given below:

$$U_0^{\overline{V}_n}(x)=1,$$

$$U_1^{\overline{V}_n}(x)=\frac{(1+2x)}{2(n-1)},$$

$$U_2^{\overline{V}_n}(x)=\frac{\left(1+3x+3nx+3x^2+3nx^2\right)}{3(n-1)^2},$$

$$U_3^{\overline{V}_n}(x)=\frac{\left(1+4x+10nx+6x^2+30nx^2+4x^3+20nx^3\right)}{4(n-1)^3},$$

$$U_4^{\overline{V}_n}(x) = \frac{1}{5(n-1)^4}\Big(1 + 5x + 25nx + 10x^2 + 125nx^2 + 15n^2x^2$$

$$+ 10x^3 + 200nx^3 + 30n^2x^3 + 5x^4 + 100nx^4 + 15n^2x^4\Big),$$

$$U_5^{\overline{V}_n}(x) = \frac{1}{6(n-1)^5}\Big(1 + 6x + 56nx + 15x^2 + 420nx^2 + 105n^2x^2$$

$$+ 20x^3 + 1120nx^3 + 420n^2x^3 + 15x^4 + 1260nx^4 + 525n^2x^4$$

$$+ 6x^5 + 504nx^5 + 210n^2x^5\Big), \tag{1.2.6}$$

$$U_6^{\overline{V}_n}(x) = \frac{1}{7(n-1)^6}\Big(1 + 7x + 119nx + 21x^2 + 1295nx^2 + 490n^2x^2$$

$$+ 35x^3 + 5215nx^3 + 3045n^2x^3 + 105n^3x^3 + 35x^4 + 9765nx^4$$

$$+ 6685n^2x^4 + 315n^3x^4 + 21x^5 + 8589nx^5 + 6195n^2x^5 + 315n^3x^5$$

$$+ 7x^6 + 2863nx^6 + 2065n^2x^6 + 105n^3x^6\Big).$$

For the Baskakov–Kantorovich operators, the central moments are given by

$$U_{n,m}^{\overline{V}_n}(x) = \left[\frac{\partial^m}{\partial\theta^m} e^{-\theta x}.\frac{(n-1)}{\theta}\left(e^{\frac{\theta}{n-1}} - 1\right)\left(1 + x\left(1 - e^{\frac{\theta}{n-1}}\right)\right)^{-n}\right]_{\theta=0} \tag{1.2.7}$$

1.2.3 Szász–Kantorovich Operators

For $f \in C[0,\infty)$ the Szász–Kantorovich operators are defined as

$$\overline{S}_n(f,x) := n\sum_{k=0}^{\infty} e^{-nx}\frac{(nx)^k}{k!}\int_{\frac{k}{n}}^{\frac{k+1}{n}} f(t)\,dt. \tag{1.2.8}$$

We have

$$\overline{S}_n(e^{\theta t},x) = \frac{1}{\theta}\,n\left(e^{\frac{\theta}{n}} - 1\right)e^{nx\left(e^{\frac{\theta}{n}}-1\right)}. \tag{1.2.9}$$

We observe that $\overline{S}_n(e^{\theta t}, x)$ is the m.g.f. of the operators $\overline{S}_n$, which may be used to obtain the moments of (1.2.8). Let

$$\mu_r^{\overline{S}_n}(x) = \overline{S}_n(e_r, x), \quad \text{where } e_r(t) = t^r, \ r \in \mathbb{N} \cup \{0\}.$$

The moments are given by

$$\begin{aligned}\mu_r^{\overline{S}_n}(x) &= \left[\frac{\partial^r}{\partial\theta^r}\overline{S}_n(e^{\theta t}, x)\right]_{\theta=0} \\ &= \left[\frac{\partial^r}{\partial\theta^r}\left\{\frac{1}{\theta}n\left(e^{\frac{\theta}{n}} - 1\right)e^{nx\left(e^{\frac{\theta}{n}}-1\right)}\right\}\right]_{\theta=0}.\end{aligned}$$

The property of change of origin of the moment generating function

$$e^{-\theta x}\frac{n}{\theta}\left(e^{\frac{\theta}{n}} - 1\right)e^{nx\left(e^{\frac{\theta}{n}}-1\right)}$$

can be used to find the central moments

$$U_{n,m}^{\overline{S}_n}(x) = \overline{S}_n((t-x)^m, x).$$

Expanding

$$e^{-\theta x}\frac{1}{\theta}n\left(e^{\frac{\theta}{n}} - 1\right)e^{nx\left(e^{\frac{\theta}{n}}-1\right)}$$

in powers of θ, we have

$$\begin{aligned}&e^{-\theta x}.\frac{1}{\theta}n\left(e^{\frac{\theta}{n}} - 1\right)e^{nx\left(e^{\frac{\theta}{n}}-1\right)} \\ &= 1 + \frac{\theta}{2n} + \frac{(1+3nx)\theta^2}{6n^2} + \frac{(1+10nx)\theta^3}{24n^3} \\ &+ \frac{1}{120n^4}\left(1 + 25nx + 15n^2x^2\right)\theta^4 + \frac{1}{720n^5}\left(1 + 56nx + 105n^2x^2\right)\theta^5 \\ &+ \frac{1}{5040n^6}\left(1 + 119nx + 490n^2x^2 + 105n^3x^3\right)\theta^6 + O[\theta]^7.\end{aligned}$$

Thus the central moments of order m of the Szász–Kantorovich operators can be obtained by collecting the coefficient of $\theta^m/m!$ in the above expansion. Some of the central moments are given below:

$$
\begin{aligned}
U_0^{\overline{S}_n}(x) &= 1,\\
U_1^{\overline{S}_n}(x) &= \frac{1}{2n},\\
U_2^{\overline{S}_n}(x) &= \frac{1+3nx}{3n^2},\\
U_3^{\overline{S}_n}(x) &= \frac{1+10nx}{4n^3},\\
U_4^{\overline{S}_n}(x) &= \frac{1+25nx+15n^2x^2}{5n^4},\\
U_5^{\overline{S}_n}(x) &= \frac{1+56nx+105n^2x^2}{6n^5},\\
U_6^{\overline{S}_n}(x) &= \frac{1+119nx+490n^2x^2+105n^3x^3}{7n^6}.
\end{aligned}
\tag{1.2.10}
$$

For the Szász–Kantorovich operators, the central moments are given by

$$
U_{n,m}^{\overline{S}_n}(x) = \left[\frac{\partial^m}{\partial\theta^m} e^{-\theta x} . \frac{n}{\theta}\left(e^{\frac{\theta}{n}}-1\right) e^{nx\left(e^{\frac{\theta}{n}}-1\right)}\right]_{\theta=0}. \tag{1.2.11}
$$

1.2.4 Lupaş–Kantorovich Operators

For $f \in C[0,\infty)$ the Lupaş–Kantorovich operators are defined as

$$
\overline{L}_n(f,x) := (n+1)\sum_{k=0}^{\infty} 2^{-nx}\frac{(nx)_k}{k!\,2^k}\int_{\frac{k}{n+1}}^{\frac{k+1}{n+1}} f(t)\,dt. \tag{1.2.12}
$$

We get the following after simple computation

$$
\begin{aligned}
\overline{L}_n(e^{\theta t},x) &= \frac{1}{\theta}(n+1)\,2^{-nx}\left(e^{\frac{\theta}{n+1}}-1\right)\left(1-\frac{e^{\frac{\theta}{n+1}}}{2}\right)^{-nx}\\
&= \frac{1}{\theta}(n+1)\left(e^{\frac{\theta}{n+1}}-1\right)\left(2-e^{\frac{\theta}{n+1}}\right)^{-nx}.
\end{aligned}
\tag{1.2.13}
$$

We observe that $\overline{L}_n(e^{\theta t},x)$ may be treated as m.g.f. of the operators $\overline{L}_n$, which may be utilized to obtain the moments of (1.2.12). Let

$$
\mu_r^{\overline{L}_n}(x) = \overline{L}_n(e_r,x), \quad \text{where } e_r(t)=t^r,\ r\in\mathbb{N}\cup\{0\}.
$$

The moments are given by

$$\begin{aligned}\mu_r^{\overline{L}_n}(x) &= \left[\frac{\partial^r}{\partial\theta^r}\overline{L}_n(e^{\theta t}, x)\right]_{\theta=0} \\ &= \left[\frac{d^r}{d\theta^r}\left\{\frac{1}{\theta}(n+1)\left(e^{\frac{\theta}{n+1}}-1\right)\left(2-e^{\frac{\theta}{n+1}}\right)^{-nx}\right\}\right]_{\theta=0}.\end{aligned}$$

The property of change of origin of the m.g.f.

$$e^{-\theta x}\frac{(n+1)}{\theta}\left(e^{\frac{\theta}{n+1}}-1\right)\left(2-e^{\frac{\theta}{n+1}}\right)^{-nx}$$

can be used to find the central moments

$$U_{n,m}^{\overline{L}_n}(x) = \overline{L}_n((t-x)^m, x)\,.$$

Expanding

$$e^{-\theta x}\frac{(n+1)}{\theta}\left(e^{\frac{\theta}{n+1}}-1\right)\left(2-e^{\frac{\theta}{n+1}}\right)^{-nx}$$

in powers of θ, we have

$$\begin{aligned}&e^{-\theta x}\frac{(n+1)}{\theta}\left(e^{\frac{\theta}{n+1}}-1\right)\left(2-e^{\frac{\theta}{n+1}}\right)^{-nx} \\ &= 1+\frac{(1-2x)\theta}{2(n+1)}+\frac{\left(1-3x+6nx+3x^2\right)\theta^2}{6(n+1)^2} \\ &+\frac{\left(1-4x+36nx+6x^2-24nx^2-4x^3\right)\theta^3}{24(n+1)^3} \\ &+\frac{1}{120(n+1)^4}\left(1-5x+210nx+10x^2-180nx^2+60n^2x^2-10x^3+60nx^3+5x^4\right)\theta^4 \\ &+\frac{1}{720(n+1)^5}\left(1-6x+1440nx+15x^2-1260nx^2+900n^2x^2-20x^3\right. \\ &\left.+540nx^3-360n^2x^3+15x^4-120nx^4-6x^5\right)\theta^5 \\ &+\frac{1}{5040(n+1)^6}\left(1-7x+11886nx+21x^2-10080nx^2\right. \\ &+10920n^2x^2-35x^3+4410nx^3-6300n^2x^3+840n^3x^3+35x^4 \\ &\left.-1260nx^4+1260n^2x^4-21x^5+210nx^5+7x^6\right)\theta^6+O[\theta]^7\end{aligned}$$

Thus the central moments of order m of the Lupaş–Kantorovich operators can be obtained by collecting the coefficient of $\theta^m/m!$ in the above expansion. Some of the central moments are given below:

$$\begin{aligned}
U_0^{\overline{L}_n}(x) &= 1,\\
U_1^{\overline{L}_n}(x) &= \frac{(1-2x)}{2(n+1)},\\
U_2^{\overline{L}_n}(x) &= \frac{\left(1-3x+6nx+3x^2\right)}{3(n+1)^2},\\
U_3^{\overline{L}_n}(x) &= \frac{\left(1-4x+36nx+6x^2-24nx^2-4x^3\right)}{4(n+1)^3},\\
U_4^{\overline{L}_n}(x) &= \frac{1-5x+210nx+10x^2-180nx^2+60n^2x^2-10x^3+60nx^3+5x^4}{5(n+1)^4},\\
U_5^{\overline{L}_n}(x) &= \frac{1}{6(n+1)^5}\Big(1-6x+1440nx+15x^2-1260nx^2+900n^2x^2-20x^3\\
&\quad+540nx^3-360n^2x^3+15x^4-120nx^4-6x^5\Big),\\
U_6^{\overline{L}_n}(x) &= \frac{1}{7(n+1)^6}\Big(1-7x+11886nx+21x^2-10080nx^2\\
&\quad+10920n^2x^2-35x^3+4410nx^3-6300n^2x^3+840n^3x^3+35x^4\\
&\quad-1260nx^4+1260n^2x^4-21x^5+210nx^5+7x^6\Big).
\end{aligned} \tag{1.2.14}$$

For the Lupaş–Kantorovich operators, the central moments are given by

$$U_{n,m}^{\overline{V}_n}(x) = \left[\frac{\partial^m}{\partial\theta^m} e^{-\theta x}.\frac{(n+1)}{\theta}\left(e^{\frac{\theta}{n+1}}-1\right)\left(2-e^{\frac{\theta}{n+1}}\right)^{-nx}\right]_{\theta=0}. \tag{1.2.15}$$

1.3 Durrmeyer Type Operators

In [126] Gupta and Rassias estimated the moments of certain Durrmeyer type operators using hypergeometric series. In this section, we use the alternative approach of moment generating function, to find moments of some operators.

1.3.1 Szász–Durrmeyer Operators

The Szász–Durrmeyer operators are defined as

$$\widetilde{S}_n(f,x) := n\sum_{k=0}^{\infty} e^{-nx}\frac{(nx)^k}{k!}\int_0^{\infty} e^{-nt}\frac{(nt)^k}{k!} f(t)\,dt. \tag{1.3.1}$$

We have

$$\widetilde{S}_n(e^{\theta t},x) = \frac{n}{n-\theta} e^{nx\theta/(n-\theta)}. \tag{1.3.2}$$

We observe that $\widetilde{S}_n(e^{\theta t},x)$ is the m.g.f. of the operators $\widetilde{S}_n$, which may be used to obtain the moments of (1.3.1). Let

$$\mu_r^{\widetilde{S}_n}(x) = \widetilde{S}_n(e_r,x), \quad \text{where } e_r(t) = t^r,\ r \in \mathbb{N}\cup\{0\}\,.$$

The moments are given by

$$\begin{aligned}\mu_r^{\widetilde{S}_n}(x) &= \left[\frac{\partial^r}{\partial\theta^r}\widetilde{S}_n(e^{\theta t},x)\right]_{\theta=0}\\ &= \left[\frac{\partial^r}{\partial\theta^r}\left\{\frac{n}{n-\theta}e^{nx\theta/(n-\theta)}\right\}\right]_{\theta=0}.\end{aligned}$$

The property of change of origin of the m.g.f.

$$e^{-\theta x}\frac{n}{n-\theta}e^{nx\theta/(n-\theta)}$$

can be used to find the central moments

$$U_{n,m}^{\widetilde{S}_n}(x) = \widetilde{S}_n((t-x)^m,x)\,.$$

Expanding

$$e^{-\theta x}\frac{n}{n-\theta}e^{nx\theta/(n-\theta)}$$

in powers of θ, we have

$$\begin{aligned}&e^{-\theta x}\frac{n}{n-\theta}e^{nx\theta/(n-\theta)}\\ &= 1 + \frac{\theta}{n} + \left(\frac{1}{n^2}+\frac{x}{n}\right)\theta^2\end{aligned}$$

$$+\left(\frac{1}{n^3}+\frac{2x}{n^2}\right)\theta^3+\left(\frac{1}{n^4}+\frac{3x}{n^3}+\frac{x^2}{2n^2}\right)\theta^4$$
$$+\left(\frac{1}{n^5}+\frac{4x}{n^4}+\frac{3x^2}{2n^3}\right)\theta^5+\left(\frac{1}{n^6}+\frac{5x}{n^5}+\frac{3x^2}{n^4}+\frac{x^3}{6n^3}\right)\theta^6+O[\theta]^7.$$

Thus the central moments of order m of the Szász–Durrmeyer operators can be obtained by collecting the coefficient of $\theta^m/m!$ in the above expansion. Some of the central moments are given below:

$$\begin{aligned}
U_0^{\widetilde{S}_n}(x) &= 1,\\
U_1^{\widetilde{S}_n}(x) &= \frac{1}{n},\\
U_2^{\widetilde{S}_n}(x) &= \left(\frac{2}{n^2}+\frac{2x}{n}\right),\\
U_3^{\widetilde{S}_n}(x) &= \left(\frac{6}{n^3}+\frac{12x}{n^2}\right),\\
U_4^{\widetilde{S}_n}(x) &= \left(\frac{24}{n^4}+\frac{72x}{n^3}+\frac{12x^2}{n^2}\right),\\
U_5^{\widetilde{S}_n}(x) &= \left(\frac{120}{n^5}+\frac{480x}{n^4}+\frac{180x^2}{n^3}\right),\\
U_6^{\widetilde{S}_n}(x) &= \left(\frac{720}{n^6}+\frac{3600x}{n^5}+\frac{2160x^2}{n^4}+\frac{120x^3}{n^3}\right).
\end{aligned} \tag{1.3.3}$$

For the Szász–Durrmeyer operators, the central moments are given by

$$U_{n,m}^{\widetilde{S}_n}(x)=\left[\frac{\partial^m}{\partial\theta^m}e^{-\theta x}.\frac{n}{n-\theta}e^{nx\theta/(n-\theta)}\right]_{\theta=0}. \tag{1.3.4}$$

1.3.2 Baskakov–Szász Operators

Gupta and Srivastava [121] considered the hybrid operators by combining the Baskakov and Szász–Mirakyan basis functions on $[0,\infty)$ viz. Baskakov–Szász operators, given as:

$$M_n(f,x)=n\sum_{k=0}^{\infty}b_{n,k}(x)\int_0^{\infty}s_{n,k}(t)\,f(t)\,dt, \tag{1.3.5}$$

where

$$b_{n,k}(x)=\binom{n+k-1}{k}\frac{x^k}{(1+x)^{n+k}} \text{ and } s_{n,k}(t)=e^{-nt}\frac{(nt)^k}{k!}.$$

Let $f(t)=e^{At}$, $A\in\mathbb{R}$, then we have

$$n\int_0^\infty e^{-nt}\frac{(nt)^k}{k!}e^{At}\,dt=\left(\frac{n}{n-A}\right)^{k+1}. \tag{1.3.6}$$

Thus, for the operators defined by (1.3.5), using (1.3.6), we have

$$\begin{aligned}M_n(e^{At},x)&=\frac{n}{(n-A)(1+x)^n(n-1)!}\sum_{k=0}^{\infty}\frac{(n+k-1)!}{k!}\left[\frac{nx}{(n-A)(1+x)}\right]^k\\&=n(n-A)^{n-1}\left[n-A(1+x)\right]^{-n}.\end{aligned} \tag{1.3.7}$$

Here, we observe that $M_n(e^{At},x)$ may be treated as m.g.f. of the operators $M_n f$, which may be utilized to obtain the moments of (1.3.5). Let

$$\mu_r(x)=M_n(e_r,x), \quad \text{where } e_r(t)=t^r,\ r\in\mathbb{N}\cup\{0\}.$$

The moments are given by

$$\begin{aligned}\mu_r(x)&=\left[\frac{d^r}{dA^r}M_n(e^{At},x)\right]_{A=0}\\&=\left[\frac{d^r}{dA^r}\left\{n(n-A)^{n-1}\left[n-A(1+x)\right]^{-n}\right\}\right]_{A=0}.\end{aligned}$$

The property of change of origin of the m.g.f.

$$e^{-Ax}\left\{n(n-A)^{n-1}\left[n-A(1+x)\right]^{-n}\right\}$$

can be used to find the central moments

$$U_{n,m}^{M_n}(x)=M_n((t-x)^m,x).$$

Expanding

$$e^{-Ax}\left\{n(n-A)^{n-1}\left[n-A(1+x)\right]^{-n}\right\}$$

in powers of A, we have

$$e^{-Ax}\left\{n(n-A)^{n-1}\left[n-A(1+x)\right]^{-n}\right\}$$

$$=1+\frac{A}{n}+\frac{\left(2+2nx+nx^2\right)A^2}{2n^2}+\frac{\left(6+12nx+9nx^2+2nx^3\right)A^3}{6n^3}$$

$$+\frac{\left(24+72nx+72nx^2+12n^2x^2+32nx^3+12n^2x^3+6nx^4+3n^2x^4\right)A^4}{24n^4}$$

$$+\frac{A^5}{120n^5}\left(120+480nx+600nx^2+180n^2x^2+400nx^3+240n^2x^3+150nx^4\right.$$

$$\left.+115n^2x^4+24nx^5+20n^2x^5\right)$$

$$+\frac{A^6}{720n^6}\left(720+3600nx+5400nx^2+2160n^2x^2+4800nx^3+3600n^2x^3\right.$$

$$+120n^3x^3+2700nx^4+2550n^2x^4+180n^3x^4+864nx^5$$

$$\left.+900n^2x^5+90n^3x^5+120nx^6+130n^2x^6+15n^3x^6\right)+O[A]^7.$$

The central moments of order m of the Baskakov–Szász operators can be obtained by collecting the coefficient of $A^m/m!$ in the above expansion. Some of the central moments are given below:

$$U_{n,0}^{M_n}(x)=1,$$

$$U_{n,1}^{M_n}(x)=\frac{1}{n},$$

$$U_{n,2}^{M_n}(x)=\frac{\left(2+2nx+nx^2\right)}{n^2},$$

$$U_{n,3}^{M_n}(x)=\frac{\left(6+12nx+9nx^2+2nx^3\right)}{n^3},$$

$$U_{n,4}^{M_n}(x)=\frac{\left(24+72nx+72nx^2+12n^2x^2+32nx^3+12n^2x^3+6nx^4+3n^2x^4\right)}{n^4},$$

$$U_{n,5}^{M_n}(x)=\frac{1}{n^5}\left(120+480nx+600nx^2+180n^2x^2+400nx^3+240n^2x^3+150nx^4\right.$$

$$\left.+115n^2x^4+24nx^5+20n^2x^5\right),$$

$$U_{n,6}^{M_n}(x) = \frac{1}{n^6}\Big(720 + 3600nx + 5400nx^2 + 2160n^2x^2 + 4800nx^3 + 3600n^2x^3$$
$$+120n^3x^3 + 2700nx^4 + 2550n^2x^4 + 180n^3x^4 + 864nx^5$$
$$+900n^2x^5 + 90n^3x^5 + 120nx^6 + 130n^2x^6 + 15n^3x^6\Big).$$

For the Baskakov–Szász operators, the central moments are given by

$$U_{\alpha,m}^{B_\alpha^\rho}(x) = \left[\frac{\partial^m}{\partial A^m} e^{-Ax}\left(1 + c\left(1 - (\alpha\rho)^\rho(-A+\alpha\rho)^{-\rho}\right)x\right)^{-\frac{\alpha}{c}}\right]_{A=0}. \quad (1.3.8)$$

1.3.3 Păltănea Type Operators

Based on the parameters $\rho > 0$ and $c \in \{0, 1\}$ Gupta and Agrawal in [123] proposed the following general operators

$$B_\alpha^\rho(f; x, c) = \sum_{k=1}^{\infty} p_{\alpha,k}(x, c)\int_0^\infty \theta_{\alpha,k}^\rho(t)f(t)dt + p_{\alpha,0}(x, c)f(0), \quad (1.3.9)$$

where

$$p_{\alpha,k}(x, c) = \frac{(\alpha/c)_k}{k!}\frac{(cx)^k}{(1+cx)^{\alpha/c+k}}, \theta_{\alpha,k}^\rho(t) = \frac{\alpha\rho}{\Gamma(k\rho)}e^{-\alpha\rho t}(\alpha\rho t)^{k\rho-1}.$$

Some of the special cases of the operators (1.3.9) are as follows:

1. If $c \to 0$, then

$$p_{\alpha,k}(x, 0) = e^{-\alpha x}\frac{(\alpha x)^k}{k!},$$

   we get the operators due to Păltănea [194]. Also, for this case if $\rho = 1$, we get the Phillips operators [89, 196].
2. If $\rho = 1, c = 1$, we get the operators studied in [19].
3. If $c \to 0, \alpha = n$ and $\rho \to \infty$, then in view of [192, Theorem 2.2], we get the Szász–Mirakjan operators.
4. If $c = 1, \alpha = n,\ f \in \overline{\Pi}$, the closure of the space of algebraic polynomials in space $C[0,\infty))$ and $\rho \to \infty$, we obtain at once Baskakov operators.

By simple computation, we have

$$B_\alpha^\rho(e^{At}; x, c) = \left[1 + cx\left\{1 - (\alpha\rho)^\rho(\alpha\rho - A)^{-\rho}\right\}\right]^{-\alpha/c}. \quad (1.3.10)$$

It may be observed that $B_\alpha^\rho(e^{At}, x)$ may be treated as m.g.f. of the operators B_α^ρ, which may be utilized to obtain the moments of (1.3.9). Let

$$\mu_r^{B_\alpha^\rho}(x) = B_\alpha^\rho(e_r, x), \quad \text{where } e_r(t) = t^r, \ r \in \mathbb{N} \cup \{0\}.$$

The moments are given by

$$B_\alpha^\rho(e_r; x, c) = \left[\frac{\partial^r}{\partial A^r}\left(1 + c\left(1 - (\alpha\rho)^\rho(-A + \alpha\rho)^{-\rho}\right)x\right)^{-\frac{\alpha}{c}}\right]_{A=0}.$$

The property of change of origin of the m.g.f.

$$e^{-Ax}\left(1 + c\left(1 - (\alpha\rho)^\rho(-A + \alpha\rho)^{-\rho}\right)x\right)^{-\frac{\alpha}{c}}$$

can be used to find the central moments

$$U_{\alpha,m}^{B_\alpha^\rho}(x) = B_\alpha^\rho((t - x)^m, x).$$

Expanding

$$e^{-Ax}\left(1 + c\left(1 - (\alpha\rho)^\rho(-A + \alpha\rho)^{-\rho}\right)x\right)^{-\frac{\alpha}{c}}$$

in powers of A, we have

$$\begin{aligned}
&e^{-Ax}\left(1 + c\left(1 - (\alpha\rho)^\rho(-A + \alpha\rho)^{-\rho}\right)x\right)^{-\frac{\alpha}{c}}\\
&= 1 + \frac{x[1 + \rho(1 + cx)]}{\alpha\rho}\frac{A^2}{2} + \frac{\left(2x + 3\rho x + \rho^2 x + 3c\rho x^2 + 3c\rho^2 x^2 + 2c^2\rho^2 x^3\right)}{\alpha^2\rho^2}\frac{A^3}{6}\\
&+ \frac{A^4}{24\alpha^3\rho^3}\Big(6x + 11\rho x + 6\rho^2 x + \rho^3 x + 11c\rho x^2 + 3\alpha\rho x^2 + 18c\rho^2 x^2\\
&+ 6\alpha\rho^2 x^2 + 7c\rho^3 x^2 + 3\alpha\rho^3 x^2 + 12c^2\rho^2 x^3 + 6c\alpha\rho^2 x^3 + 12c^2\rho^3 x^3\\
&+ 6c\alpha\rho^3 x^3 + 6c^3\rho^3 x^4 + 3c^2\alpha\rho^3 x^4\Big)\\
&+ \frac{A^5}{120\alpha^4\rho^4}\Big(24x + 50\rho x + 35\rho^2 x + 10\rho^3 x + \rho^4 x + 50c\rho x^2\\
&+ 20\alpha\rho x^2 + 105c\rho^2 x^2 + 50\alpha\rho^2 x^2 + 70c\rho^3 x^2 + 40\alpha\rho^3 x^2\\
&+ 15c\rho^4 x^2 + 10\alpha\rho^4 x^2 + 70c^2\rho^2 x^3 + 50c\alpha\rho^2 x^3 + 120c^2\rho^3 x^3\\
&+ 90c\alpha\rho^3 x^3 + 50c^2\rho^4 x^3 + 40c\alpha\rho^4 x^3 + 60c^3\rho^3 x^4\\
&+ 50c^2\alpha\rho^3 x^4 + 60c^3\rho^4 x^4 + 50c^2\alpha\rho^4 x^4 + 24c^4\rho^4 x^5 + 20c^3\alpha\rho^4 x^5\Big),
\end{aligned}$$

$$+\frac{A^6}{720\alpha^5\rho^5}\Big(120x+274\rho x+225\rho^2x+85\rho^3x+15\rho^4x+\rho^5x$$
$$+274c\rho x^2+130\alpha\rho x^2+675c\rho^2x^2+375\alpha\rho^2x^2+595c\rho^3x^2$$
$$+385\alpha\rho^3x^2+225c\rho^4x^2+165\alpha\rho^4x^2+31c\rho^5x^2+25\alpha\rho^5x^2$$
$$+450c^2\rho^2x^3+375c\alpha\rho^2x^3+15\alpha^2\rho^2x^3+1020c^2\rho^3x^3+900c\alpha\rho^3x^3$$
$$+45\alpha^2\rho^3x^3+750c^2\rho^4x^3+705c\alpha\rho^4x^3+45\alpha^2\rho^4x^3+180c^2\rho^5x^3$$
$$+180c\alpha\rho^5x^3+15\alpha^2\rho^5x^3+510c^3\rho^3x^4+515c^2\alpha\rho^3x^4+45c\alpha^2\rho^3x^4$$
$$+900c^3\rho^4x^4+930c^2\alpha\rho^4x^4+90c\alpha^2\rho^4x^4+390c^3\rho^5x^4+415c^2\alpha\rho^5x^4$$
$$+45c\alpha^2\rho^5x^4+360c^4\rho^4x^5+390c^3\alpha\rho^4x^5+45c^2\alpha^2\rho^4x^5+360c^4\rho^5x^5$$
$$+390c^3\alpha\rho^5x^5+45c^2\alpha^2\rho^5x^5+120c^5\rho^5x^6+130c^4\alpha\rho^5x^6+15c^3\alpha^2\rho^5x^6\Big).$$

The central moments of order m of the Păltănea type operators can be obtained by collecting the coefficient of $A^m/m!$ in the above expansion. Some of the central moments are given below:

$$U^{B^\rho_\alpha}_{\alpha,0}(x)=1,$$

$$U^{B^\rho_\alpha}_{\alpha,1}(x)=0,$$

$$U^{B^\rho_\alpha}_{\alpha,2}(x)=\frac{x[1+\rho(1+cx)]}{\alpha\rho},$$

$$U^{B^\rho_\alpha}_{\alpha,3}(x)=\frac{\left(2x+3\rho x+\rho^2x+3c\rho x^2+3c\rho^2x^2+2c^2\rho^2x^3\right)}{\alpha^2\rho^2},$$

$$U^{B^\rho_\alpha}_{\alpha,4}(x)=\frac{1}{\alpha^3\rho^3}\Big(6x+11\rho x+6\rho^2x+\rho^3x+11c\rho x^2+3\alpha\rho x^2+18c\rho^2x^2$$
$$+6\alpha\rho^2x^2+7c\rho^3x^2+3\alpha\rho^3x^2+12c^2\rho^2x^3+6c\alpha\rho^2x^3+12c^2\rho^3x^3$$
$$+6c\alpha\rho^3x^3+6c^3\rho^3x^4+3c^2\alpha\rho^3x^4\Big)$$

$$U^{B^\rho_\alpha}_{\alpha,5}(x)=\frac{1}{\alpha^4\rho^4}\Big(24x+50\rho x+35\rho^2x+10\rho^3x+\rho^4x+50c\rho x^2$$
$$+20\alpha\rho x^2+105c\rho^2x^2+50\alpha\rho^2x^2+70c\rho^3x^2+40\alpha\rho^3x^2$$
$$+15c\rho^4x^2+10\alpha\rho^4x^2+70c^2\rho^2x^3+50c\alpha\rho^2x^3+120c^2\rho^3x^3$$
$$+90c\alpha\rho^3x^3+50c^2\rho^4x^3+40c\alpha\rho^4x^3+60c^3\rho^3x^4$$
$$+50c^2\alpha\rho^3x^4+60c^3\rho^4x^4+50c^2\alpha\rho^4x^4+24c^4\rho^4x^5+20c^3\alpha\rho^4x^5\Big),$$

$$
\begin{aligned}
U_{\alpha,6}^{B_\alpha^\rho}(x) = \frac{1}{\alpha^5\rho^5}\Big(& 120x + 274\rho x + 225\rho^2 x + 85\rho^3 x + 15\rho^4 x + \rho^5 x \\
&+274c\rho x^2 + 130\alpha\rho x^2 + 675c\rho^2 x^2 + 375\alpha\rho^2 x^2 + 595c\rho^3 x^2 \\
&+385\alpha\rho^3 x^2 + 225c\rho^4 x^2 + 165\alpha\rho^4 x^2 + 31c\rho^5 x^2 + 25\alpha\rho^5 x^2 \\
&+450c^2\rho^2 x^3 + 375c\alpha\rho^2 x^3 + 15\alpha^2\rho^2 x^3 + 1020c^2\rho^3 x^3 + 900c\alpha\rho^3 x^3 \\
&+45\alpha^2\rho^3 x^3 + 750c^2\rho^4 x^3 + 705c\alpha\rho^4 x^3 + 45\alpha^2\rho^4 x^3 + 180c^2\rho^5 x^3 \\
&+180c\alpha\rho^5 x^3 + 15\alpha^2\rho^5 x^3 + 510c^3\rho^3 x^4 + 515c^2\alpha\rho^3 x^4 + 45c\alpha^2\rho^3 x^4 \\
&+900c^3\rho^4 x^4 + 930c^2\alpha\rho^4 x^4 + 90c\alpha^2\rho^4 x^4 + 390c^3\rho^5 x^4 + 415c^2\alpha\rho^5 x^4 \\
&+45c\alpha^2\rho^5 x^4 + 360c^4\rho^4 x^5 + 390c^3\alpha\rho^4 x^5 + 45c^2\alpha^2\rho^4 x^5 + 360c^4\rho^5 x^5 \\
&+390c^3\alpha\rho^5 x^5 + 45c^2\alpha^2\rho^5 x^5 + 120c^5\rho^5 x^6 + 130c^4\alpha\rho^5 x^6 + 15c^3\alpha^2\rho^5 x^6\Big).
\end{aligned}
$$

For the Păltănea type operators, the central moments are given by

$$
U_{\alpha,m}^{B_\alpha^\rho}(x) = \left[\frac{\partial^m}{\partial A^m} e^{-Ax}\left(1 + c\left(1 - (\alpha\rho)^\rho(-A+\alpha\rho)^{-\rho}\right)x\right)^{-\frac{\alpha}{c}}\right]_{A=0} \tag{1.3.11}
$$

Chapter 2
Quantitative Estimates

The well-known theorem due to Bohman and Korovkin (cf. [62, 152]) states that if $\{L_n\}$ is a sequence of positive linear operators on the space $C[a, b]$, then $L_n f \to f$ for every $f \in C[a, b]$, provided $L_n(e^r, x) \to e_r(x), r = 0, 1, 2$ with $e_r(t) = t^r$ for n sufficiently large. Efforts have been made by several researchers to enlarge the domain of approximation operators and to include bounded or unbounded functions. A systematic study on Korovkin-type theorems was done by Altomare in [27], who provided applications concerning the approximation of continuous functions (as well as of L^p-functions), by means of linear positive operators. Also, Boyanov and Veselinov [63] established the uniform convergence of any sequence of positive linear operators. Suppose $C^*[0, \infty)$ denotes the subspace of all real-valued continuous functions possessing a finite limit at infinity and equipped with the uniform norm. Boyanov and Veselinov proved the following theorem for the general sequence of linear positive operators:

Theorem 2.1 *The sequence* $L_n : C^*[0, \infty) \to C^*[0, \infty)$ *of positive linear operators satisfies the conditions*

$$\lim_{n\to\infty} L_n(e^{-kt}, x) = e^{-kx}, k = 0, 1, 2$$

uniformly in $[0, \infty)$ *if and only if*

$$\lim_{n\to\infty} L_n(f, x) = f(x)$$

uniformly on $[0, \infty)$, *for all* $f \in C^*[0, \infty)$.

Motivated by the work of Boyanov and Veselinov [63], Holhoş in [139] obtained the quantitative error estimate for general sequence of linear positive operators and also mentioned the results with some examples. The quantitative estimate developed in [139] is the following theorem:

V. Gupta et al., *Recent Advances in Constructive Approximation Theory*,
Springer Optimization and Its Applications 138,
https://doi.org/10.1007/978-3-319-92165-5_2

Theorem 2.2 ([139]) *If a sequence of linear positive operators $L_n : C^*[0, \infty) \to C^*[0, \infty)$ satisfies the equalities*

$$||L_n e_0 - 1||_{[0,\infty)} = \alpha_n$$

$$||L_n(e^{-t}) - e^{-x}||_{[0,\infty)} = \beta_n$$

$$||L_n(e^{-2t}) - e^{-2x}||_{[0,\infty)} = \gamma_n$$

then

$$||L_n f - f||_{[0,\infty)} \leq \alpha_n ||f||_{[0,\infty)} + (2+\alpha_n)\omega^*(f, \sqrt{\alpha_n + 2\beta_n + \gamma_n}),\ f \in C^*[0, \infty).$$

The modulus of continuity used in the above theorem is defined as:

$$\omega^*(f, \delta) := \sup_{\substack{|e^{-x}-e^{-t}|\leq\delta \\ x,t>0}} |f(t) - f(x)|.$$

2.1 Discrete Operators

In this section, we discuss some approximation properties of modified forms of the discretely defined operators.

2.1.1 Modified Szász–Mirakyan Operators

For the Szász–Mirakyan operators $S_n : C^*[0, \infty) \to C^*[0, \infty)$ defined by

$$S_n(f, x) = e^{-nx} \sum_{k=0}^{\infty} \frac{(nx)^k}{k!} f\left(\frac{k}{n}\right)$$

Holhoş [139] obtained the following

$$||S_n f - f||_{[0,\infty)} \leq 2\omega^*\left(f, n^{-1/2}\right), n \geq 1.$$

Recently Acar–Aral–Gonska [8] considered the modified form of Szász–Mirakyan operators, which preserve e^{2ax}. The operators take the following form:

$$S_n^1(f,x) = e^{-\frac{2ax}{(e^{2a/n}-1)}} \sum_{k=0}^{\infty} \frac{(2ax)^k}{k!\,(e^{2a/n}-1)^k} f\left(\frac{k}{n}\right). \tag{2.1.1}$$

Relevant to the Theorem 2.2 Acar et al. obtained the following theorem:

Theorem 2.3 ([8]) *If $f \in C^*[0,\infty)$ we have*

$$||S_n^1 f - f||_{[0,\infty)} \leq 2\omega^*(f, \sqrt{2\beta_n + \gamma_n}),$$

where

$$||S_n^1(e^{-t}) - e^{-x}||_{[0,\infty)} = \beta_n$$

$$||S_n^1(e^{-2t}) - e^{-2x}||_{[0,\infty)} = \gamma_n.$$

Moreover, β_n and γ_n tend to zero as n goes to infinity, so S_n^1 converges uniformly to f.

Theorem 2.4 ([8]) *Let $f, f'' \in C^*[0,\infty)$, then, for $x \in [0,\infty)$, the following inequality holds:*

$$\left| n\,[S_n^1(f,x) - f(x)] - axf'(x) - \frac{x}{2}\frac{f''(x)}{2} \right|$$
$$\leqslant |p_n(x)|\,|f'| + |q_n(x)|\,|f''| + 2\,(2\,q_n(x) + x + r_n(x))\;\omega^*(f'', 1/\sqrt{n}),$$

where

$$p_n(x) = n\;S_n^1(\phi_x^1(t),x) - x,\;\; q_n(x) = \frac{1}{2}\left(S_n^1(\phi_x^2(t),x) - x\right)$$

and

$$r_n(x) = n^2\sqrt{S_n^1\left((e^{-x} - e^{-t})^4, x\right)}\sqrt{S_n^1\left((t-x)^4, x\right)}.$$

Recently Gupta and Malik [129] considered another modification of Szász–Mirakyan operators preserving e^{-2x} and they were able to obtain a better approximation. Acar et al. in [9] introduced another family of linear positive operators having Szász–Mirakyan basis functions that reproduce the functions e^{ax} and e^{2ax}, $a > 0$. They started with the following form

$$S_n^2(f,x) = e^{-na_n(x)} \sum_{k=0}^{\infty} \frac{(nb_n(x))^k}{k!} f\left(\frac{k}{n}\right). \tag{2.1.2}$$

Thus with this choice, one has

$$S_n^2(e^{at}, x) = e^{ax} \text{ and } S_n^2(e^{2at}, x) = e^{2ax}.$$

By simple computation the values of $a_n(x)$ and $b_n(x)$ defined in (2.1.2) were obtained as

$$a_n(x) = \frac{ax(2 - e^{a/n})}{n(e^{a/n} - 1)}, \quad b_n(x) = \frac{ax}{ne^{a/n}(e^{a/n} - 1)}. \tag{2.1.3}$$

In [9] with weight $\varphi(x) = 1 + e^{2ax}$, $x \in \mathbb{R}^+$ the spaces are defined as

$$B_\varphi(\mathbb{R}^+) = \{f : \mathbb{R}^+ \to \mathbb{R} : |f(x)| \leq M_f \varphi(x), x \geq 0\},$$

$$C_\varphi(\mathbb{R}^+) = C(\mathbb{R}^+) \bigcap B_\varphi(\mathbb{R}^+)$$

and

$$C_\varphi^k(\mathbb{R}^+) = \{f \in C_\varphi(\mathbb{R}^+) : \lim_{x \to \infty} \frac{f(x)}{\varphi(x)} = k_f \text{ exists and it is finite}\}.$$

The following main results were discussed in [9]:

Theorem 2.5 ([9]) *For each function $f \in C_\varphi^k(\mathbb{R}^+)$ we have*

$$\lim_{n \to \infty} ||S_n^2(f) - f||_\varphi = 0,$$

where

$$||f||_\varphi = \sup_{x \in \mathbb{R}^+} \frac{|f(x)|}{1 + e^{2ax}}.$$

Theorem 2.6 ([9]) *For $f \in C_\varphi^k(\mathbb{R}^+)$ we have*

$$\lim_{n \to \infty} ||[S_n^2(f) - f||_{5a/2} \leq \frac{a}{e}\frac{1}{n}||f||_\alpha + C\tilde{\omega}(f, n^{-1/2}),$$

where

$$||f||_\alpha = \sup_{x \in \mathbb{R}^+} \frac{|f(x)|}{e^{\alpha x}}$$

and the weighted modulus of continuity is

$$\tilde{\omega}(f, \delta) = \sup_{|t-x| \leq \delta, x \geq 0} \frac{|f(t) - f(x)|}{e^{at} + e^{ax}}.$$

Theorem 2.7 ([9]) *Let $f \in C_\varphi(\mathbb{R}^+)$. If f is twice differentiable in $x \in \mathbb{R}^+$ and f'' is continuous at x, then the following limit holds:*

$$\lim_{n\to\infty} n[S_n^2(f,x) - f(x)] = a^2xf(x) - \frac{3}{2}axf'(x) + \frac{x}{2}f''(x).$$

2.1.2 Lupaş-Type Modified Operators

Abel and Ivan [2] considered an important general form of the discrete operators and established the complete asymptotic expansion of these operators. The operators discussed in [2] for $x \in [0,\infty)$ are defined by

$$L_n^c(f,x) = \sum_{k=0}^{\infty} \left(\frac{c}{1+c}\right)^{ncx} \frac{(ncx)_k}{k!(1+c)^k} f\left(\frac{k}{n}\right) \tag{2.1.4}$$

These operators provide some of the well-known operators as special cases. The following are some examples of these operators:

If $c \to \infty$ and $x \in [0,\infty)$, we get the well-known Szász–Mirakyan operator

$$L_n^\infty(f,x) = \sum_{k=0}^{\infty} e^{-nx} \frac{(nx)^k}{k!} f\left(\frac{k}{n}\right).$$

Also if $c = 1$ and $x \in [0,\infty)$, we get Lupaş operators [158] defined as

$$L_n^1(f,x) = \sum_{k=0}^{\infty} 2^{-nx} \frac{(nx)_k}{k!2^k} f\left(\frac{k}{n}\right).$$

For $f \in C[0,\infty)$ and taking the right side of (2.1.4) to be absolutely convergent, we consider the operators as

$$G_n^c(f,x) = \left(\frac{c}{1+c}\right)^{nca_n(x)} \sum_{k=0}^{\infty} \frac{(nca_n(x))_k}{k!(1+c)^k} f\left(\frac{k}{n}\right), x \geq 0, n \in \mathbb{N}, \tag{2.1.5}$$

such that the condition $G_n^c(e^{-t},x) = e^{-x}$ holds for all x and n. Thus using (2.1.5) and the well-known binomial series

$$\sum_{k=0}^{\infty} \frac{(a)_k}{k!} z^k = (1-z)^{-a}, |z| < 1,$$

we have

$$
\begin{aligned}
e^{-x} &= \left(\frac{c}{1+c}\right)^{nca_n(x)} \sum_{k=0}^{\infty} \frac{(nca_n(x))_k}{k!(1+c)^k} e^{-k/n} \\
&= \left(\frac{c}{1+c}\right)^{nca_n(x)} \left(1 - \frac{e^{-1/n}}{1+c}\right)^{-nca_n(x)} \\
&= c^{nca_n(x)} (1 + c - e^{-1/n})^{-nca_n(x)},
\end{aligned}
$$

which implies that

$$
a_n(x) = \frac{-x}{nc\left(\ln c - \ln(1 + c - e^{-1/n})\right)}. \tag{2.1.6}
$$

Thus the operators (2.1.5) take the form

$$
\begin{aligned}
G_n^c(f, x) &= \left(\frac{c}{1+c}\right)^{-x/\left(\ln c - \ln(1+c-e^{-1/n})\right)} \\
&\quad \sum_{k=0}^{\infty} \frac{\left(-x/\left(\ln c - \ln(1 + c - e^{-1/n})\right)\right)_k}{k!(1+c)^k} f\left(\frac{k}{n}\right) \\
&= L_n(f, \varphi_n(x)),
\end{aligned} \tag{2.1.7}
$$

where $\varphi_n(x) = ((L_n(e^{-t}, x))^{-1} \circ e^{-x}$.

Actually by considering the operators in the form (2.1.5) (which preserve e^{-x}) is more appropriate than those which preserve e^{2at}, $a > 0$, as discussed in [8]. Here we study the modified form (2.1.5), and establish quantitative asymptotic estimates for these operators.

In the sequel, we need the following lemmas:

Lemma 2.1 *For $a > 0$, we have*

$$
\begin{aligned}
G_n^c(e^{at}, x) &= c^{nc\alpha_n(x)}(1 + c - e^{a/n})^{-nc\alpha_n(x)} \\
&= c^{-x/\left(\ln c - \ln(1+c-e^{-1/n})\right)}(1 + c - e^{a/n})^{x/\left(\ln c - \ln(1+c-e^{-1/n})\right)} \\
&= \left(\frac{1 + c - e^{a/n}}{c}\right)^{x/\ln[c/(1+c-e^{-1/n})]}.
\end{aligned}
$$

Lemma 2.2 *If we set $e_r(t) = t^r$, $r = 0, 1, 2, \ldots$, then we have*

$$
\begin{aligned}
G_n^c(e_0, x) &= 1, \\
G_n^c(e_1, x) &= a_n(x) \\
G_n^c(e_2, x) &= a_n^2(x) + \frac{(1+c)a_n(x)}{cn}
\end{aligned}
$$

$$G_n^c(e_3,x) = a_n^3(x) + \frac{3(1+c)a_n^2(x)}{nc} + \frac{(1+c)(2+c)a_n(x)}{n^2c^2}$$

$$G_n^c(e_4,x) = a_n^4(x) + \frac{6(1+c)a_n^3(x)}{nc} + \frac{(11+18c+7c^2)a_n^2(x)}{n^2c^2} + \frac{(6+12c+7c^2+c^3)a_n(x)}{n^3c^3}.$$

Lemma 2.3 *If we denote $\mu_{n,m}^c(x) := G_n^c((t-x)^m, x)$, then by Lemma 2.2, we have*

$$\mu_{n,1}^c(x) = a_n(x) - x$$

$$\mu_{n,2}^c(x) = (a_n(x)-x)^2 + \frac{(1+c)a_n(x)}{cn}.$$

Moreover, by simple computation

$$\lim_{n\to\infty} nx\left(\frac{-1}{nc\left(\log c - \log\left(1+c-e^{\frac{-1}{n}}\right)\right)} - 1\right) = \frac{(1+c)x}{2c},$$

$$\lim_{n\to\infty} n\left[\left(\frac{-x}{nc\left(\log c - \log\left(1+c-e^{\frac{-1}{n}}\right)\right)} - x\right)^2 - \frac{(1+c)x}{n^2c^2\left(\log c - \log\left(1+c-e^{\frac{-1}{n}}\right)\right)}\right]$$
$$= \frac{(1+c)x}{c}.$$

In this section, we obtain the following uniform convergence estimate corresponding to Theorem 2.2 for the operators (2.1.5).

Theorem 2.8 *For $f \in C^*[0,\infty)$, we have*

$$||G_n^c f - f||_{[0,\infty)} \le 2\omega^*(f, \sqrt{\gamma_n}),$$

where

$$\gamma_n = ||G_n^c(e^{-2t}) - e^{-2x}||_{[0,\infty)}$$
$$= \left\|\frac{(1+c)e^{-2x}x}{cn} + \frac{(1+c)e^{-2x}x(-3-c+x+cx)}{2c^2n^2} + O(n^{-3})\right\|_{[0,\infty)}.$$

Proof The operators G_n^c preserve constant as well as e^{-x} thus $\alpha_n = \beta_n = 0$. We only have to evaluate γ_n. In view of Lemma 2.1, we have

$$G_n^c(e^{-2t}, x) = \left(\frac{1+c-e^{-2/n}}{c}\right)^{x/\ln[c/(1+c-e^{-1/n})]},$$

Thus, using the software Mathematica, we get

$$G_n^c(e^{-2t}, x) = e^{-2x} + \frac{(1+c)e^{-2x}x}{cn} + \frac{(1+c)e^{-2x}x(-3-c+x+cx)}{2c^2n^2} + O(n^{-3}).$$

This completes the proof of the theorem. ∎

Theorem 2.9 *Let $f, f'' \in C^*[0,\infty)$, then for any $x \in [0,\infty)$, we have*

$$\left| n[G_n^c(f,x) - f(x)] - \frac{(1+c)x}{2c}[f'(x) + f''(x)] \right|$$
$$\leq |p_n(x)||f'(x)| + |q_n(x)||f''(x)| + 2\left(2q_n(x) + \frac{(1+c)x}{c} + r_n(x)\right)\omega^*(f'', n^{-1/2}),$$

where

$$p_n(x) = n\mu_{n,1}^c(x) - \frac{(1+c)x}{2c}$$
$$q_n(x) = \frac{1}{2}\left[n\mu_{n,2}^c(x) - \frac{(1+c)x}{c}\right]$$
$$r_n(x) = n^2\sqrt{G_n^c((e^{-x}-e^{-t})^4, x)}\sqrt{G_n^c((t-x)^4, x)}.$$

Proof By the Taylor's formula, we can write

$$f(t) = f(x) + f'(x)(t-x) + \frac{f''(x)}{2}(t-x)^2 + h(t,x)(t-x)^2$$

where

$$h(t,x) := [f''(\eta) - f''(x)]/2,$$

η lies between x and t and h is a continuous function which vanishes at zero. Applying the operator $G_n^c(f,x)$ to both sides of above equality, we obtain

$$G_n^c(f,x) - f(x) = f'(x)\mu_{n,1}^c(x) + \frac{f''(x)}{2}\mu_{n,2}^c(x) + G_n^c\left(h(t,x)(t-x)^2, x\right).$$

We can also write that

$$\left| n\left[G_n^c(f,x) - f(x)\right] - \frac{(1+c)x}{2c}[f'(x) + f''(x)] \right|$$
$$\leq |f'(x)|.\left| n\mu_{n,1}^c(x) - \frac{(1+c)x}{2c} \right| + \frac{|f''(x)|}{2}\left| n\mu_{n,2}^c(x) - \frac{(1+c)x}{c} \right|$$
$$+ \left| nG_n^c\left(h(t,x)(t-x)^2, x\right) \right|.$$

To estimate the last inequality following the methods of [139] and applying the Cauchy–Schwarz inequality we obtain

$$\left|nG_n^c\left(h(t,x)(t-x)^2,x\right)\right| \le nG_n^c\left(|h(t,x)|(t-x)^2,x\right)$$

$$\le 2n\omega^*\left(f'',\delta\right)\mu_{n,2}^c(x)+\frac{2n}{\delta^2}\omega^*\left(f'',\delta\right)\sqrt{G_n^c\left(\left(e^{-x}-e^{-t}\right)^4,x\right)}\sqrt{\mu_{n,4}^c(x)}.$$

Choosing $\delta=\frac{1}{\sqrt{n}}$ we have

$$nG_n^c\left(|h(t,x)|(t-x)^2,x\right)\le 2\omega^*\left(f'',\frac{1}{\sqrt{n}}\right)\left[n\mu_{n,2}^c(x)+r_n(x)\right],$$

where

$$r_n(x)=\sqrt{n^2G_n^c\left(\left(e^{-x}-e^{-t}\right)^4,x\right)}\sqrt{n^2\mu_{n,4}^c(x)}.$$

which was our claim. ■

Corollary 2.1 *Let $f, f''\in C^*[0,\infty)$, then for $x\in[0,\infty)$ we have*

$$\lim_{n\to\infty} n[G_n^c(f,x)-f(x)]=\frac{(1+c)x}{2c}[f'(x)+f''(x)].$$

2.1.3 Modified Baskakov Type Operators

The Baskakov operators are defined as

$$\widetilde{V}_n f(x)=\sum_{k=0}^{\infty}\binom{n+k-1}{k}\frac{x^k}{(1+x)^{n+k}}f\left(\frac{k}{n}\right) \tag{2.1.8}$$

Then, we have

$$\widetilde{V}_n(e^{Ax},x)=(1+x(1-e^{A/n}))^{-n}. \tag{2.1.9}$$

In [139] for $f\in C^*[0,\infty)$ the following estimate was established

$$||\widetilde{V}_n f-f||_\infty\le 2\omega^*\left(f,\frac{5}{2\sqrt{n}}\right),\ n\ge 2.$$

We observe from (2.1.9) that its right side is the moment generating function of Baskakov operators, which may be utilized to find the moments of Baskakov operators as

$$\mu_r' = \left[\frac{d^r}{dA^r}(1 + x(1 - e^{A/n}))^{-n}\right]_{A=0},$$

where

$$\mu_r' = \sum_{k=0}^{\infty} \binom{n+k-1}{k} \frac{x^k}{(1+x)^{n+k}} \left(\frac{k}{n}\right)^r$$

Chen [72] gave a generalization of the Baskakov operators based on certain parameter $\alpha > 0$ as

$$\widetilde{V}_{n,\alpha} f(x) = \sum_{k=0}^{\infty} \frac{n(n+\alpha)\cdots[n+(k-1)\alpha]}{k!} \frac{x^k}{(1+\alpha x)^{\frac{n}{\alpha}+k}} f\left(\frac{k}{n}\right), \tag{2.1.10}$$

For functions $f \in C[0, \infty)$, we consider the following form of operators such that these operators are absolutely convergent

$$V_{n,\alpha} f(x) = \sum_{k=0}^{\infty} \frac{n(n+\alpha)\cdots[n+(k-1)\alpha]}{k!} \frac{(b_n(x))^k}{(1+\alpha b_n(x))^{\frac{n}{\alpha}+k}} f\left(\frac{k}{n}\right), \tag{2.1.11}$$

$x \geq 0, n \in \mathbb{N}$, with the condition

$$V_{n,\alpha}(e^{-t}, x) = e^{-x},$$

being satisfied for all x and n. Thus, we get

$$b_n(x) = \frac{(1 - e^{\alpha x/n})e^{1/n}}{\alpha(1 - e^{1/n})}. \tag{2.1.12}$$

Therefore in view of (2.1.12), the operators (2.1.11) take the form

$$\begin{aligned} V_{n,\alpha}(f, x) = &\left(\frac{1 - e^{(\alpha x+1)/n}}{1 - e^{1/n}}\right)^{-n/\alpha} \\ &\sum_{k=0}^{\infty} \frac{\left(\frac{n}{\alpha}\right)_k}{k!} \left(\frac{(1 - e^{\alpha x/n})e^{1/n}}{1 - e^{(\alpha x+1)/n}}\right)^k f\left(\frac{k}{n}\right) \\ = &\, V_{n,\alpha}(f, \varphi_n(x)), \end{aligned} \tag{2.1.13}$$

where $\varphi_n(x) = ((V_{n,\alpha}(e^{-t}, x))^{-1} \circ e^{-x}$. In the present section, we discuss approximation properties of the operators $V_{n,\alpha}$. We shall need the following basic results:

Lemma 2.4 *We have*

$$\begin{aligned} V_{n,\alpha}(e^{At},x) &= [1+\alpha b_n(x)(1-e^{A/n})]^{-n/\alpha} \\ &= \left[1+\frac{(1-e^{\alpha x/n})e^{1/n}}{1-e^{1/n}}(1-e^{A/n})\right]^{-n/\alpha} \\ &= \left[\frac{1+e^{(A+1)/n}+e^{(\alpha x+1)/n}-e^{(\alpha x+A+1)/n}}{1-e^{1/n}}\right]^{-n/\alpha}. \end{aligned}$$

Lemma 2.5 *The moments $T_{n,m}(x)=V_{n,\alpha}(t^m,x), m=1,2,\ldots$ are given by*

$$\begin{aligned} T_{n,1}(x) &= b_n(x) \\ T_{n,2}(x) &= b_n^2(x)+\frac{b_n(x)(1+\alpha b_n(x))}{n}, \end{aligned}$$

where $\beta_n(x)$ is defined in (2.1.12).

Lemma 2.6 *Applying Lemma 2.5, the central moments*

$$\mu_{n,m}(x)=V_{n,\alpha}((t-x)^m,x), m=1,2,\ldots$$

are given by

$$\begin{aligned} \mu_{n,1}(x) &= b_n(x)-x \\ \mu_{n,2}(x) &= (b_n(x)-x)^2+\frac{b_n(x)(1+\alpha b_n(x))}{n}. \end{aligned}$$

We have the following limits:

$$\lim_{n\to\infty} n\left(\frac{(1-e^{\alpha x/n})e^{1/n}}{\alpha(1-e^{1/n})}-x\right)=\frac{x(1+\alpha x)}{2},$$

$$\lim_{n\to\infty} n\left(\left(\frac{(1-e^{\alpha x/n})e^{1/n}}{\alpha(1-e^{1/n})}-x\right)^2+\frac{(1-e^{\alpha x/n})(1-e^{(\alpha x+1)/n})e^{1/n}}{n\alpha(1-e^{/n})^2}\right)=x(1+\alpha x).$$

We obtain the following quantitative estimate corresponding to Theorem 2.2 for the operators $V_{n,\alpha}$:

Theorem 2.10 *For $f\in C^*[0,\infty)$, we have*

$$||V_{n,\alpha}f-f||_{[0,\infty)}\le 2\omega^*\left(f,\sqrt{\left|\frac{2\alpha x^3+(\alpha-2)x^2+x}{n}+O(n^{-2})\right|}\right).$$

Proof In view of Lemma 2.4, we obtain

$$\begin{aligned} V_{n,\alpha}(e^{-2t}, x) &= [1 + \alpha b_n(x)(1 - e^{-2/n})]^{-n/\alpha} \\ &= \left[1 + \frac{(1 - e^{\alpha x/n})e^{1/n}}{(1 - e^{1/n})}(1 - e^{-2/n})\right]^{-n/\alpha} \\ &= \left[1 + (e^{\alpha x/n} - 1)(1 + e^{-1/n})\right]^{-n/\alpha}. \end{aligned}$$

Consider

$$\begin{aligned} u_n(x) &= (e^{\alpha x/n} - 1)(1 + e^{-1/n}) \\ &= \left(\frac{\alpha x}{n} + \frac{\alpha^2 x^2}{2!n^2} + \frac{\alpha^3 x^3}{3!n^3} + \cdots\right)\left(2 - \frac{1}{n} + \frac{1}{2n^2} - \frac{1}{3!n^3} + \cdots\right) \\ &= \frac{2\alpha x}{n} + \frac{\alpha^2 x^2 - \alpha x}{n^2} + O(n^{-3}). \end{aligned}$$

Thus, we get

$$\begin{aligned} V_{n,\alpha}(e^{-2t}, x) &= 1 - \frac{n}{\alpha}u_n(x) + \frac{n}{\alpha}\left(\frac{n}{\alpha} + 1\right)\frac{u_n^2(x)}{2!} + O(n^{-3}) \\ &= 1 - 2x + \frac{4x^2}{2!} + \cdots + \frac{g(x)}{n} + O(n^{-2}) \\ &= e^{-2x} + \frac{g(x)}{n} + O(n^{-2}). \end{aligned}$$

Hence,

$$|V_{n,\alpha}(e^{-2t}, x) - e^{-2x}| = \frac{g(x)}{n} + O(n^{-2}),$$

where $g(x) = 2\alpha x^3 + (\alpha - 2)x^2 + x$. This completes the proof of theorem. ■

Remark 2.1 For the usual Baskakov operators (2.1.10), which preserve the constant and linear functions, using Mathematica software, we obtain the following theorem:

Theorem 2.11 *For $f \in C^*[0, \infty)$, we have*

$$||\widetilde{V}_{n,\alpha}f - f||_{[0,\infty)} \leq 2\omega^*(f, \sqrt{2\beta_n + \gamma_n}),$$

where

$$\beta_n = \left|\left|\frac{e^{-x}x(1 + \alpha x)}{2n} + \frac{e^{-x}x(-4 + 3x - 12\alpha x + 6\alpha x^2 - 8\alpha^2 x^2 + 3\alpha^2 x^3)}{24n^2} + O(n^{-3})\right|\right|_{[0,\infty)}$$

$$\gamma_n = \left\| \frac{2e^{-2x}x(1+\alpha x)}{2n} + \frac{2e^{-2x}x(-2+3x-6\alpha x+6\alpha x^2-4\alpha^2x^2+3\alpha^2x^3)}{3n^2} + O(n^{-3}) \right\|_{[0,\infty)}.$$

Remark 2.2 We may remark that the complicated analysis which was applied in proving [139, Corollary 3.3] is not required to prove the above theorems. We used Mathematica.

The following estimate is a quantitative asymptotic formula:

Theorem 2.12 *Let $f, f'' \in C^*[0,\infty)$, then for any $x \in [0,\infty)$, we obtain*

$$\left| n[V_{n,\alpha}(f,x) - f(x)] - \frac{x(1+\alpha x)}{2}[f'(x) + f''(x)] \right|$$
$$\leq |p_n(x)||f'(x)| + |q_n(x)||f''(x)| + 2(2q_n(x) + x + r_n(x))\omega^*(f'', n^{-1/2}),$$

where

$$p_n(x) = n\mu_{n,1}(x) - \frac{x(1+\alpha x)}{2}$$
$$q_n(x) = \frac{1}{2}(n\mu_{n,2}(x) - x(1+\alpha x))$$
$$r_n(x) = n^2\sqrt{V_n((e^{-x}-e^{-t})^4, x)}\sqrt{\mu_{n,4}(x)}.$$

Corollary 2.2 *Let $f, f'' \in C^*[0,\infty)$, then for $x \in [0,\infty)$ we have*

$$\lim_{n\to\infty} n[V_{n,\alpha}(f,x) - f(x)] = \frac{x(1+\alpha x)}{2}\left[f'(x) + f''(x)\right].$$

Also very recently Yılmaz, Gupta, and Aral in [228] considered the usual Baskakov operators (1.1.5), preserving e^{2ax}. They introduced the operators for $x \in [0,\infty)$, as

$$V_{n,\beta}(f,x) = \sum_{k=0}^{\infty} f\left(\frac{k}{n}\right)\binom{n+k-1}{k}\frac{(b_n(x))^k}{(1+b_n(x))^{n+k}}$$
$$= V_n(f, b_n(x)), \tag{2.1.14}$$

where

$$b_n(x) = \left(V_n(e^{2at}, x)\right)^{-1} \circ e^{2ax}.$$

and

$$b_n(x) = \frac{e^{\frac{-2ax}{n}} - 1}{\left(1 - e^{\frac{2a}{n}}\right)}.$$

The following direct results were established in [228]:

Theorem 2.13 *For $f \in C^*[0,\infty)$, we obtain*

$$\left\| V_{n,\beta} f - f \right\|_{[0,\infty)} \leq 2\omega^* \left(f, \sqrt{2\beta_n + \gamma_n} \right),$$

where

$$\left\| V_{n,\beta}\left(e^{-t}, x\right) - e^{-x} \right\|_{[0,\infty)} = \beta_n,$$

$$\left\| V_{n,\beta}\left(e^{-2t}, x\right) - e^{-2x} \right\|_{[0,\infty)} = \gamma_n.$$

Here when $n \longrightarrow \infty$, b_n and c_n tend to zero and $\left\{V_{n,\beta} f\right\}$ converges uniformly to the function f.

Theorem 2.14 ([228]) *Let $f,\ f'' \in C^*[0,\infty)$. Then we get*

$$\left| n\left[V_{n,\beta}(f,x) - f(x)\right] + ax(x+1)f'(x) - \frac{x(x+1)}{2} f''(x) \right|$$

$$\leq |p_n(x)| \left|f'(x)\right| + |q_n(x)| \left|f''(x)\right| + 2\left(2q_n + x(x+1) + r_n(x)\right)\omega^*\left(f'', \frac{1}{\sqrt{n}}\right),$$

where $V_{n,\beta}\left(e_r^x(t), x\right)$ are the r-th central moments of (2.1.14) and

$$p_n(x) = nV_{n,\beta}\left(e_1^x(t), x\right) + ax(x+1),$$

$$q_n(x) = \frac{1}{2}\left(nV_{n,\beta}\left(e_2^x(t), x\right) - x(x+1)\right),$$

$$r_n(x) = n^2 \sqrt{V_{n,\beta}\left(\left(e^{-x} - e^{-t}\right)^4, x\right)} \sqrt{V_{n,\beta}\left(e_4^x(t), x\right)}.$$

Theorem 2.15 ([228]) *Let $x \in [0,\infty)$ and $f,\ f'' \in C^*[0,\infty)$, we have*

$$\lim_{n \longrightarrow \infty} n\left[V_{n,\beta}(f,x) - f(x)\right] = -ax(x+1)f'(x) + \frac{x(x+1)}{2} f''(x).$$

Theorem 2.16 ([228]) *If the function $f \in C^2[0,\infty)$ is strictly φ-convex with respect to $\tau(x) = e^{2ax}$, $a > 0$, then for all $x \geq 0$ there exists $n_0 = n_0(x) \in \mathbb{N}$ such that for $n \geq n_0$ it holds*

$$f(x) \leq V_{n,\beta}(f,x)$$

Theorem 2.17 ([228]) *Let $f \in C[0,\infty)$ be a decreasing and convex function. Then we have*

$$f(x) \leq V_{n+1,\beta}(f,x) \leq V_{n,\beta}(f,x).$$

2.2 Some Integral Operators

This section deals with quantitative estimates of certain integral type operators, which preserve different exponential type functions.

2.2.1 *Post-Widder Operators*

Recently Gupta and Maheshwari in [133] considered the modified form of Post-Widder operators (1.1.17) which preserve the test function e^{-x}. The modified form is defined as

$$\widetilde{P}_n(f,x) := \frac{1}{n!}\left(e^{x/(n+1)}-1\right)^{-(n+1)} \int_0^\infty t^n\, e^{-\frac{t}{(e^{x/(n+1)}-1)}} f(t)\,dt, \quad (2.2.1)$$

$x \in (0,\infty)$ and $\widetilde{P}_n(f,0) = f(0)$, which preserve constant and the test function e^{-x}.

Lemma 2.7 *We have for $\theta > 0$ that*

$$\widetilde{P}_n(e^{\theta t},x) = \left(1-(e^{x/(n+1)}-1)\theta\right)^{-(n+1)}.$$

One of the main results discussed in [133] is the following estimate:

Theorem 2.18 *The sequence of modified Post-Widder operators*

$$\widetilde{P}_n : C^*[0,\infty) \to C^*[0,\infty)$$

satisfies

$$||\widetilde{P}_n f - f||_{[0,\infty)} \leq 2\omega^*(f,\sqrt{\gamma_n}),\ f \in C^*[0,\infty).$$

Here the convergence takes place if n is sufficiently large.

Proof The operators $\widetilde{P}_n$ preserve constant functions as well as e^{-x} so by Theorem A, $\alpha_n = \beta_n = 0$. We only have to evaluate γ_n. In view of Lemma 2.7, we have

$$\widetilde{P}_n\left(e^{-2t};x\right)=\left(1+2(e^{x/(n+1)}-1)\right)^{-(n+1)}.$$

Let

$$f_n(x)=\left(2e^{x/(n+1)}-1\right)^{-(n+1)}-e^{-2x}$$

Since $f_n(0)=f_n(\infty)=0$, there exists a point $\xi_n\in(0,\infty)$ such that

$$\|f_n\|_\infty=f_n(\xi_n).$$

It follows that $f_n'(\xi_n)=0$, i.e.

$$e^{-2\xi_n}=e^{\frac{\xi_n}{1+n}}\left(-1+2e^{\frac{\xi_n}{1+n}}\right)^{-n-2}$$

and

$$f_n(\xi_n)=\left(2e^{\frac{\xi_n}{1+n}}-1\right)^{-n-1}-e^{-2\xi_n}=\left(e^{\frac{\xi_n}{1+n}}-1\right)\left(2e^{\frac{\xi_n}{1+n}}-1\right)^{-2-n}.$$

Let $x_n:=e^{\frac{\xi_n}{1+n}}-1>0$. It follows that

$$f_n(\xi_n)=\frac{x_n}{(2x_n+1)^{n+2}}\le\min\left\{x_n,\frac{1}{(2x_n+1)^{n+1}}\right\}\to 0\ \text{as}\ n\to\infty.$$

This completes the proof of the theorem. ■

2.2.2 Lupaş–Kantorovich Type Modified Operators

In 1995 Lupaş [158] proposed the following discrete operators:

$$L_n(f,x)=\sum_{k=0}^{\infty}\frac{2^{-nx}(nx)_k}{k!2^k}f\left(\frac{k}{n}\right),$$

where $(nx)_k$ is the rising factorial given by

$$(nx)_k=nx(nx+1)(nx+2)\cdots(nx+k-1),\quad (nx)_0=1.$$

Agratini [18] introduced the Kantorovich type modification of the operators L_n. Erençin and Taşdelen in [82] and Gupta et al. in [134] considered a generalization of the operators discussed in [18] based on some parameters, and established some approximation properties. In [136] Gupta–Rassias–Agrawal considered the Kantorovich variant of Lupaş operators, defined by

$$\overline{L}_n(f,x) = n\sum_{k=0}^{\infty}\frac{2^{-na_n(x)}(na_n(x))_k}{k!2^k}\int_{k/n}^{(k+1)/n} f(t)dt \tag{2.2.2}$$

with the hypothesis that these operators preserve the function e^{-x}. Then, we can write

$$\begin{aligned} e^{-x} &= n\sum_{k=0}^{\infty}\frac{2^{-na_n(x)}(na_n(x))_k}{k!2^k}\int_{k/n}^{(k+1)/n} e^{-t}dt \\ &= n\sum_{k=0}^{\infty}\frac{2^{-na_n(x)}(na_n(x))_k}{k!2^k}e^{-k/n}(1-e^{-1/n}) \\ &= n(1-e^{-1/n})(2-e^{-1/n})^{-na_n(x)}, \end{aligned}$$

where we have used the identity

$$\sum_{k=0}^{\infty}\frac{(a)_k}{k!}z^k = (1-z)^{-a},\ |z|<1,$$

which concludes

$$a_n(x) = \frac{x+\ln\left(n(1-e^{-1/n})\right)}{n\ln\left(2-e^{-1/n}\right)}. \tag{2.2.3}$$

Therefore the operators defined by (2.2.2) take the following alternative form

$$\overline{L}_n(f,x) = \sum_{k=0}^{\infty}\frac{n}{k!2^k}2^{-\frac{x+\ln\left(n(1-e^{-1/n})\right)}{\ln\left(2-e^{-1/n}\right)}}\left(\frac{x+\ln\left(n(1-e^{-1/n})\right)}{\ln\left(2-e^{-1/n}\right)}\right)_k\int_{k/n}^{(k+1)/n} f(t)dt.$$

These operators preserve constant and the function e^{-x}. In order to prove the main results, the following lemmas were proved in [136]:

Lemma 2.8 *The following representation holds*

$$\overline{L}_n(e^{At},x) = \frac{n(e^{A/n}-1)}{A}(2-e^{A/n})^{-na_n(x)}.$$

Lemma 2.9 *If $e_r(t)=t^r, r\in\mathbb{Z}$ (the set of whole numbers), then the moments of the operators (2.2.2) are given as follows:*

$$\begin{aligned}
\overline{L}_n(e_0, x) &= 1,\\
\overline{L}_n(e_1, x) &= a_n(x) + \frac{1}{2n},\\
\overline{L}_n(e_2, x) &= (a_n(x))^2 + \frac{3a_n(x)}{n} + \frac{1}{3n^2},\\
\overline{L}_n(e_3, x) &= (a_n(x))^3 + \frac{15(a_n(x))^2}{2n} + \frac{10a_n(x)}{n^2} + \frac{1}{4n^3},\\
\overline{L}_n(e_4, x) &= (a_n(x))^4 + \frac{14(a_n(x))^3}{n} + \frac{50(a_n(x))^2}{n^2} + \frac{53a_n(x)}{n^3} + \frac{1}{5n^4}.
\end{aligned}$$

Lemma 2.10 *If $\mu_{n,m}(x) = \overline{L}_n\left((t-x)^m, x\right)$, then by using Lemma 2.9, we have*

$$\begin{aligned}
\mu_{n,0}(x) &= 1,\\
\mu_{n,1}(x) &= a_n(x) + \frac{1}{2n} - x,\\
\mu_{n,2}(x) &= (a_n(x) - x)^2 + \frac{3a_n(x)}{n} - \frac{x}{n} + \frac{1}{3n^2},\\
\mu_{n,4}(x) &= (a_n(x) - x)^4 + \frac{14(a_n(x))^3 - 30x(a_n(x))^2 + 18x^2a_n(x) - 2x^3}{n}\\
&\quad + \frac{50(a_n(x))^2 - 40xa_n(x) + 2x^2}{n^2} + \frac{53a_n(x) - x}{n^3} + \frac{1}{5n^4}.
\end{aligned}$$

Furthermore,

$$\lim_{n\to\infty} n\left[\frac{x + \ln\left(n(1 - e^{-1/n})\right)}{n\ln\left(2 - e^{-1/n}\right)} + \frac{1}{2n} - x\right] = x$$

and

$$\lim_{n\to\infty} n\left[\left(\frac{x + \ln\left(n(1 - e^{-1/n})\right)}{n\ln\left(2 - e^{-1/n}\right)} - x\right)^2 + \frac{3\left[x + \ln\left(n(1 - e^{-1/n})\right)\right]}{n^2\ln\left(2 - e^{-1/n}\right)} - \frac{x}{n} + \frac{1}{3n^2}\right] = 2x.$$

In [136] the following quantitative estimate corresponding to Theorem 2.2 was calculated for the operators $\overline{L}_n$.

Theorem 2.19 *For $f \in C^*[0,\infty)$, we have*

$$||\overline{L}_n f - f||_{[0,\infty)} \le 2\omega^*\left(f, \sqrt{\gamma_n}\right),$$

where

$$\gamma_n = ||\overline{L}_n(e^{-2t}) - e^{-2x}||_{[0,\infty)} = \left\|\frac{2xe^{-2x}}{n} + \frac{(24x^2 - 48x - 11)e^{-2x}}{12n^2} + O\left(\frac{1}{n^3}\right)\right\|_{[0,\infty)}.$$

Theorem 2.20 ([136]) *Let $f, f'' \in C^*[0,\infty)$. Then the inequality*

$$\begin{aligned}&\left|n\left[\overline{L}_n(f,x)-f(x)\right]-x[f'(x)+f''(x)]\right|\\ &\leq |p_n(x)||f'|+|q_n(x)||f''|+2\left(2q_n(x)+2x+r_n(x)\right)\omega^*\left(f'',n^{-1/2}\right),\end{aligned}$$

holds for any $x \in [0,\infty)$, where

$$\begin{aligned}p_n(x)&=n\mu_{n,1}(x)-x,\\ q_n(x)&=\frac{1}{2}\left(n\mu_{n,2}(x)-2x\right),\\ r_n(x)&=n^2\sqrt{\overline{L}_n\left((e^{-x}-e^{-t})^4,x\right)}\sqrt{\mu_{n,4}(x)},\end{aligned}$$

and $\mu_{n,1}(x)$, $\mu_{n,2}(x)$, and $\mu_{n,4}(x)$ are given in Lemma 2.10.

Remark 2.3 ([136]) From the Lemma 2.10, $p_n(x) \to 0, q_n(x) \to 0$ as $n \to \infty$ and using Mathematica, we get

$$\lim_{n\to\infty} n^2\mu_{n,4}(x)=12x^2.$$

Furthermore

$$\lim_{n\to\infty} n^2\overline{L}_n\left((e^{-t}-e^{-x})^4,x\right)=12e^{-4x}x^2.$$

Thus in the above Theorem 2.20, convergence occurs for sufficiently large n.

Corollary 2.3 ([136]) *Let $f, f'' \in C^*[0,\infty)$. Then, the inequality*

$$\lim_{n\to\infty} n\left[\overline{L}_n(f,x)-f(x)\right]=x[f'(x)+f''(x)]$$

holds for any $x \in [0,\infty)$.

Remark 2.4 ([136]) If the operators (2.2.2) preserve the function e^{-2x}, then by Lemma 2.8, we have

$$e^{-2x}=\frac{n(1-e^{-2/n})}{2}\left(2-e^{-2/n}\right)^{-na_n(x)},$$

which implies

$$a_n(x)=\frac{2x+\ln\left(\frac{n(1-e^{-2/n})}{2}\right)}{n\ln(2-e^{-2/n})} \tag{2.2.4}$$

Additionally, one has

$$\lim_{n\to\infty} n\left[\frac{2x+\ln\left(\frac{n(1-e^{-2/n})}{2}\right)}{n\ln(2-e^{-2/n})}+\frac{1}{2n}-x\right]=2x$$

and

$$\lim_{n\to\infty} n\left[\left(\frac{2x+\ln\left(\frac{n(1-e^{-2/n})}{2}\right)}{n\ln(2-e^{-2/n})}-x\right)^2+\frac{3(2x+\ln\left(\frac{n(1-e^{-2/n}))}{2}\right)}{n^2\ln(2-e^{-2/n})}-\frac{x}{n}+\frac{1}{3n^2}\right]=2x.$$

Thus, Theorems 2.19, 2.20 and Corollary 2.3 obtain the following forms:

Theorem 2.21 ([136]) *For $f\in C^*[0,\infty)$, in the operators (2.2.2) with $a_n(x)$ given by (2.2.4), we have*

$$||\overline{L}_n f-f||_{[0,\infty)}\le 2\omega^*\left(f,\sqrt{2\beta_n}\right),$$

where

$$\beta_n=||\overline{L}_n(e^{-t})-e^{-x}||_{[0,\infty)}=\left\|\frac{-xe^{-x}}{n}+\frac{(12x^2+24x+11)e^{-x}}{24n^2}+O\left(\frac{1}{n^3}\right)\right\|_{[0,\infty)}.$$

Theorem 2.22 ([136]) *Let $f, f''\in C^*[0,\infty)$. Then in the operators (2.2.2) with $a_n(x)$ given by (2.2.4), the inequality*

$$\begin{aligned}&\left|n\left[\overline{L}_n(f,x)-f(x)\right]-x[2f'(x)+f''(x)]\right|\\&\le|\hat{p}_n(x)||f'|+|\hat{q}_n(x)||f''|+2\left(2\hat{q}_n(x)+2x+\hat{r}_n(x)\right)\omega^*\left(f'',n^{-1/2}\right)\end{aligned}$$

holds for any $x\in[0,\infty)$, where

$$\begin{aligned}\hat{p}_n(x)&=n\mu_{n,1}(x)-x,\\\hat{q}_n(x)&=\frac{1}{2}\left(n\mu_{n,2}(x)-4x\right),\\\hat{r}_n(x)&=n^2\sqrt{\overline{L}_n\left((e^{-x}-e^{-t})^4,x\right)}\sqrt{\mu_{n,4}(x)}.\end{aligned}$$

and $\mu_{n,1}(x)$, $\mu_{n,2}(x)$ and $\mu_{n,4}(x)$ are given in Lemma 2.10, with the values of $a_n(x)$, given by (2.2.4).

Corollary 2.4 ([136]) *Let $f, f'' \in C^*[0,\infty)$. Then, for the operators (2.2.2), with $a_n(x)$ given by (2.2.4) the inequality*

$$\lim_{n\to\infty} n\left[\overline{L}_n(f,x) - f(x)\right] = x[2f'(x) + f''(x)]$$

holds for any $x \in [0,\infty)$.

2.2.3 Modified Szász–Kantorovich Operators

Suppose the Szász–Kantorovich operators defined by (1.2.8) preserve e^{-x}, then the modified form of Szász–Kantorovich operators takes the following form

$$\overline{S}_n^M(f,x) = n\sum_{k=0}^{\infty}\frac{(-1)^k}{k!}e^{\frac{x+\ln\left(n(1-e^{-1/n})\right)}{(e^{-1/n}-1)}}\left(\frac{x+\ln\left(n(1-e^{-1/n})\right)}{\left(e^{-1/n}-1\right)}\right)^k\int_{k/n}^{(k+1)/n} f(t)dt.$$

Gupta and Aral [132] established the following modified form:

Theorem 2.23 *For $f \in C^*[0,\infty)$, we have*

$$||\overline{S}_n^M f - f||_{[0,\infty)} \le 2\omega^*(f,\sqrt{\gamma_n}),$$

where γ_n tends to zero as n goes to infinity, so that $\overline{S}_n^M f$ converges uniformly to f.

Proof The operators $\overline{S}_n^M$ preserve constant as well as e^{-x} thus $\alpha_n = \beta_n = 0$. We only have to evaluate γ_n. We have

$$\begin{aligned}\overline{S}_n^M(e^{-2t},x) &= n\sum_{k=0}^{\infty}\frac{e^{-na_n(x)}(na_n(x))^k}{k!}\int_{k/n}^{(k+1)/n} e^{-2t}dt\\ &= \frac{n(1-e^{-2/n})}{2}e^{na_n(x)[e^{-2/n}-1]},\end{aligned}$$

where $a_n(x)$ is given by

$$a_n(x) = \frac{-x-\ln\left(n(1-e^{-1/n})\right)}{n\left(e^{-1/n}-1\right)}.$$

Thus, using the software Mathematica, we get

$$\overline{S}_n^M(e^{-2t}, x) = \frac{n(1-e^{-2/n})}{2} e^{na_n(x)[e^{-2/n}-1]}$$
$$= \frac{n(1-e^{-2/n})}{2} e^{\left[-x-\ln\left(n(1-e^{-1/n})\right)\right][e^{-1/n}+1]}$$
$$= e^{-2x} + \frac{xe^{-2x}}{n} + \frac{e^{-2x}(6x^2-6x-5)}{n^2} + O(n^{-3}).$$

Since

$$\sup_{x\in[0,\infty)} xe^{-2x} = \frac{1}{2e}, \quad \sup_{x\in[0,\infty)} x^2e^{-2x} = \frac{1}{4e},$$

we get

$$\gamma_n = ||\overline{S}_n^M(e^{-2t}) - e^{-2x}||_{[0,\infty)}$$
$$= \sup_{x\in[0,\infty)} \left|\overline{S}_n^M(e^{-2t}) - e^{-2x}\right| \le \frac{1}{2ne} + \frac{1}{n^2}\left(\frac{9}{2e}+5\right) + O(n^{-3})$$
$$\le O(n^{-1})$$

■

Remark 2.5 ([132]) For the original Szász–Kantorovich operators $\overline{S}_n$ under the conditions of Theorem 2.23, we have

$$||\overline{S}_n f - f||_{[0,\infty)} \le 2\omega^*(f, \sqrt{2\beta_n + \gamma_n}),$$

where

$$\beta_n = \sup_{x\in[0,\infty)} |S_n(e^{-t}) - e^{-x}|$$
$$= \sup_{x\in[0,\infty)} |n(1-e^{-1/n})e^{nx[e^{-1/n}-1]} - e^{-x}|$$
$$= \sup_{x\in[0,\infty)} e^{-x}\left|n(1-e^{-1/n})e^{x\left(n[e^{-1/n}-1]+1\right)} - 1\right|$$
$$\le \left(n(1-e^{-1/n}) - 1\right) \quad (\text{for } x=0).$$

From the mean value theorem:

$$\frac{e^{-1/n}-1}{1/n} = -e^{x_0} \le -1 \Rightarrow n[e^{-1/n}-1]+1 \le 0$$

$$\sup_{x\in[0,\infty)} e^{-x} = 1, \quad \sup_{x\in[0,\infty)} e^{ax} = 1, \ a<0$$

and

$$\begin{aligned}
\gamma_n &= \sup_{x\in[0,\infty)} |\overline{S}_n(e^{-2t}) - e^{-2x}| \\
&= \sup_{x\in[0,\infty)} |\frac{n(1-e^{-2/n})}{2} e^{nx[e^{-2/n}-1]} - e^{-2x}| \\
&= \sup_{x\in[0,\infty)} e^{-2x} \left| n(1-e^{-2/n}) e^{2x(\frac{n}{2}[e^{-2/n}-1]+1)} - 1 \right| \\
&\leq n(1-e^{-2/n}) - 1 \quad (\text{for } x = 0)\,.
\end{aligned}$$

From the mean value theorem

$$\frac{e^{-2/n}-1}{2/n} = -e^{x_0} \leq -1 \Rightarrow \frac{n}{2}[e^{-2/n}-1]+1 \leq 0$$

$$\sup_{x\in[0,\infty)} e^{-2x} = 1, \quad \sup_{x\in[0,\infty)} e^{2ax} = 1, \ \ a < 0$$

Thus Theorem 2.23 provides better approximation for the modified Szász–Kantorovich operators $\overline{S}_n^M$, than the usual Szász–Kantorovich operators S_n.

Along with the above result in [132], Gupta and Aral also estimated the quantitative asymptotic formula.

2.2.4 Modified Szász–Durrmeyer Operators

The modified form of Szász–Durrmeyer operators (1.3.1) for $n > 2a$, preserving the function e^{2ax}, $a > 0$, was considered in [76]. Set

$$\widetilde{S}_n^* (f; x) = n \sum_{k=0}^{\infty} e^{-n\alpha_n(x)} \frac{(n\alpha_n(x))^k}{k!} \int_0^{\infty} e^{-nt} \frac{(nt)^k}{k!} f(t)\, dt, \tag{2.2.5}$$

i.e., $\widetilde{S}_n^* \left(e^{2at}; x\right) = e^{2ax}$. Thus by simple computation, we immediately get

$$e^{2ax} = e^{\frac{2an\alpha_n(x)}{n-2a}} \frac{n}{n-2a}.$$

implying

$$\alpha_n(x) = \frac{n-2a}{2an}\left[2ax - \ln\left(\frac{n}{n-2a}\right)\right], n > 2a. \tag{2.2.6}$$

By direct estimates Deniz et al. proved in [76] the following:

Theorem 2.24 *For each function $f \in C^*[0,\infty)$ and $n > 2a$, the following relation holds true*

$$\left\|\widetilde{S}_n^* f - f\right\|_{[0,\infty)} \leq 2\omega^*\left(f; \sqrt{2\beta_n + \gamma_n}\right),$$

where

$$\beta_n = \left\|\widetilde{S}_n^*\left(e^{-t};x\right) - e^{-x}\right\|_{[0,\infty)},$$
$$\gamma_n = \left\|\widetilde{S}_n^*\left(e^{-2t};x\right) - e^{-2x}\right\|_{[0,\infty)}.$$

Here, β_n and γ_n tend to zero as $n \to \infty$.

Proof Using the inequality

$$\frac{u-v}{\ln u - \ln v} < \frac{u+v}{2} \text{ for } 0 < v < u,$$

and consider $u = e^{-xu_n}$ and $v = e^{-x}$, we have

$$e^{-xu_n} - e^{-x} < \frac{1-u_n}{2}\left(xe^{-xu_n} + xe^{-x}\right).$$

Moreover,

$$\max_{x>0} xe^{-bx} = \frac{1}{eb} \text{ for every } b > 0.$$

Thus, we obtain

$$e^{-xu_n} - e^{-x} < \frac{1-u_n}{2}\left(\frac{1}{eu_n} + \frac{1}{e}\right) = \frac{1-u_n^2}{2eu_n}. \tag{2.2.7}$$

Similarly we obtain

$$e^{-xv_n} - e^{-2x} < \frac{4-v_n^2}{4ev_n}. \tag{2.2.8}$$

On the other hand, by simple calculation we obtain

$$\widetilde{S}_n^*\left(e^{-t};x\right) = e^{-\frac{n\alpha_n(x)}{n+1}}\frac{n}{n+1}$$
$$= e^{-\frac{(n-2)\left[2x-\ln\left(\frac{n}{n-2}\right)\right]}{2(n+1)}}\frac{n}{n+1}$$

$$= a_n e^{-\frac{x(n-2)}{(n+1)}},$$

$$\widetilde{S}_n^*\left(e^{-2t}; x\right) = e^{-\frac{2n\alpha_n(x)}{n+2}} \frac{n}{n+2}$$

$$= e^{-\frac{(n-2)\left[2x-\ln\left(\frac{n}{n-2}\right)\right]}{n+2}} \frac{n}{n+2}$$

$$= b_n e^{-\frac{x(2n-4)}{n+2}},$$

where

$$a_n = \left(1+\frac{2}{n-2}\right)^{\frac{n-2}{2(n+1)}} \frac{n}{n+1} \text{ and } b_n = \left(1+\frac{2}{n-2}\right)^{\frac{n-2}{n+2}} \frac{n}{n+2}.$$

In the inequalities (2.2.7) and (2.2.8), choosing $u_n = \frac{(n-2)}{(n+1)}$ and $v_n = \frac{(2n-4)}{n+2}$ we get

$$\left\|\widetilde{S}_n^*\left(e^{-t}; x\right) - e^{-x}\right\|_{[0,\infty)} = \left\|a_n\left(e^{-\frac{x(n-2)}{(n+1)}} - e^{-x}\right) + e^{-x}\left(a_n - 1\right)\right\|_{[0,\infty)}$$

$$= \beta_n < a_n \frac{1-u_n^2}{2eu_n} + (a_n - 1) \to 0,$$

$$\left\|\widetilde{S}_n^*\left(e^{-2t}; x\right) - e^{-2x}\right\|_{[0,\infty)} = \left\|b_n\left(e^{-\frac{x(2n-4)}{n+2}} - e^{-2x}\right) + e^{-2x}\left(b_n - 1\right)\right\|_{[0,\infty)}$$

$$= \gamma_n < b_n \frac{4-v_n^2}{4ev_n} + (b_n - 1) \to 0$$

as $n \to \infty$. This completes the proof of the theorem. ■

2.2.5 *Modifications of Phillips Operators*

The well-known Phillips operators (see [89, 196]) are defined as

$$P_n(f; x) = n \sum_{k=1}^{\infty} s_{n,k}(x) \int_0^{\infty} s_{n,k-1}(t) f(t)\, dt + e^{-nx} f(0),$$

where $s_{n,k}(x) = e^{-nx}\frac{(nx)^k}{k!}$. These operators preserve constant as well as linear functions. Very recently Gupta and Tachev in [131] considered a modification of the Phillips operators P_n, which preserve the test function e^{-x}. The modified form takes the following form

$$H_n(f;x) = n\sum_{k=1}^{\infty} s_{n+1,k}(x)\int_0^{\infty} s_{n,k-1}(t)f(t)\,dt + e^{-x(n+1)}f(0) \tag{2.2.9}$$

Below, we mention some of the results, discussed in [131].

Theorem 2.25 ([131]) *For $f \in C^*[0,\infty)$, we have*

$$||H_n f - f||_{[0,\infty)} \leq 2\omega^*(f, \sqrt{\gamma_n}),$$

where

$$\gamma_n = ||H_n(e^{-2t}) - e^{-2x}||_{[0,\infty)} = \left(1 - \frac{1}{n+2}\right)^{n+2}\frac{1}{n+1}.$$

Remark 2.6 For the original Phillips operators under the conditions of Theorem 5.34, we have

$$||P_n f - f||_{[0,\infty)} \leq 2\omega^*(f, \sqrt{2\beta_n + \gamma_n}),$$

where

$$\beta_n = ||P_n(e^{-t}) - e^{-x}||_{[0,\infty)} = \left(1 - \frac{1}{n+1}\right)^{n+1}\frac{1}{n}$$

and

$$\gamma_n = ||P_n(e^{-2t}) - e^{-2x}||_{[0,\infty)} = \left(1 - \frac{2}{n+2}\right)^{(n+2)/2}\frac{2}{n}.$$

Thus Theorem 2.25 provides better approximation for the modified Phillips operators H_n, than the usual Phillips operators P_n.

Theorem 2.26 ([131]) *For all $f \in C[0,\infty)$ the following holds true:*

$$H_n(f(t);x) = P_{n+1}\left(f\left(\frac{(n+1)t}{n}\right);x\right),$$

where P_n is the Phillips operators (see [196]).

Theorem 2.27 ([131]) *For every $f \in C_B[0,\infty)$ and $n > 0$ the following inequality holds true:*

$$||H_n f - f||_{[0,\infty)} \leq C\omega_\varphi^2(f; n^{-1/2}) + \omega(f; n^{-1}) \tag{2.2.10}$$

where the Ditzian–Totik modulus of continuity $\omega_\varphi^2(f;\delta)$ is defined in [140, (27),(26)].

The following asymptotic-type result was also established by Gupta and Tachev in [131].

Theorem 2.28 ([131]) *Let $f, f'' \in C^*[0,\infty)$, then for any $x \in [0,\infty)$, we have*

$$\begin{aligned}&\left|n[H_n(f,x)-f(x)]-x[f'(x)+f''(x)]\right|\\ \le\ & \frac{x(x+2)}{2n}.|f''(x)|+2\left[\frac{x^2}{n}+2x+\frac{2x}{n}+r_n(x)\right].\omega^*(f''(x),n^{-1/2}),\end{aligned}$$

where

$$r_n(x)=n^2[H_n((e^{-x}-e^{-t})^4,x).\mu_{n,4}(x)]^{1/2},$$

and $\mu_{n,4}(x)$ is the fourth order central moment of H_n.

Also, in [131] Gupta and Tachev considered another modification of Phillips operator such that the exponential function e^{At}, $A \in \mathbb{R}$, is reproduced. In that case the modified Phillips operators takes the following form

$$H_n^A(f;x)=n\sum_{k=1}^{\infty}s_{n-A,k}(x)\int_0^{\infty}s_{n,k-1}(x)f(t)\,dt+e^{-x(n-A)}f(0).$$

The main result from [224] states the following:

Theorem 2.29 *Let E be a subspace of $C[0,\infty)$ which contains the polynomials and suppose $L_n : E \to C[0,\infty)$ be sequence of linear positive operators, preserving the linear functions. We suppose that for each constant $A > 0$ and fixed $x \in [0,\infty)$ the operators L_n satisfy*

$$(i)\ L_n\left((t-x)^2e^{At};x\right)\le C(A,x)\cdot\mu_{n,2}^{L_n}(x).$$

If in addition $f \in C^2[0,\infty)\cap E$ and $f'' \in Lip(\alpha, A)$, $0<\alpha\le 1$ then we have for $x \in [0,\infty)$

$$\begin{aligned}&\left|L_n(f,x)-f(x)-\frac{1}{2}f''(x)\mu_{n,2}^{L_n}(x)\right|\\ \le\ & [e^{Ax}+\frac{C(A,x)}{2}+\frac{\sqrt{C(2A,x)}}{2}]\cdot\mu_{n,2}^{L_n}(x)\cdot\omega_1\left(f'',\sqrt{\frac{\mu_{n,4}^{L_n}(x)}{\mu_{n,2}^{L_n}(x)}},A\right).\quad(2.2.11)\end{aligned}$$

For this form, the following result as an application of Theorem 2.29 was established in [131].

Theorem 2.30 *If* $f \in E := \{f \in C[0,\infty); ||f||_A < \infty, f \in C^2[0,\infty) \bigcap E$ *and* $f'' \in Lip(\alpha, A), 0 < \alpha \leq 1$, *then for* $n > 2A$, $x \in [0,\infty)$ *we have*

$$\left| H_n^A(f,x) - f(x) + \frac{Ax}{n} f'(x) - \left(\frac{x^2A^2}{n^2} - \frac{2xA}{n^2} + \frac{2x}{n} \right) \frac{1}{2} f''(x) \right|$$

$$\leq [e^{Ax} + \frac{C(A,x)}{2} + \frac{\sqrt{C(2A,x)}}{2}] \cdot \mu_{n,2}^{H_n^A}(x) \cdot \omega_1 \left(f'', \sqrt{\frac{\mu_{n,4}^{P_n^A}(x)}{\mu_{n,2}^{H_n^A}(x)}}, A \right),$$

where $\mu_{n,r}^{H_n^A}(x)$ *denotes the* r*-th central moment of the operators* H_n^A.

2.3 Better Approximation by Certain Positive Linear Operators

In this section we deal with the modified positive linear operators that present a better degree of approximation than the original ones. Starting from the classical Bernstein operators B_n, King [150] proposed the following sequence of positive linear operators defined for $f \in C[0,1]$ as $f \to (B_n f) \circ r_n$, where r_n is a sequence of continuous functions defined on $[0,1]$ with $0 \leq r_n(x) \leq 1$ for each $x \in [0,1]$ and $n \in \{1, 2, \dots\}$. The modified Bernstein operators hold fixed the functions e_0 and e_2 and approximate each continuous function on $[0,1]$ with an order of approximation at least as good as that of classical Bernstein operators for a certain subinterval of $[0,1]$. Using the same type of technique introduced by King or more recent methods many authors published new results dealing with this subject (cf. [4, 109]).

2.3.1 *Bernstein Type Operators*

Let τ be a continuous strictly increasing function defined on $[0,1]$ with $\tau(0) = 0$ and $\tau(1) = 1$. Gonska et al. [109] studied the sequence $V_n^\tau : C[0,1] \to C[0,1]$ defined by

$$V_n^\tau f := (B_n f) \circ (B_n \tau)^{-1} \circ \tau,$$

where B_n are the Bernstein operators. Note that the operators V_n^τ preserve e_0 and τ. In [109] these operators were studied with respect to uniform convergence, global smoothness preservation, the approximation of decreasing, and convex functions. In [69], the authors considered the sequence of linear Bernstein-type operators defined for $f \in C[0,1]$ by $B_n^\tau f := B_n(f \circ \tau^{-1}) \circ \tau$, τ as being any function that is continuously differentiable infinitely many times on [0, 1], such that $\tau(0) = 0$, $\tau(1) = 1$ and $\tau'(x) > 0$ for $x \in [0, 1]$.

The next result gives some basic properties of these operators which can be obtained from the properties of B_n.

Lemma 2.11 ([69]) *For the modified Bernstein operators it holds*

$$B_n^\tau e_0 = e_0,\ B_n^\tau \tau = \tau,\ B_n^\tau \tau^2 = \left(1 - \frac{1}{n}\right)\tau^2 + \frac{\tau}{n}.$$

The following propositions provide some aspects of the asymptotic behavior, monotonic convergence, and saturation of the sequence B_n^τ. An important role is played by the notion of convexity with respect to a function.

Definition 2.1 A function $f \in C[0, 1]$ is convex with respect to τ if

$$\begin{vmatrix} 1 & 1 & 1 \\ \tau(x_0) & \tau(x_1) & \tau(x_2) \\ f(x_0) & f(x_1) & f(x_2) \end{vmatrix} \geq 0,\ \ 0 \leq x_0 < x_1 < x_2 \leq 1.$$

Proposition 2.1 ([69]) *Suppose that we have* $f \in C[0, 1]$ *and* $x \in (0, 1)$ *such that* $f''(x)$ *exists. Then*

$$\lim_{n\to\infty} 2n(B_n^\tau(f; x) - f(x)) = \tau(x)(1 - \tau(x))\left(-\frac{\tau''(x)f'(x)}{\tau'(x)^3} + \frac{f''(x)}{\tau'(x)^2}\right).$$

Proposition 2.2 ([69]) *Suppose that* $f \in C[0, 1]$. *Then the following items are equivalent:*

i) f *is convex with respect to* τ;
ii) $B_n^\tau f \geq f$ *for* $n \in \mathbb{N}$;
iii) $B_n^\tau f \geq B_{n+1}^\tau f$ *for* $n \in \mathbb{N}$.

In [69] the operators V_n^τ and B_n^τ are compared with each other and with the Bernstein operators B_n. In the following we present some of these results.

Theorem 2.31 ([69]) *Let* $f \in C[0, 1]$ *be increasing and convex with respect to* τ. *Assume also that* τ *is convex. Then*

$$f(x) \leq V_n^\tau(f; x) \leq B_n(f; x),\ \ 0 \leq x \leq 1.$$

Theorem 2.32 ([69]) *Let* $f \in C^2[0, 1]$. *Suppose that there exists* $n_0 \in \mathbb{N}$ *such that*

$$f(x) \leq B_n^\tau(f; x) \leq B_n(f; x),\ \textit{for all } n \geq n_0,\ x \in (0, 1).$$

Then

$$f''(x) \geq \frac{\tau''(x)}{\tau'(x)} f'(x) \geq \left(1 - \frac{x(1-x)\tau'(x)^2}{\tau(x)(1-\tau(x))}\right) f''(x), \quad x \in (0,1). \tag{2.3.1}$$

In particular, $f''(x) \geq 0$.

Conversely, if (2.3.1) holds with strict inequalities at a given point $x_0 \in (0,1)$, *then there exists* $n_0 \in \mathbb{N}$ *such that for* $n \geq n_0$

$$f(x_0) < B_n^\tau(f; x_0) < B_n(f; x_0).$$

Theorem 2.33 ([69]) *Let* $f \in C^2[0,1]$. *Suppose that there exists* $n_0 \in \mathbb{N}$ *such that*

$$V_n^\tau(f; x) \leq B_n^\tau(f; x), \ n \geq n_0, \ x \in (0,1).$$

Then for all $x \in (0,1)$ *the following inequality holds:*

$$\left(1 - \frac{x(1-x)\tau'(x)^2}{\tau(x)(1-\tau(x))}\right) f''(x) \leq \left(\frac{\tau''(x)}{\tau'(x)} - \frac{\tau'(x)\tau''(x)x(1-x)}{\tau(x)(1-\tau(x))}\right) f'(x). \tag{2.3.2}$$

Conversely, if (2.3.2) is satisfied with strict inequalities at a given point $x_0 \in (0,1)$, *then there exists* $n_0 \in \mathbb{N}$ *such that for* $n \geq n_0$

$$V_n^\tau(f; x_0) < B_n^\tau(f; x_0).$$

In [69], the authors considered a special case of function τ and proved that the modified Bernstein operators represent a good shape preserving approximation process making a comparison with Bernstein operators.

Corollary 2.5 ([69]) *Let* $f \in C^2[0,1]$ *be increasing and strictly convex. Then there exists* $\alpha > 0$ *such that the operators* B_n^τ *and* V_n^τ *for the function*

$$\tau(x) = \frac{x^2 + \alpha x}{1 + \alpha}$$

satisfy the following properties:

i) For each $x \in [0,1]$, $f(x) \leq V_n^\tau(f; x) \leq B_n(f; x)$.
ii) For each $x \in (0,1)$ *there exists* $n_0 \in \mathbb{N}$ *such that*

$$V_n^\tau(f; x) < B_n^\tau(f; x), \ n \geq n_0.$$

iii) For each $x \in \left(\frac{1 - 2\alpha + \sqrt{4\alpha^2 + 8\alpha + 1}}{6}, 1\right]$, *there exists* $n_0 \in \mathbb{N}$ *such that* $f(x) < B_n^\tau(f; x) < B_n(f; x), \ n \geq n_0$.

2.3.2 Bernstein–Durrmeyer Type Operators

Recently Aral–Acar [35, pp. 1–15] considered a generalization of the Bernstein–Durrmeyer operators. Also, Acar et al. in [4] introduced another new type of Bernstein–Durrmeyer operators based on a function τ which is continuously differentiable infinitely many times on the interval $[0, 1]$, such that $\tau(0) = 0$, $\tau(1) = 1$ and $\tau'(x) > 0$ for $x \in [0, 1]$. The modified Bernstein–Durrmeyer operators were introduced as follows:

$$\tilde{B}_n^{\tau}(f;x) = (n+1)\sum_{k=0}^{n} p_{n,k}^{\tau}(x)\int_0^1 \left(f\circ\tau^{-1}\right)(t)\,p_{n,k}(t)\,dt,$$

where

$$p_{n,k}^{\tau}(x) := \binom{n}{k}\tau^k(x)\,(1-\tau(x))^{n-k}$$

and

$$p_{n,k}(x) := \binom{n}{k}x^k(1-x)^{n-k}.$$

The following results are similar to the corresponding results for the Bernstein–Durrmeyer operators and can be verified by taking $\tau = e_1$.

Lemma 2.12 ([4]) *The modified Bernstein–Durrmeyer operators satisfy*

$$\tilde{B}_n^{\tau}e_0 = e_0,\quad \tilde{B}_n^{\tau}\tau = \frac{1+\tau n}{n+2},\quad \tilde{B}_n^{\tau}\tau^2 = \frac{\tau^2 n(n-1)+4n\tau+2}{(n+2)(n+3)}.$$

Lemma 2.13 ([4]) *If we define the central moment operator by*

$$\begin{aligned}\mu_{n,m}^{\tau}(x) &= \tilde{B}_n^{\tau}\left((\tau(t)-\tau(x))^m;x\right)\\ &= (n+1)\sum_{k=0}^{n} p_{n,k}^{\tau}(x)\int_0^1 (t-\tau(x))^m\,p_{n,k}(t)\,dt,\ m\in\mathbb{N},\end{aligned}$$

then we have

$$\mu_{n,0}^{\tau}(x) = 1,\ \mu_{n,1}^{\tau}(x) = \frac{1-2\tau(x)}{n+2}$$

$$\mu_{n,2}^{\tau}(x) = \frac{\tau(x)(1-\tau(x))(2n-6)+2}{(n+2)(n+3)}.$$

for all $n, m \in \mathbb{N}$.

Let $L_n : C[0,1] \to C[0,1]$, $n \geq 1$, be positive linear operators and $L_n e_0 = e_0$. Consider the operator $K_n : C[0,1] \to C[0,1]$ defined as follows

$$K_n g := \left(L_n\left(g \circ \tau^{-1}\right)\right) \circ \tau, \quad n \geq 1,$$

where $\tau \in C^2[0,1]$ such that $\tau(1) = 1$, $\tau'(x) > 0$, $x \in [0,1]$ and $g \in C[0,1]$. It is obvious that K_n are linear positive operators and $K_n e_0 = e_0$.

A Voronovskaja type asymptotic formula for the modified operators K_n was proved in [4].

Theorem 2.34 ([4]) *Let $f \in C[0,1]$ with $f''(x)$ finite for $x \in [0,1]$. If there exist $\alpha, \beta \in C[0,1]$ such that*

$$\lim_{n\to\infty} n\left(L_n(f,x) - f(x)\right) = \alpha(x) f''(x) + \beta(x) f'(x),$$

then we have

$$\lim_{n\to\infty} n\left(K_n(g,t) - g(t)\right) = \frac{\alpha(\tau(t))}{\tau'(t)^2} g''(t) + \left(\frac{\beta(\tau(t))}{\tau'(t)} - \frac{\alpha(\tau(t))\tau''(t)}{\tau'(t)^3}\right) g'(t)$$

for $g \in C[0,1]$ with $g''(x)$ finite for $x \in [0,1]$.

Using Theorem 2.34, the following Voronovskaja type asymptotic formula for the modified Bernstein–Durrmeyer operators can be obtained:

Theorem 2.35 ([4]) *If $f \in C^2[0,1]$, then*

$$\lim_{n\to\infty} n\left[\tilde{B}_n^{\tau}(f;x) - f(x)\right] = \frac{\alpha(\tau(x))}{(\tau'(x))^2} f''(x) + \left(\frac{\beta(\tau(x))}{\tau'(x)} - \frac{\alpha(\tau(x))\tau''(x)}{(\tau'(x))^3}\right) f'(x)$$

uniformly on $[0,1]$, with α and β defined above.

In [4], the authors study local approximation properties of $\tilde{B}_n^{\tau}$ in quantitative form using an appropriate K-functional. For $\eta > 0$ and

$$\mathcal{W}^2 = \left\{g \in C[0,1] : g', g'' \in C[0,1]\right\},$$

the Peetre's K-functional [195] is defined by

$$K(f,\eta) = \inf_{g\in\mathcal{W}^2} \left\{\|f-g\| + \eta \|g\|_{\mathcal{W}^2}\right\}, \tag{2.3.3}$$

where

$$\|f\|_{\mathcal{W}^2} = \|f\| + \left\|f'\right\| + \left\|f''\right\|.$$

Theorem 2.36 ([4]) *For the operator $\tilde{B}_n^\tau f$, there exist absolute constants $C, C_f > 0$ (C is independent of f and n, C_f is depend only on f) such that*

$$\left|\tilde{B}_n^\tau (f; x) - f(x)\right| \leq CK\left(f, \frac{\delta_{n,\tau}^2(x)}{(n+2)}\right) + \omega\left(f; \frac{|1-2\tau(x)|}{(n+2)a}\right),$$

where $\inf\limits_{x\in[0,1]} \tau'(x) \geq a,\ a \in \mathbb{R}^+$.

2.3.3 Bézier Variant of the Bernstein–Durrmeyer Type Operators

In the last two decades Bézier variants of several operators have been introduced and the rate of convergence has been established. We mention here some of the papers in this direction due to Gupta [116], Srivastava–Gupta [219], Abel–Gupta [1], Zeng–Gupta [229], etc. For any function τ being infinitely many times continuously differentiable on [0, 1], such that $\tau(0) = 0$, $\tau(1) = 1$ and $\tau'(x) > 0$ for $x \in [0, 1]$, Acar et al. [6] defined the Bézier-variant of the Bernstein–Durrmeyer operators as follows:

$$\tilde{B}_n^{\tau,\theta}(f; x) = (n+1)\sum_{k=0}^{n} Q_{n,k}^{\tau,\theta}(x)\int_0^1 (f\circ\tau^{-1})(t)p_{n,k}(t)dt,$$

where

$$Q_{n,k}^{\tau,\theta}(x) = \left[I_{n,k}^\tau(x)\right]^\theta - \left[I_{n,k+1}^\tau(x)\right]^\theta,\quad \theta \geq 1$$

with $I_{n,k}^\tau(x) = \sum\limits_{j=k}^{n} p_{n,k}^\tau(x)$, when $k \leq n$ and 0 otherwise.

The degree of approximation in terms of the modulus of continuity was studied in [6]:

Theorem 2.37 ([6]) *For $f \in C[0, 1]$ and $x \in [0, 1]$, there holds*

$$\left|\tilde{B}_n^{\tau,\theta}(f; x) - f(x)\right| \leq \left\{1 + \sqrt{2\theta\left(\varphi_\tau^2(x) + \frac{1}{n+3}\right)}\right\}\omega\left(f; \sqrt{\frac{1}{n}}\right),$$

where $\varphi_\tau^2(x) := \tau(x)(1-\tau(x))$, $x \in [0, 1]$ and $\omega(f; \delta)$ is the usual modulus of continuity.

In [6], Acar et al. proved a global approximation theorem and a quantitative Voronovskaja type theorem for the operator $\tilde{B}_n^{\tau,\theta}$ using the first order Ditzian–Totik modulus of smoothness which is given for $f \in C[0, 1]$ by

$$\omega_{\varphi_\tau}(f;t) = \sup_{0<h\leq t}\left\{\left|f\left(x+\frac{h\varphi_\tau(x)}{2}\right)-f\left(x-\frac{h\varphi_\tau(x)}{2}\right)\right|, x\pm\frac{h\varphi_\tau(x)}{2}\in[0,1]\right\}, \tag{2.3.4}$$

where $\varphi_\tau(x) := \sqrt{\tau(x)(1-\tau(x))}$.

Further, the corresponding K-functional to (2.3.4) is defined by

$$K_{\varphi_\tau}(f;t) = \inf_{g\in W_{\varphi_\tau}[0,1]}\{||f-g|| + t||\varphi_\tau g'||\}\ (t>0),$$

where

$$W_{\varphi_\tau}[0,1] = \{g : g \in AC_{loc}[0,1], \|\varphi_\tau g'\| < \infty\}$$

and $g \in AC_{loc}[0, 1]$ means that g is absolutely continuous on every interval $[a, b] \subset (0, 1)$.

Theorem 2.38 ([6]) *Let $f \in C[0, 1]$. Then for every $x \in (0, 1)$, we have*

$$\left|\tilde{B}_n^{\tau,\theta}(f;x) - f(x)\right| \leq C(\theta)\omega_{\varphi_\tau}\left(f;\frac{1}{a}\sqrt{\frac{\theta}{n+2}\left(1+\frac{1}{(n+3)\varphi_\tau^2(x)}\right)}\right),$$

where $C(\theta)$ is a positive constant and $\inf_{x\in[0,1]}\tau'(x) \geq a$, $a \in \mathbb{R}^+$.

Theorem 2.39 ([6]) *For any $f \in C^2[0, 1]$ and $x \in [0, 1]$ the following inequalities hold*

$$\left|\sqrt{n}\left(\tilde{B}_n^{\tau,\theta}(f;x) - f(x)\right)\right| \leq \sqrt{2\theta\left\{\varphi_\tau^2(x)+\frac{1}{n+3}\right\}}\|(f\circ\tau^{-1})'\|$$

$$+\|(f\circ\tau^{-1})''\|\frac{\theta}{\sqrt{n}}\varphi_\tau^2(x)+\frac{C}{\sqrt{n}}\omega_{\varphi_\tau}\left((f\circ\tau^{-1})'';\frac{2\sqrt{6}}{an^{1/2}}\varphi_\tau(x)\right)+o(n^{-1});$$

$$\left|\sqrt{n}\left(\tilde{B}_n^{\tau,\theta}(f;x) - f(x)\right)\right| \leq \sqrt{2\theta\left\{\varphi_\tau^2(x)+\frac{1}{n+3}\right\}}\|(f\circ\tau^{-1})'\|$$

$$+\|(f\circ\tau^{-1})''\|\frac{\theta}{\sqrt{n}}\varphi_\tau^2(x)+\frac{C}{\sqrt{n}}\varphi_\tau(x)\omega_{\varphi_\tau}\left((f\circ\tau^{-1})'';\frac{2\sqrt{6}}{an^{1/2}}\right)+o(n^{-1}),$$

where C is a constant depending on θ.

2.3.4 *Bernstein–Stancu Type Operators*

In 1968, Stancu [221] proposed the sequence of positive linear operators $S_n^{<\alpha>}$: $C[0,1] \to C[0,1]$, depending on a nonnegative parameter α given by

$$S_n^{<\alpha>}(f;x) = \sum_{k=0}^{n} f\left(\frac{k}{n}\right) p_{n,k}^{<\alpha>}(x),\ x \in [0,1] \tag{2.3.5}$$

where

$$p_{n,k}^{<\alpha>}(x) = \binom{n}{k} \frac{x^{[k,-\alpha]}(1-x)^{[n-k,-\alpha]}}{1^{[n,-\alpha]}}$$

and

$$t^{[n,h]} := t(t-h)\cdots(t-\overline{n-1}h)$$

is the nth factorial power of t with increment h.

The following form of Bernstein operators using the divided difference is well known

$$B_n(f;x) = \sum_{k=0}^{n} \frac{k!}{n^k}\binom{n}{k}\left[0, \frac{1}{n}, \ldots, \frac{k}{n}; f\right] x^k. \tag{2.3.6}$$

Starting with the form (2.3.6) of the Bernstein operators, the following Stancu type operators are constructed in [73, 74]:

$$C_n : C[0,1] \to \Pi_n$$

$$C_n(f;x) = \sum_{k=0}^{n} \frac{k!}{n^k}\binom{n}{k} \mathbf{m}_{k,n}\left[0, \frac{1}{n}, \ldots, \frac{k}{n}; f\right] x^k, \quad f \in C[0,1], \tag{2.3.7}$$

where the real numbers $\left(m_{k,n}\right)_{k=0}^{\infty}$ are selected in order to preserve some important properties of Bernstein operators and Π_n is the linear space of all real polynomials of degree $\leq n$.

Let $\mathbf{m}_{0,n} = 1$, $\lim_{n\to\infty} \mathbf{m}_{1,n} = 1$ and $\mathbf{m}_{k,n} = \frac{(a_n)_k}{k!}$, $a_n \in (0,1]$.

For this special case of real sequence $(m_{k,n})_{k=0}^{\infty}$, the Bernstein–Stancu operators C_n were written in the Bernstein basis as follows (see [74], Theorem 10):

$$C_n(f;x) = \sum_{k=0}^{n} p_{n,k}(x) C_{k,n}[f], \tag{2.3.8}$$

where

$$C_{k,n}[f] = \frac{1}{k!}\sum_{j=0}^{k}\binom{k}{j} f\left(\frac{j}{n}\right)(a_n)_j(1-a_n)_{k-j}.$$

We remark that $a_n \in (0, 1]$ leads to C_n linear positive operators.

The coefficients $C_{k,n}[f]$ can be written in the form

$$C_{k,n}[f] = \sum_{j=0}^{k} p_{k,j}^{<1>}(a_n) f\left(\frac{j}{n}\right).$$

Therefore,

$$C_{k,n}[f] = S_k^{<1>}(\tilde{f}; a_n), \text{ where } \tilde{f}(t) = f\left(t\frac{k}{n}\right).$$

In [153] the modified Bernstein–Stancu operators were introduced as follows:

$$C_n^{\tau}(f;x) = \sum_{k=0}^{n} p_{n,k}^{\tau}(x)\sum_{j=0}^{k} p_{k,j}^{<1>}(a_n)\left(f\circ\tau^{-1}\right)\left(\frac{j}{n}\right), \tag{2.3.9}$$

where

$$p_{n,k}^{\tau}(x) = \binom{n}{k}\tau(x)^k\left(1-\tau(x)\right)^{n-k}$$

and τ is any function that is continuously differentiable infinitely many times on $[0, 1]$, such that $\tau(0) = 0$, $\tau(1) = 1$, and $\tau'(x) > 0$ for $x \in [0, 1]$.

Note that these operators are positive and linear and for the case $\tau(x) = x$, these operators (2.3.9) reduce to the Bernstein–Stancu operators defined by Cleciu [73, 74].

Lemma 2.14 ([153]) *The modified operators C_n^{τ} verify*

i) $C_n^{\tau} e_0 = 1$,
ii) $C_n^{\tau}\tau = a_n\tau$,
iii) $C_n^{\tau}\tau^2 = \tau^2 + \dfrac{\tau(1-\tau)}{n}a_n + \dfrac{1-a_n}{2}\left(\dfrac{a_n}{n} - (2+a_n)\right)\tau^2.$

Let

$$\begin{aligned}\mu_{n,m}^{\tau}(x) &= C_n^{\tau}\left((\tau(t)-\tau(x))^m; x\right)\\ &= \sum_{k=0}^{n} p_{n,k}^{\tau}(x)\sum_{j=0}^{k} p_{k,j}^{<1>}(a_n)\left(\frac{j}{n}-\tau(x)\right)^m, \quad n, m \in \mathbb{N}.\end{aligned}$$

be the central moment operators.

Lemma 2.15 ([153]) *The central moment operators verify*

i) $\mu_{n,0}^{\tau}(x) = 1,$
ii) $\mu_{n,1}^{\tau}(x) = (a_n - 1)\tau(x),$
iii) $\mu_{n,2}^{\tau}(x) = \dfrac{\tau(x)(1-\tau(x))}{n} a_n + \tau(x)^2(1-a_n)\left(\dfrac{2-a_n}{2} + \dfrac{a_n}{2n}\right),$
iv) $\mu_{n,4}(x) = \tau(x)^4 + \left[\dfrac{6(a_n)_3}{n^4}\binom{n}{3} - \dfrac{12(a_n)_2}{n^3}\binom{n}{2} + \dfrac{6a_n}{n}\right]\tau(x)^3$
$+\left[\dfrac{7(a_n)_2}{n^4}\binom{n}{2} - \dfrac{4a_n}{n^2}\right]\tau(x)^2 + \dfrac{a_n}{n^3}\tau(x) + \dfrac{(a_n)_4}{n^4}\binom{n}{4}$
$-\dfrac{4(a_n)_3}{n^3}\binom{n}{3} + \dfrac{6(a_n)_2}{n^2}\binom{n}{2} - 4a_n.$

Lemma 2.16 ([153]) *For all $n \in \mathbb{N}$ we have*

$$\mu_{n,2}^{\tau}(x) \le \delta_{n,\tau}^2(x), \text{ for all } x \in [0,1],$$

where

$$\delta_{n,\tau}^2(x) := \frac{a_n}{n}\varphi_\tau^2(x) + (1-a_n)\,.$$

$$\varphi_\tau^2(x) := \tau(x)(1-\tau(x)).$$

Theorem 2.40 ([153]) *Let $f \in C[0,1]$, $a_n \in (0,1]$ and $\lim_{n\to\infty} a_n = 1$, then $C_n^{\tau} f$ converges to f as n tends to infinity, uniformly on* $[0,1]$.

Proposition 2.3 ([153]) *Let $f \in C[0,1]$ with modulus of continuity $\omega(f,\cdot)$. Then*

$$\left|C_n^{\tau}(f;x) - f(x)\right| \le \left(1 + \frac{\mu_{n,2}^{\tau}(x)}{\delta^2}\right)\omega(f,\delta),$$

for $\delta > 0$ and $x \in [0,1]$.

Example 2.1 ([153]) If we choose $\tau(x) = \dfrac{x^2+x}{2}$, we have

$$\tau(x)(1-\tau(x)) \le x(1-x), \quad \text{for all } x \in [0,1/2]$$

and this inequality leads to $\mu_{n,2}^{\tau}(x) \le \mu_{n,2}(x)$. Therefore, the modified operators C_n^{τ} present an order of approximation better than C_n in that interval.

Example 2.2 ([153]) Now using a graphical example we try to illustrate these approximation processes. Let

$$f(x) = \sin(9x), \ \tau(x) = \frac{x^2+x}{2} \text{ and } a_n = 1/2\,.$$

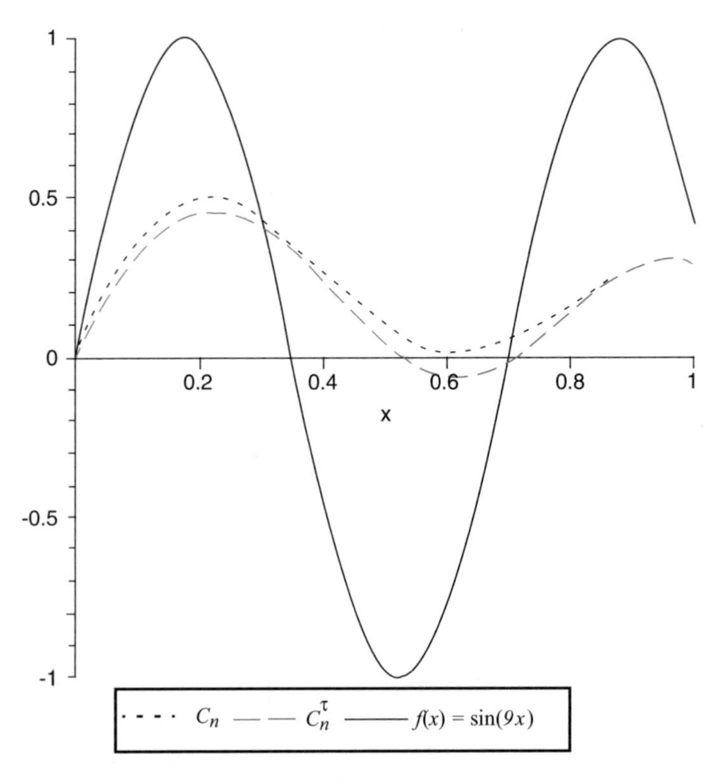

Fig. 2.1 Approximation process by C_n and C_n^τ

For $n = 20$, the approximation to the function f by C_n and C_n^τ is shown in Figure 2.1.
Also, the approximation to the function $f(x) = \log(x + 1)$ by C_n and C_n^τ is shown in Figure 2.2.

In [153], Kwun et al. obtained a Voronovskaya type theorem for C_n^τ and a direct approximation theorem by means of Ditzian–Totik modulus of smoothness.

Theorem 2.41 ([153]) *Let* $f \in C^2[0, 1]$. *If* $a_n \in (0, 1)$, $\lim_{n\to\infty} a_n = 1$ *and* $L := \lim_{n\to\infty} n(1-a_n)$ *exists, then*

$$\lim_{n\to\infty} n\left(C_n^\tau(f, x) - f(x)\right) = \frac{\alpha(\tau(x))}{\tau'(x)^2} f''(x) + \left(\frac{\beta(\tau(x))}{\tau'(x)} - \frac{\alpha(\tau(x))\tau''(x)}{\tau'(x)^3}\right) f'(x)$$

uniformly on $[0, 1]$, *with*

$$\alpha(x) = -\frac{x(1-x)}{2} - \frac{x^2}{4} L$$

and $\beta(x) = xL$.

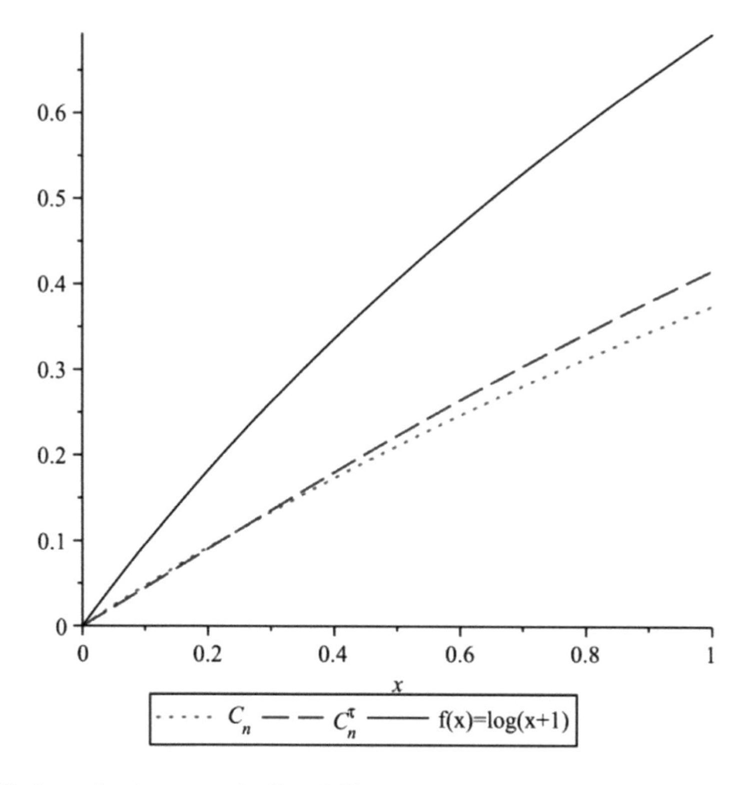

Fig. 2.2 Approximation process by C_n and C_n^τ

Theorem 2.42 ([153]) *Let* $f \in C[0,1]$ *and* $\varphi_\tau(x) = \sqrt{\tau(x)(1-\tau(x))}$ *then for every* $x \in (0,1)$, *we have*

$$\left|C_n^\tau(f;x) - f(x)\right| \leq \tilde{C}\omega_{\varphi_\tau}\left(f; \frac{\delta_{n,\tau}(x)}{\varphi_\tau(x)}\right),$$

where $\tilde{C}$ *is a constant independent of* n *and* x.

2.3.5 Lupaş Operators Based on Pólya Distribution

In [159], Lupaş and Lupaş introduced a special case of the operators (2.3.5) as follows:

$$P_n^{<\frac{1}{n}>}(f;x) = \frac{2n!}{(2n)!}\sum_{k=0}^{n}\binom{n}{k} f\left(\frac{k}{n}\right)(nx)_k(n-nx)_{n-k}. \tag{2.3.10}$$

Let τ be continuously differentiable infinitely many times on $[0, 1]$, such that $\tau(0) = 0$, $\tau(1) = 1$, and $\tau'(x) > 0$ for $x \in [0, 1]$. In [17] the sequence of Lupaş type operators for $f \in C[0, 1]$ was introduced as

$$P_n^{<\frac{1}{n},\tau>}(f;x) = \sum_{k=0}^{n} p_{n,k}^{<\frac{1}{n},\tau>}(x)(f \circ \tau^{-1})\left(\frac{k}{n}\right), \; x \in [0, 1], \tag{2.3.11}$$

where

$$p_{n,k}^{<\frac{1}{n},\tau>}(x) = \frac{2n!}{(2n)!}\binom{n}{k}(n\tau(x))_k\,(n - n\tau(x))_{n-k}\,.$$

Example 2.3 Let

$$\tau_1(x) = \frac{x^2 + x}{2}, \;\; \tau_2(x) = \sin\frac{\pi}{2}x \text{ and } f(x) = \cos(10x), \;\; x \in [0, 1]\,.$$

For $n = 40$, the approximation to the function f by the modified Lupaş operators and the classical ones is illustrated in Figure 2.3. The error of approximation for $P_n^{<\frac{1}{n}>}$, $P_n^{<\frac{1}{n},\tau_1>}$ and $P_n^{<\frac{1}{n},\tau_2>}$ at certain points from $[0, 1]$ is computed in Table 2.1. Therefore, depending on the choice of the function τ, the modified operator $P_n^{<\frac{1}{n},\tau>}$ presents a better order of approximation than $P_n^{<\frac{1}{n}>}$ on a certain interval.

Lemma 2.17 ([17]) *The modified Lupaş operators satisfy*

$$P_n^{<\frac{1}{n},\tau>}e_0 = e_0, \; P_n^{<\frac{1}{n},\tau>}\tau = \tau, \; P_n^{<\frac{1}{n},\tau>}\tau^2 = \tau^2 + \frac{2\tau(1-\tau)}{n+1}.$$

Let

$$\mu_{n,m}^{\tau}(x) = P_n^{<\frac{1}{n},\tau>}\left((\tau(t) - \tau(x))^m; x\right) = \sum_{k=0}^{n} p_{n,k}^{<\frac{1}{n},\tau>}(x)\left(\frac{k}{n} - \tau(x)\right)^m$$

be the central moment operator.

Lemma 2.18 ([17]) *The central moment operator satisfies:*

i) $\mu_{n,2}^{\tau}(x) = \dfrac{2}{n+1}\varphi_\tau^2(x)$;

ii) $\mu_{n,4}^{\tau}(x) = \dfrac{12(n^2 - 7n)\varphi_\tau^2(x) + (26n - 2)}{n(n+1)(n+2)(n+3)}\varphi_\tau^2(x)$,

where $\varphi_\tau^2(x) := \tau(x)(1 - \tau(x))$.

Lemma 2.19 ([17]) *If* $f \in C[0, 1]$, *then* $\|P_n^{<\frac{1}{n},\tau>}f\| \le \|f\|$, *where* $\|\cdot\|$ *is the uniform norm on* $C[0, 1]$.

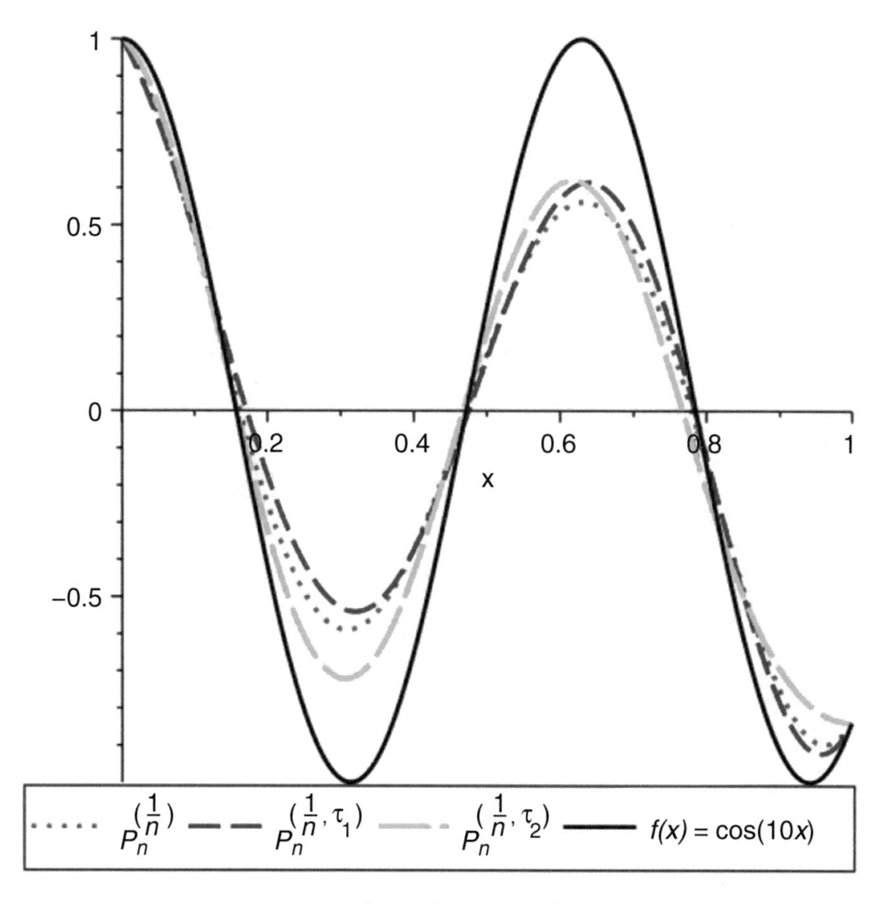

Fig. 2.3 Approximation process by $P_n^{<\frac{1}{n}>}$, $P_n^{<\frac{1}{n},\tau_1>}$ and $P_n^{<\frac{1}{n},\tau_2>}$

Theorem 2.43 ([17]) *If $f \in C[0,1]$, then $P_n^{<\frac{1}{n},\tau>} f$ converges to f as n tends to infinity, uniformly on* $[0, 1]$.

Example 2.4 ([17]) We consider

$$f : [0,1] \to \mathbb{R},\ f(x) = \cos(10x) \text{ and } \tau(x) = \frac{x^2 + x}{2}.$$

The convergence of the modified Lupaş operator to the function f is illustrated in Figure 2.4. We remark that as the values of n increase, the error in the approximation of the function by the operator becomes smaller.

Table 2.1 Error of approximation for $P_n^{<\frac{1}{n}>}$, $P_n^{<\frac{1}{n},\tau_1>}$, and $P_n^{<\frac{1}{n},\tau_2>}$

x	$\lvert P_n^{<\frac{1}{n}>}(f;x)-f(x)\rvert$	$\lvert P_n^{<\frac{1}{n},\tau_1>}(f;x)-f(x)\rvert$	$\lvert P_n^{<\frac{1}{n},\tau_2>}(f;x)-f(x)\rvert$
0.04	0.0719078440	0.0918804542	0.0491775787
0.08	0.0899902045	0.0949187558	0.0656722344
0.12	0.0472944810	0.0215916952	0.0431323566
0.16	0.0464266575	0.1024165160	0.0144204290
0.20	0.1684314306	0.2437174913	0.0943043728
0.24	0.2889352546	0.3677734502	0.1783441641
0.28	0.3780208231	0.4453240720	0.2468493195
0.32	0.4121574493	0.4575642665	0.2826616675
0.36	0.3791213446	0.3992005780	0.2745554027
0.40	0.2804560850	0.2789088716	0.2194175892
0.44	0.1310964673	0.1171877712	0.1228721695
0.48	0.0436865547	0.0579307255	0.0017044307
0.52	0.2133696908	0.2164713841	0.1355396341
0.56	0.3482733387	0.3323367044	0.2577115929
0.60	0.4254746884	0.3882544972	0.3486510319
0.64	0.4333817144	0.3789143428	0.3932519814
0.68	0.3740607726	0.3116949702	0.3831315495
0.72	0.2628264094	0.2048387962	0.3179584077
0.76	0.1251131158	0.0834187551	0.2061763439
0.80	0.0088333903	0.0261223418	0.0655133299
0.84	0.1104739425	0.1017791431	0.0769549422
0.88	0.1594595824	0.1309512530	0.1865335751
0.92	0.1487803128	0.1135004422	0.2263598837
0.96	0.0874521271	0.0621541233	0.1673068131

Using the result of Shisha and Mond [214] we have

$$\left| P_n^{<\frac{1}{n},\tau>}(f;x) - f(x) \right| \leq \left(1 + \frac{\mu_{n,2}^{\tau}(x)}{\delta^2} \right) \omega(f,\delta), \text{ for } \delta > 0,$$

where $\omega(f;\delta)$ is the usual modulus of continuity of $f \in C[0,1]$.

Example 2.5 ([17]) The rates of convergence of the modified operators depend on the selection of the function τ. If we choose $\tau(x) = \left(\sin\frac{\pi x}{2}\right)^2$, we have $\tau(x)(1-\tau(x)) \leq x(1-x)$, for all $x \in [0,1]$ and this inequality leads to $\mu_{n,2}^{\tau}(x) \leq \mu_{n,2}(x)$. Therefore, the modified operator $P_n^{<\frac{1}{n},\tau>}$ presents a better order of approximation than $P_n^{<\frac{1}{n}>}$.

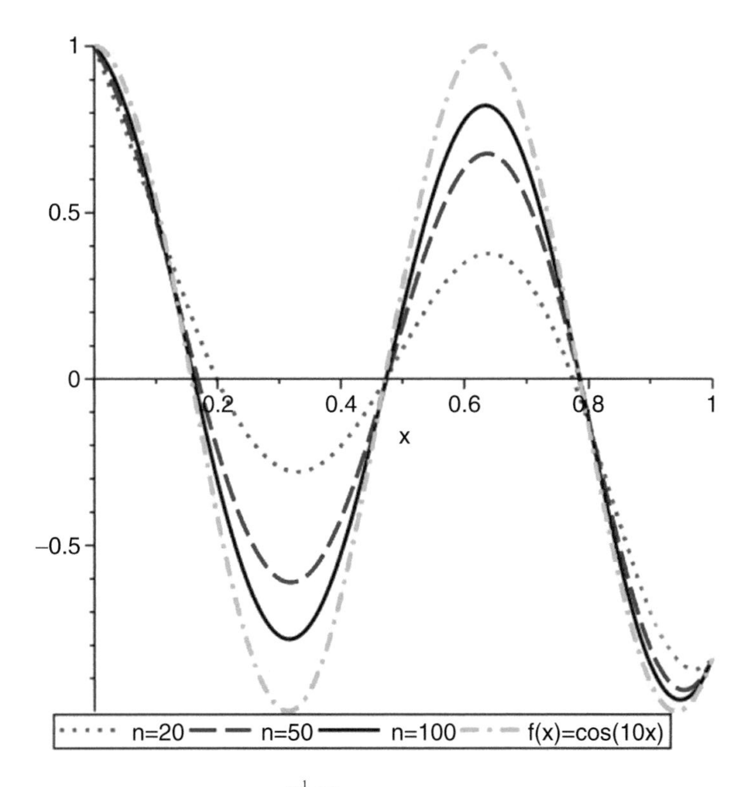

Fig. 2.4 Approximation process by $P_n^{<\frac{1}{n},\tau>}$ for $n \in \{20, 50, 100\}$

Let ω_k be the classical k^{th} order modulus of smoothness given in (5.2.2). The next result presents approximation properties for modified Lupaş operators in terms of the modulus of smoothness. We assume that

$$\inf_{x\in[0,1]} \tau'(x) \geq a, a \in \mathbb{R}^+ .$$

Theorem 2.44 ([17]) *If* $f \in C[0, 1]$, *then the operators* $P_n^{<\frac{1}{n},\tau>}$ *satisfy the following inequality*

$$\left| P_n^{<\frac{1}{n},\tau>}(f; x) - f(x) \right| \leq \frac{3}{2}\left(1 + \frac{1}{a^2}\right)\omega_2\left(f; \frac{\varphi_\tau(x)}{\sqrt{n+1}}\right) + \frac{5\varphi_\tau(x)\|\tau''\|}{a^3\sqrt{n+1}}\omega_1\left(f; \frac{\varphi_\tau(x)}{\sqrt{n+1}}\right).$$

Estimates for continuous functions and for twice continuously differentiable functions, dealing with the Voronovskaja type theorem for Lupaş operators were considered in [171] as follows:

Theorem 2.45 ([171]) *For any $f \in C^2[0,1]$ and $x \in [0,1]$ the following inequalities hold*

i) $\left| P_n^{<\frac{1}{n}>}(f;x) - f(x) \right| \le \frac{3}{2}\omega_1\left(f; \frac{1}{\sqrt{n}}\right)$,

ii) $n\left| P_n^{<\frac{1}{n}>}(f;x) - f(x) - \frac{x(1-x)}{n+1} f''(x) \right| \le \frac{5}{8}\omega_1\left(f''; \frac{1}{\sqrt{n}}\right)$.

A global approximation theorem and a quantitative Voronovskaja type theorem for the operator $P_n^{<\frac{1}{n},\tau>}$ using the first order Ditzian–Totik modulus of smoothness were established in [17].

Theorem 2.46 ([17]) *Let $f \in C[0,1]$. Then for every $x \in (0,1)$, we have*

$$\left| P_n^{<\frac{1}{n},\tau>}(f;x) - f(x) \right| \le C\omega_{\varphi_\tau}\left(f; \frac{2}{a\sqrt{n+1}}\right),$$

where $C > 0$ is a constant.

Theorem 2.47 ([17]) *For any $f \in C^2[0,1]$ and $x \in (0,1)$ the following inequalities hold*

i) $n\left| P_n^{<\frac{1}{n},\tau>}(f;x) - f(x) - \frac{1}{n+1} \cdot \frac{1}{[\tau'(x)]^2}\left[f''(x) - f'(x)\frac{\tau''(x)}{\tau'(x)}\right]\varphi_\tau^2(x) \right|$
$\le C\omega_{\varphi_\tau}\left((f \circ \tau^{-1})''; \frac{\varphi_\tau(x)}{a} u_n^\tau(x)\right)$,

ii) $n\left| P_n^{<\frac{1}{n},\tau>}(f;x) - f(x) - \frac{1}{n+1} \cdot \frac{1}{[\tau'(x)]^2}\left[f''(x) - f'(x)\frac{\tau''(x)}{\tau'(x)}\right]\varphi_\tau^2(x) \right|$
$\le C\varphi_\tau(x)\omega_{\varphi_\tau}\left((f \circ \tau^{-1})''; \frac{u_n^\tau(x)}{a}\right)$,

where $C > 0$ is a constant and

$$u_n^\tau(x) = 2\sqrt{\frac{2(n^2-7n)\varphi_\tau^2(x) + 13n - 1}{n(n+2)(n+3)}}.$$

Remark 2.7 ([17]) If we set $\tau(x) = x$ in Theorem 2.46 and in Theorem 2.47, we get

i) $\left| P_n^{<\frac{1}{n}>}(f;x) - f(x) \right| \le C_1\omega_\varphi\left(f; \frac{2}{\sqrt{n+1}}\right)$, for $f \in C[0,1]$

ii) $n\left| P_n^{<\frac{1}{n}>}(f;x) - f(x) - \frac{\varphi^2(x)}{n+1} f''(x) \right| \le C_2\omega_\varphi\left(f''; \varphi(x)u_n(x)\right)$ for $f \in C^2[0,1]$,

where C_1, C_2 are positive constants and

$$u_n^\tau(x) = 2\sqrt{\frac{2(n^2-7n)\varphi^2(x) + 13n - 1}{n(n+2)(n+3)}}.$$

Chapter 3
Basics of Post-quantum Calculus

3.1 Introduction

Quantum calculus is an old area of research and has various applications in mathematics, physics, and engineering sciences. As far as its role in mathematics is concerned, it was widely used by researchers in number theory, hypergeometric functions, special functions [173], etc. We refer the readers to the two important books on the basics on q-calculus by Ernst [83] and Kac and Cheung [147]. In the field of approximation theory the first paper came into existence in 1987, when the Romanian mathematician Lupaş [157] proposed the q analogue of Bernstein polynomials. He studied some properties of the q variant of Bernstein polynomials. But unfortunately there was limited interest among the researchers of the time towards the applications of quantum-calculus in approximation theory. After a gap of almost 10 years Phillips [197] proposed another q variant of the Bernstein polynomials and thereafter some researchers started studying in this direction. The initial contribution on q-Bernstein polynomials was due to Sofiya Ostrovka [185–187], etc. Later several other q generalizations of Bernstein–Jacobi polynomials and Bernstein–Durrmeyer operators were respectively proposed by Derriennic [77] and Gupta [117]. In this direction, we mention the work due to Aral and Gupta [33]

Furthermore it is possible to extend q-calculus to post-quantum calculus, namely the (p, q)-calculus. Actually such an extension of quantum calculus cannot be obtained directly by substituting q by q/p in q-calculus. But there is a link between q-calculus and (p, q)-calculus. The q calculus may be obtained by substituting $p = 1$ in (p, q)-calculus.

V. Gupta et al., *Recent Advances in Constructive Approximation Theory*,
Springer Optimization and Its Applications 138,
https://doi.org/10.1007/978-3-319-92165-5_3

3.2 Some Notations of q-Calculus

In this section we mention some basic notations of q calculus, which may also be found in the books [39, 83, 147] and references therein.

For $q > 0$ the q-integer $[n]_q$ is defined by

$$[n]_q = \begin{cases} \frac{1-q^n}{1-q}, & q \neq 1 \\ n, & q = 1 \end{cases},$$

for $n \in \mathbb{N}$. Also, the q-factorial $[n]_q!$ is defined as

$$[n]_q! = \begin{cases} [n]_q [n-1]_q \cdots [1]_q, & n = 1, 2, \ldots \\ 1 & n = 0. \end{cases},$$

for $n \in \mathbb{N}$.

The q-binomial coefficients are given by

$$\begin{bmatrix} n \\ k \end{bmatrix}_q = \frac{[n]_q!}{[k]_q! [n-k]_q!}, \quad 0 \leq k \leq n,$$

for $n, k \in \mathbb{N}$.

We use the notation $(1+x)_q^n$ as

$$(1+x)_q^n := \begin{cases} (1+x)(1+qx)\ldots\left(1+q^{n-1}x\right) & n = 1, 2, \ldots \\ 1 & n = 0. \end{cases}$$

Furthermore, we have the Gauss binomial formula:

$$(x+a)_q^n = \sum_{j=0}^{n} \begin{bmatrix} n \\ j \end{bmatrix}_q q^{j(j-1)/2} a^j x^{n-j}.$$

For $f \in C[0, 1]$ and $0 < q < 1$, the q-Bernstein polynomials proposed by Phillips [197] are defined as

$$\mathcal{B}_{n,q}(f, x) = \sum_{k=0}^{n} b_{n,k}^q(x) f\left(\frac{[k]_q}{[n]_q}\right), \tag{3.2.1}$$

where the q-Bernstein basis function is given by

$$b_{n,k}^q(x) = \begin{bmatrix} n \\ k \end{bmatrix}_q x^k (1-x)_q^{n-k}, x \in [0, 1]$$

and

$$(a-b)_q^n = \prod_{s=0}^{n-1}(a-q^s b), \quad a, b \in \mathbf{R}.$$

The q-derivative $D_q f$ of a function f is given by

$$\left(D_q f\right)(x) = \frac{f(x) - f(qx)}{(1-q)x}, \quad \text{if } x \neq 0.$$

The two q-analogues of classical exponential function e^x are given by

$$e_q(x) = \sum_{k=0}^{\infty} \frac{x^k}{[k]_q!}$$

and

$$E_q(x) = \sum_{k=0}^{\infty} q^{\frac{k(k-1)}{2}} \frac{x^k}{[k]_q!}.$$

The Jackson definite integral of function f is defined by (see [144]):

$$\int_0^a f(x)\, d_q x = (1-q)\, a \sum_{n=0}^{\infty} f\left(aq^n\right) q^n, \ a \in \mathbb{R}.$$

For $m, n > 0$ the q Beta function of first and second kinds (see [147]) are respectively defined as

$$B_q(m,n) = \int_0^1 t^{m-1}(1-qt)_q^{n-1} d_q t.$$

$$B_q(t,s) = K(A,t) \int_0^{\infty/A} \frac{x^{t-1}}{(1+x)_q^{t+s}} d_q x,$$

The following definitions of q-gamma functions are considered as

$$\Gamma_q(t) = \int_0^{1/1-q} x^{t-1} E_q(-qx)\, d_q x, \quad t > 0$$

$$\Gamma_q(t) = K(A,t) \int_0^{\infty/A(1-q)} x^{t-1} e_q(-x) d_q x,$$

where

$$K(x,t)=\frac{1}{x+1}x^{t}\left(1+\frac{1}{x}\right)_{q}^{t}(1+x)_{q}^{1-t}.$$

The q-gamma and q-beta functions are connected with the following relation:

$$B_{q}(t,s)=\frac{\Gamma_{q}(t)\,\Gamma_{q}(s)}{\Gamma_{q}(t+s)}.$$

The function $K(x,t)$ is a q-constant, i.e., $K(qx,t)=K(x,t)$. In particular for any positive integer n, it holds

$$K(x,n)=q^{\frac{n(n-1)}{2}},\quad K(x,0)=1.$$

The Riemann type q-integral of f over the interval $[a,b]$ $(0\leq a<b; 0<q<1)$ is given by

$$\int_{a}^{b}f(u)d_{q}^{R}u=(1-q)(b-a)\sum_{j=0}^{\infty}q^{j}f(a+(b-a)q^{j}).\tag{3.2.2}$$

(see [93, 168]).

3.3 Certain Definitions of (p,q)-Calculus

Some basic definitions of (p,q)-Calculus are given in [209–211] and the references therein:

Definition 3.1 The (p,q)-numbers for $p\neq q$ are defined as

$$[n]_{p,q}:=p^{n-1}+p^{n-2}q+p^{n-3}q^{2}+\cdots+pq^{n-2}+q^{n-1}=\frac{p^{n}-q^{n}}{p-q}.$$

Some basic identities of (p,q) numbers are

- $[n+m]_{p,q}=q^{m}[n]_{p,q}+p^{n}[m]_{p,q}$
- $[n+m]_{p,q}=p^{m}[n]_{p,q}+q^{n}[m]_{p,q}$
- $[n]_{p,q}=[2]_{p,q}[n-1]_{p,q}-pq[n-2]_{p,q}$.

Obviously, it may be seen that $[n]_{p,q}=p^{n-1}[n]_{q/p}$.

Definition 3.2 The (p, q)-factorial is defined by

$$[n]_{p,q}! = \prod_{k=1}^{n} [k]_{p,q}\,, n \geq 1, [0]_{p,q}! = 1.$$

Definition 3.3 The (p, q)-binomial coefficient is given by

$$\begin{bmatrix} n \\ k \end{bmatrix}_{p,q} = \frac{[n]_{p,q}!}{[n-k]_{p,q}!\,[k]_{p,q}!}, 0 \leq k \leq n.$$

As a special case when $p = q = 1$ the (p, q)-binomial coefficient $\begin{bmatrix} n \\ k \end{bmatrix}_{p,q}$ reduces to the usual binomial coefficient $\binom{n}{k}$. The (p, q)-binomial coefficients satisfy the following relations:

- $\begin{bmatrix} n \\ k \end{bmatrix}_{p,q} = \begin{bmatrix} n \\ n-k \end{bmatrix}_{p,q} = p^{k(n-k)} \begin{bmatrix} n \\ k \end{bmatrix}_{q/p}$.
- $\begin{bmatrix} n+1 \\ k \end{bmatrix}_{p,q} = p^{k} \begin{bmatrix} n \\ k \end{bmatrix}_{p,q} + q^{n+1-k} \begin{bmatrix} n \\ k-1 \end{bmatrix}_{p,q}$.

Definition 3.4 The (p, q)-power basis is defined below and it also has a link with q-power basis as

$$(a \oplus b)^{n}_{p,q} = (a + b)(pa + qb)(p^{2}a + q^{2}b) \cdots (p^{n-1}a + q^{n-1}b).$$

$$(a \ominus b)^{n}_{p,q} = (a - b)(pa - qb)(p^{2}a - q^{2}b) \cdots (p^{n-1}a - q^{n-1}b).$$

For m and n nonnegative integers, we have

$$(a \ominus b)^{m+n}_{p,q} = (a \ominus b)^{m}_{p,q}(p^{m}a \ominus q^{m}b)^{n}_{p,q}.$$

Theorem 3.1 *The* (p, q)*-binomial theorem for* $0 \leq k \leq n, n \in \mathbb{N}$ *is defined as*

$$(a \ominus b)^{n}_{p,q} = \sum_{k=0}^{n} (-1)^{k} \begin{bmatrix} n \\ k \end{bmatrix}_{p,q} p^{(n-k)(n-k-1)/2} q^{k(k-1)/2} a^{n-k} b^{k}.$$

Definition 3.5 The (p, q)-derivative of the function f is defined as

$$D_{p,q} f(x) = \frac{f(px) - f(qx)}{(p-q)x}, x \neq 0.$$

Note also that for $p = 1$, the (p, q)-derivative reduces to the q-derivative.

The (p, q)-derivative satisfies the following product rules

$$D_{p,q}(f(x)g(x)) = f(px)D_{p,q}g(x) + g(qx)D_{p,q}f(x),$$

$$D_{p,q}(f(x)g(x)) = g(px)D_{p,q}f(x) + f(qx)D_{p,q}g(x).$$

The following quotient rule holds for (p, q)-derivatives:

$$D_{p,q}\left(\frac{f(x)}{g(x)}\right) = \frac{g(qx)D_{p,q}f(x) - f(qx)D_{p,q}g(x)}{g(px)g(qx)},$$

$$D_{p,q}\left(\frac{f(x)}{g(x)}\right) = \frac{g(px)D_{p,q}f(x) - f(px)D_{p,q}g(x)}{g(px)g(qx)}.$$

Example 3.1 ([209]) For any integer n, we have

$$D_{p,q}(x \ominus a)^n_{p,q} = [n]_{p,q}(px \ominus a)^{n-1}_{p,q}$$

and

$$D_{p,q}(a \ominus x)^n_{p,q} = -[n]_{p,q}(a \ominus qx)^{n-1}_{p,q}.$$

In general for $0 \le k \le n, n \in \mathbb{N}$, we have

$$D^k_{p,q}(x \ominus a)^n_{p,q} = p^{k(k-1)/2}\frac{[n]_{p,q}!}{[n-k]_{p,q}!}(p^k x \ominus a)^{n-k}_{p,q}$$

and

$$D^k_{p,q}(a \ominus x)^n_{p,q} = (-1)^k q^{k(k-1)/2}\frac{[n]_{p,q}!}{[n-k]_{p,q}!}(a \ominus q^k x)^{n-k}_{p,q}$$

Definition 3.6 Let f be an arbitrary function and $a \in \mathbb{R}$. The (p, q)-integral of $f(x)$ on $[0, a]$ (see [209]) is defined as

$$\int_0^a f(x)\, d_{p,q}x = (q-p)\, a \sum_{k=0}^{\infty} \frac{p^k}{q^{k+1}} f\left(\frac{p^k}{q^{k+1}}a\right) \quad \text{if} \quad \left|\frac{p}{q}\right| < 1$$

and

$$\int_0^a f(x)\, d_{p,q}x = (p-q)\, a \sum_{k=0}^{\infty} \frac{q^k}{p^{k+1}} f\left(\frac{q^k}{p^{k+1}}a\right) \quad \text{if} \quad \left|\frac{q}{p}\right| < 1.$$

The formula of (p, q)-integration by parts is given by

$$\int_a^b f(px) D_{p,q} g(x) D_{p,q} x \ 2 = f(b) g(b) - f(a) g(a) - \int_a^b g(qx) D_{p,q} f(x) D_{p,q} x \qquad (3.3.1)$$

Recently Acar and Aral [5] proposed the following modified (p, q)-integral as

$$\int_a^b f(x) d_{p,q} x = (q-p)(b-a) \sum_{k=0}^{\infty} \frac{p^k}{q^{k+1}} f\left(a + (b-a)\frac{p^k}{q^{k+1}}\right) \quad \text{if} \quad \left|\frac{p}{q}\right| < 1$$

and

$$\int_a^b f(x) d_{p,q} x = (p-q)(b-a) \sum_{k=0}^{\infty} \frac{q^k}{p^{k+1}} f\left(a + (b-a)\frac{q^k}{p^{k+1}}\right) \quad \text{if} \quad \left|\frac{q}{p}\right| < 1.$$

Definition 3.7 Let n is a nonnegative integer, we define the (p, q)-Gamma function as

$$\Gamma_{p,q}(n+1) = \frac{(p \ominus q)^n_{p,q}}{(p-q)^n} = [n]_{p,q}!, \quad 0 < q < p,$$

where

$$(a \ominus b)^n_{p,q} = \prod_{i=0}^{n-1} (p^i a - q^i b).$$

3.3.1 (p, q)-Beta Function of First Kind

Recently Gupta and Aral in [128] defined (p, q)-Beta integral $B_{p,q}(m, n)$ for $m, n \in \mathbb{N}$, in the following way

$$B_{p,q}(m, n) = \int_0^1 (px)^{m-1} (p \ominus pqx)^{n-1}_{p,q} d_{p,q} x. \qquad (3.3.2)$$

Theorem 3.2 ([128]) *The (p, q)-Gamma and (p, q)-Beta functions fulfil the following fundamental relation*

$$B_{p,q}(m, n) = p^{[n(2m+n-2)+n-2]/2} \frac{\Gamma_{p,q}(m) \Gamma_{p,q}(n)}{\Gamma_{p,q}(m+n)},$$

where $m, n \in \mathbb{N}$.

Proof For any $m, n \in \mathbb{N}$, since

$$B_{p,q}(m, n) = \int_0^1 (px)^{m-1} (p \ominus pqx)_{p,q}^{n-1} d_{p,q}x,$$

using (p, q)-integration by parts

$$\int_a^b f(px) D_{p,q} g(x) d_{p,q}x = f(b)g(b) - f(a)g(a) - \int_a^b g(qx) D_{p,q} f(x) d_{p,q}x$$

and consider

$$f(x) = x^{m-1}, \quad g(x) = -\frac{(p \ominus px)_{p,q}^n}{p[n]_{p,q}}$$

with the relations

$$D_{p,q} x^{m-1} = [m-1]_{p,q} x^{m-2} \text{ and } D_{p,q} (p \ominus px)^n = -[n]_{p,q}\, p\, (p \ominus pqx)^{n-1},$$

we have

$$B_{p,q}(m, n) = \frac{[m-1]_{p,q}}{p^{m-1}[n]_{p,q}} B_{p,q}(m-1, n+1). \tag{3.3.3}$$

Also, we can write for positive integer n the following

$$\begin{aligned} B_{p,q}(m, n+1) &= \int_0^1 (px)^{m-1} (p \ominus pqx)_{p,q}^n d_{p,q}x \\ &= \int_0^1 (px)^{m-1} (p \ominus pqx)_{p,q}^{n-1} \left(p^n - pq^n x\right) d_{p,q}x \\ &= p^n B_{p,q}(m, n) - q^n B_{p,q}(m+1, n). \end{aligned}$$

Using (3.3.3), we have

$$B_{p,q}(m, n+1) = p^n B_{p,q}(m, n) - q^n \frac{[m]_{p,q}}{p^m [n]_{p,q}} B_{p,q}(m, n+1),$$

which implies that

$$B_{p,q}(m, n+1) = p^{n+m} \frac{p^n - q^n}{p^{n+m} - q^{n+m}} B_{p,q}(m, n).$$

Also, by (p, q)-integration

$$B_{p,q}(m, 1) = \frac{p^{m-1}}{[m]_{p,q}},$$

thus, with the above relation, with repeated applications, we immediately have

$$\begin{aligned}
B_{p,q}(m, n) &= p^{n+m-1} \frac{p^{n-1} - q^{n-1}}{p^{n+m-1} - q^{n+m-1}} B_{p,q}(m, n-1) \\
&= p^{n+m-1} \frac{p^{n-1} - q^{n-1}}{p^{n+m-1} - q^{n+m-1}} p^{n+m-2} \frac{p^{n-2} - q^{n-2}}{p^{n+m-2} - q^{n+m-2}} B_{p,q}(m, n-2) \\
&= p^{n+m-1} \frac{p^{n-1} - q^{n-1}}{p^{n+m-1} - q^{n+m-1}} p^{n+m-2} \frac{p^{n-2} - q^{n-2}}{p^{n+m-2} - q^{n+m-2}} \cdots \\
&\quad \cdots p^{m+1} \frac{p - q}{p^{m+1} - q^{m+1}} B_{p,q}(m, 1) \\
&= \frac{p^{(m-1)+m+(m+1)+\cdots+(m+n-1)}}{p^m} \frac{(p \ominus q)_{p,q}^{n-1}}{(p^m \ominus q^m)_{p,q}^n} (p - q),
\end{aligned}$$

i.e.,

$$B_{p,q}(m, n) = p^s \frac{(p \ominus q)_{p,q}^{n-1}}{(p^m \ominus q^m)_{p,q}^n} (p - q), \tag{3.3.4}$$

where

$$s = [n(2m + n - 2) + n - 2]/2.$$

Following [210], we have

$$(a \ominus b)_{p,q}^{n+m} = (a \ominus b)_{p,q}^n (ap^n \ominus bq^n)_{p,q}^m$$

thus (3.3.4) leads to

$$
\begin{aligned}
B_{p,q}(m,n) &= p^{s}\frac{(p\ominus q)_{p,q}^{n-1}}{(p^{m}\ominus q^{m})_{p,q}^{n}}(p-q)\\
&= p^{s}\frac{(p\ominus q)_{p,q}^{n-1}}{(p-q)^{n-1}}.\frac{(p\ominus q)_{p,q}^{m-1}}{(p-q)^{m-1}}.\frac{(p-q)^{m-1}(p-q)^{n-1}}{(p\ominus q)_{p,q}^{m-1}(p^{m}\ominus q^{m})_{p,q}^{n}}(p-q)\\
&= p^{s}\frac{(p\ominus q)_{p,q}^{n-1}}{(p-q)^{n-1}}.\frac{(p\ominus q)_{p,q}^{m-1}}{(p-q)^{m-1}}.\frac{(p-q)^{m+n-1}}{(p\ominus q)_{p,q}^{m+n-1}} = p^{s}\frac{\Gamma_{p,q}(m)\,\Gamma_{p,q}(n)}{\Gamma_{p,q}(m+n)}.
\end{aligned}
$$

This completes the proof of the theorem. ■

Recently Milovanović et al. in [175] considered a slightly different form of (p,q)-Beta functions as:

$$
\hat{B}_{p,q}(m,n) = \int_{0}^{1} x^{m-1}(1\ominus qx)_{p,q}^{n-1}\,d_{p,q}x. \tag{3.3.5}
$$

Theorem 3.3 ([175]) *The (p,q)-Gamma and (p,q)-Beta functions satisfy the following fundamental relation*

$$
\hat{B}_{p,q}(m,n) = p^{(n-1)(2m+n-2)/2}\frac{\Gamma_{p,q}(m)\,\Gamma_{p,q}(n)}{\Gamma_{p,q}(m+n)},
$$

where $m,n\in\mathbb{N}$.

The following observations have been made for (p,q)-Beta functions:

- For $m,n\in\mathbb{N}$, we have

$$
\hat{B}_{p,q}(m,n+1) = p^{n-1}\hat{B}_{p,q}(m,n) - q^{n}\hat{B}_{p,q}(m+1,n).
$$

- The (p,q)-Beta integrals defined by (3.3.2) and (3.3.5) are not commutative. In order to make them commutative, one may consider the following form

$$
\widetilde{B}_{p,q}(m,n) = \int_{0}^{1} p^{m(m-1)/2}x^{m-1}(1\ominus qx)_{p,q}^{n-1}\,D_{p,q}x.
$$

For this form, (p,q)-Gamma and (p,q)-Beta functions fulfil the following fundamental relation

$$
\widetilde{B}_{p,q}(m,n) = p^{(2mn+m^{2}+n^{2}-3m-3n+2)/2}\frac{\Gamma_{p,q}(m)\,\Gamma_{p,q}(n)}{\Gamma_{p,q}(m+n)}, \tag{3.3.6}
$$

where $m,n\in\mathbb{N}$. Obviously for the form (3.3.6), we get

$$
\widetilde{B}_{p,q}(m,n) = \widetilde{B}_{p,q}(n,m).
$$

3.3.2 (p, q)-Beta Function of Second Kind

Let $m, n \in \mathbb{N}$, Aral and Gupta [36] defined (p, q)-Beta function of second kind as

$$B^{s}_{p,q}(m, n) = \int_0^\infty \frac{x^{m-1}}{(1 \oplus px)^{m+n}_{p,q}} d_{p,q}x$$

Theorem 3.4 ([36]) *Let $m, n \in \mathbb{N}$. We have the following relation between (p, q)-Beta and (p, q)-Gamma function:*

$$B^{s}_{p,q}(m, n) = q^{[2-m(m-1)]/2} p^{-m(m+1)/2} \frac{\Gamma_{p,q}(m)\,\Gamma_{p,q}(n)}{\Gamma_{p,q}(m+n)}.$$

Proof We know that

$$D_{p,q} \frac{1}{(1 \oplus x)^{n}_{p,q}} = -\frac{p\,[n]_{p,q}}{(1 \oplus px)^{n+1}_{p,q}}$$

If we choose

$$f(x) = x^m \text{ and } g(x) = -\frac{1}{p\,[m+n]_{p,q}\,(1 \oplus x)^{m+n}_{p,q}}$$

and use (3.3.1) we have

$$\begin{aligned}
B^{s}_{p,q}(m+1, n) &= \int_0^\infty \frac{x^{m}}{(1 \oplus px)^{m+n+1}_{p,q}} d_{p,q}x \\
&= -\frac{p^{-m}}{p\,[m+n]_{p,q}} \int_0^\infty (px)^m\, D_{p,q} \frac{1}{(1 \oplus x)^{m+n}_{p,q}} d_{p,q}x \\
&= \frac{p^{-m}}{p\,[m+n]_{p,q}} \int_0^\infty D_{p,q}x^m \frac{1}{(1 \oplus qx)^{m+n}_{p,q}} d_{p,q}x \\
&= \frac{p^{-m}\,[m]_{p,q}}{p\,[m+n]_{p,q}} \int_0^\infty x^{m-1} \frac{1}{(1 \oplus qx)^{m+n}_{p,q}} d_{p,q}x \\
&= \frac{p^{-m-1}\,[m]_{p,q}}{q^{m-1}\,[m+n]_{p,q}} \int_0^\infty (qx)^{m-1} \frac{1}{(1 \oplus qx)^{m+n}_{p,q}} d_{p,q}x \\
&= \frac{p^{-1}\,[m]_{p,q}}{(pq)^m\,[m+n]_{p,q}} \int_0^\infty (x)^{m-1} \frac{1}{(1 \oplus x)^{m+n}_{p,q}} d_{p,q}x \\
&= \frac{p^{-1}\,[m]_{p,q}}{(pq)^m\,[m+n]_{p,q}} B^{s}_{p,q}(m, n)
\end{aligned}$$

$$B^{s}_{p,q}(1, n) = \int_0^\infty \frac{1}{(1 \oplus px)^{n+1}_{p,q}} d_{p,q}x = -\frac{1}{p\,[n]_{p,q}} \int_0^\infty D_{p,q} \frac{1}{(1 \oplus x)^{n}_{p,q}} d_{p,q}x = \frac{1}{p\,[n]_{p,q}}$$

$$\begin{aligned}B^s_{p,q}(m,n) &= \frac{p^{-1}[m-1]_{p,q}}{(pq)^{m-1}[m+n-1]_{p,q}}B^s_{p,q}(m-1,n)\\ &= \frac{p^{-1}[m-1]_{p,q}}{(pq)^{m-1}[m+n-1]_{p,q}}\frac{p^{-1}[m-2]_{p,q}}{(pq)^{m-2}[m+n-2]_{p,q}}B^s_{p,q}(m-2,n)\\ &= \frac{p^{-1}[m-1]_{p,q}}{(pq)^{m-1}[m+n-1]_{p,q}}\frac{p^{-1}[m-2]_{p,q}}{(pq)^{m-2}[m+n-2]_{p,q}}\cdots\frac{p^{-1}}{pq[n+1]_{p,q}}B^s_{p,q}(1,n)\\ &= \frac{p^{-1}[m-1]_{p,q}}{(pq)^{m-1}[m+n-1]_{p,q}}\frac{p^{-1}[m-2]_{p,q}}{(pq)^{m-2}[m+n-2]_{p,q}}\cdots\frac{p^{-1}}{pq[n+1]_{p,q}}\frac{q}{pq[n]_{p,q}}\\ &= \frac{qp^{-m}}{(pq)^{(m-1)m/2}}\frac{\Gamma_{p,q}(m)\Gamma_{p,q}(n)}{\Gamma_{p,q}(m+n)}\end{aligned}$$

This completes the proof of the theorem. ■

3.3.3 (p,q)-Exponential and Gamma Function

Two different (p,q) analogues of exponential function namely $E_{p,q}$ and $e_{p,q}$ are given as follows:

$$e_{p,q}(x) = \sum_{n=0}^{\infty}\frac{p^{n(n-1)/2}}{[n]_{p,q}!}x^n$$

and

$$E_{p,q}(x) = \sum_{n=0}^{\infty}\frac{q^{n(n-1)/2}}{[n]_{p,q}!}x^n.$$

We know that is the following relation between (p,q)-exponential functions

$$e_{p,q}(x)E_{p,q}(-x) = 1 \tag{3.3.7}$$

holds. It was observed in [37] that these (p,q)-analogues of the classical exponential functions are valid for $0 < q < p \le 1$. Moreover $E_{p,q}(x)$ and $e_{p,q}(x)$ tend to e^x as $p \to 1^-$ and $q \to 1^-$. By simple computation, one has

$$D_{p,q}E_{p,q}(x) = E_{p,q}(qx),\; D_{p,q}E_{p,q}(ax) = aE_{p,q}(aqx)$$

Definition 3.8 For any $n \in \mathbb{N}$, the (p,q)-Gamma function considered in [37] is defined as

$$\Gamma_{p,q}(n) = \int_0^{\infty} p^{(n-1)(n-2)/2}x^{n-1}E_{p,q}(-qx)\,d_{p,q}x.$$

Lemma 3.1 ([37]) *For any $n \in \mathbb{N}$, we have*

$$\Gamma_{p,q}(n+1) = [n]_{p,q}!.$$

3.4 Some Discrete (p, q) Operators

Throughout the present section and in the next chapter, we consider $0 < q < p \leq 1$. In 2015 and 2016, Mursaleen et al. in [178, 180] introduced the (p, q) Bernstein operators for $x \in [0, 1]$ as

$$B_{n,p,q}(f; x) = \sum_{k=0}^{n} b_{n,k}^{p,q}(1, x) f\left(\frac{p^{n-k}[k]_{p,q}}{[n]_{p,q}}\right), \tag{3.4.1}$$

where

$$b_{n,k}^{p,q}(1, x) = \begin{bmatrix} n \\ k \end{bmatrix}_{p,q} p^{[k(k-1)-n(n-1)]/2} x^k (1 \ominus x)_{p,q}^{n-k}.$$

Also, it was observed by Gupta and Aral in [128] that the above representation can be obtained easily by replacing q by q/p in the q-Bernstein polynomials defined by (3.2.1) and using the identities which link the quantum calculus with post-quantum calculus. The following relations were shown in [128]:

$$\begin{aligned}
\begin{bmatrix} n \\ k \end{bmatrix}_{q/p} &= \frac{[n]_{q/p}!}{[k]_{q/p}![n-k]_{q/p}!} \\
&= \frac{[n]_{q/p}[n-1]_{q/p} \cdots [2]_{q/p}.1}{([k]_{q/p}[k-1]_{q/p} \cdots [2]_{q/p}.1)([n-k]_{q/p}[n-k-1]_{q/p} \cdots [2]_{q/p}.1)} \\
&= \frac{p^{k(k-1)/2} p^{(n-k)(n-k-1)/2} [n]_{p,q}[n-1]_{p,q} \cdots [2]_{p,q}.1}{p^{n(n-1)/2}([k]_{p,q}[k-1]_{p,q} \cdots [2]_{p,q}.1)([n-k]_{p,q}[n-k-1]_{p,q} \cdots [2]_{p,q}.1)} \\
&= \frac{p^{k(k-1)/2-n(n-1)/2} p^{(n-k)(n-k-1)/2} [n]_{p,q}!}{[k]_{p,q}![n-k]_{p,q}!} = p^{k(k-n)} \begin{bmatrix} n \\ k \end{bmatrix}_{p,q}.
\end{aligned}$$

Also, we have (p, q)-power basis as

$$\begin{aligned}
(x \ominus a)_{p,q}^n &= (x-a)(px-qa)(p^2x-q^2a) \cdots (p^{n-1}x-q^{n-1}a) \\
&= 1 \cdot p \cdot p^2 \cdots p^{n-1}(x-a)\left(x - a\frac{q}{p}\right)\left(x - a\frac{q^2}{p^2}\right) \cdots \left(x - a\frac{q^{n-1}}{p^{n-1}}\right) \\
&= p^{n(n-1)/2}(x-a)_{q/p}^n.
\end{aligned}$$

The following recurrence formula for moments holds true for (p, q) Bernstein polynomials:

Lemma 3.2 ([119]) *If we define*

$$U_{n,m}^{p,q}(x) := B_{n,p,q}(e_m, x) = \sum_{k=0}^{n} b_{n,k}^{p,q}(1, x)\left(\frac{p^{n-k}[k]_{p,q}}{[n]_{p,q}}\right)^m,$$

where $e_i = t^i, i = 0, 1, 2, \cdots$, *then for* $m \geq 1$, *we have the following recurrence relation:*

$$[n]_{p,q}\, U_{n,m+1}^{p,q}(px) = p^n x(1-px)D_{p,q}[U_{n,m}^{p,q}(x)] + [n]_{p,q}\, px U_{n,m}^{p,q}(px).$$

In particular, we have

$$B_{n,p,q}(e_0, x) = 1,\ B_{n,p,q}(e_1, x) = x,\ B_{n,p,q}(e_2, x) = x^2 + \frac{p^{n-1}x(1-x)}{[n]_{p,q}}.$$

For $0 < q < p \leq 1$, $f \in C[0, 1]$, $x \in [0, 1]$ and $n = 1, 2, \ldots$, Finta [88] considered the following slightly modified form of the operators (3.4.1) as

$$B_n^{p,q}(f; x) = \sum_{k=0}^{n} \begin{bmatrix} n \\ k \end{bmatrix}_{p,q} p^{[k(k-1)-n(n-1)]/2} x^k (1 \ominus x)_{p,q}^{n-k} f\left(p^n \frac{[k]_{p,q}}{[n]_{p,q}}\right) \tag{3.4.2}$$

The following direct estimates for the operators (3.4.2) were established in [88]:

Theorem 3.5 ([88]) *If the sequences* (p_n) *and* (q_n) *satisfy* $0 < q_n < p_n \leq 1$ *for* $n = 1, 2, \cdots$ *and* $p_n \to 1, q_n \to 1, p_n^n \to 1$ *as* $n \to \infty$, *then*

$$|B_n^{p_n,q_n}(f; x) - f(x)| \leq 2\omega\left(f, \left(2(1-p_n^n)x^2 + \frac{x(1-x)}{[n]_{q_n/p_n}}\right)^{1/2}\right)$$

for all $f \in C[0, 1]$ *and* $x \in [0, 1]$.

Theorem 3.6 ([88]) *If the sequences* (p_n) *and* (q_n) *satisfy* $0 < q_n < p_n \leq 1$ *for* $n = 1, 2, \ldots$ *and* $p_n \to 1, q_n \to 1, p_n^n \to 1$ *as* $n \to \infty$, *then*

$$|B_n^{p_n,q_n}(f; x) - B_n^{q_n/p_n}(f; x)| \leq 2\omega(f, (1-p_n^n)),$$

for all $f \in C[0, 1]$ *and* $x \in [0, 1]$.

It was pointed out by Finta [88] that the two different generalizations considered in [178, 181] of the q-Bernstein polynomials involve (p, q)-integers. The first [178] does not preserve even the constant functions, and the second [181] is a (q/p)-Bernstein polynomial. The (p, q)-Bernstein polynomials defined by (3.4.2) are different from those discussed in [178] and [181] and allows one to introduce the limit (p, q)-Bernstein operator.

One can remark here that in discrete operators like Bernstein and Baskakov operators the (p,q) analogue can easily be defined by using the link identities between quantum and post-quantum calculus.

The (p,q)-variant of Baskakov operators for $x \in [0,\infty)$ may be defined as

$$V_{n,p,q}(f,x) = \sum_{k=0}^{\infty} b_{n,k}^{p,q}(x) f\left(\frac{p^{n-1}[k]_{p,q}}{q^{k-1}[n]_{p,q}}\right), \tag{3.4.3}$$

where

$$b_{n,k}^{p,q}(x) = \begin{bmatrix} n+k-1 \\ k \end{bmatrix}_{p,q} p^{k+n(n-1)/2} q^{k(k-1)/2} \frac{x^k}{(1 \oplus x)_{p,q}^{n+k}}.$$

In case $p=1$, we get the q-Baskakov operators [34]. Also, if $p=q=1$, we obtain at once the well-known Baskakov operators.

As far as the Szász operators are concerned, they cannot easily be obtained by a q analogue of Szász operators. Acar [3] considered the (p,q)-analogue of Szász operators for $x \in [0,\infty)$, which are defined in the following way

$$S_{n,p,q}(f;x) = \sum_{k=0}^{n} s_{n,k}^{p,q}(x) f\left(\frac{[k]_{p,q}}{q^{k-2}[n]_{p,q}}\right), \tag{3.4.4}$$

where

$$s_{n,k}^{p,q}(x) = \frac{1}{E_{p,q}\left([n]_{p,q}x\right)} \frac{q^{k(k-1)/2}}{[k]_{p,q}!} \left([n]_{p,q}x\right)^k.$$

If $p=q=1$, we get at once the well-known Szász operators. Also in [3] some direct results were established. Acar et al. [10] extended the studies and they proposed the following modification of the operators (3.4.4) preserving the test function x^2 by

$$S_{n,p,q}^*(f;x) := \frac{1}{E_{p,q}\left([n]_{p,q} r_n(x)\right)} \sum_{k=0}^{\infty} f\left(\frac{[k]_{p,q}}{q^{k-2}[n]_{p,q}}\right) q^{\frac{k(k-1)}{2}} \frac{[n]_{p,q}^k r_n^k(x)}{[k]_{p,q}!}, \tag{3.4.5}$$

where $q \in (0,1)$, $p \in (q,1]$, $x \in [0,\infty)$ and

$$r_n(x) = \frac{-q^2 + \sqrt{q^4 + 4pq[n]_{p,q}^2 x^2}}{2pq[n]_{p,q}}.$$

The order of approximation of the operators (3.4.5) via the Peetre $\mathcal{K}$-functional, weighted approximation properties, and approximation for functions in a Lipschitz space were discussed in [10].

By $C_B[0,\infty)$, we denote the space of real-valued uniformly continuous and bounded functions f defined on the interval $[0,\infty)$. The norm $\|\cdot\|$ on the space $C_B[0,\infty)$ is given by

$$\|f\| = \sup_{0\leq x<\infty} |f(x)|.$$

The Peetre's K-functional is defined by

$$K_2(f,\delta) = \inf_{g\in W^2} \{\|f-g\| + \delta\|g''\|\},$$

where

$$W^2 = \{g \in C_B[0,\infty) : g', g'' \in C_B[0,\infty)\}.$$

By [78, p. 177, Theorem 2.4], there exists a positive constant $C>0$ such that

$$K_2(f,\delta) \leq C\omega_2(f,\sqrt{\delta}),\ \delta>0,$$

where

$$\omega_2(f,\sqrt{\delta}) = \sup_{0<h<\sqrt{\delta},x\in[0,\infty)} |f(x+2h) - 2f(x+h) + f(x)|$$

is the second order modulus of continuity of the function $f \in C_B[0,\infty)$.

Also, for $f \in C_B[0,\infty)$ the first order modulus of continuity is given by

$$\omega(f,\sqrt{\delta}) = \sup_{0<h<\sqrt{\delta},x\in[0,\infty)} |f(x+h) - f(x)|.$$

Let $H_m[0,\infty)$ be the set of all functions f defined on $[0,\infty)$ satisfying

$$|f(x)| \leq M_f\left(1+x^m\right),$$

where M_f is a certain constant depending only on f. By $C_{x^m}[0,\infty)$, we denote the subspace of all continuous functions belonging to $H_m[0,\infty)$. Also, let $C^*_{x^m}[0,\infty)$ be the subspace of all functions $f \in C_{x^m}[0,\infty)$, for which $\lim\limits_{|x|\to\infty} \frac{f(x)}{1+x^m}$ is finite. The norm on $C^*_{x^m}[0,\infty)$ is

$$\|f\|_{x^m} = \sup_{x\in[0,\infty)} |f(x)|(1+x^m)^{-1}.$$

Theorem 3.7 ([10]) *Let $q = q_n \in (0, 1)$, $p = p_n \in (q, 1]$ such that $q_n \to 1$, $p_n \to 1$ as $n \to \infty$. Then for each function $f \in C_2^* [0, \infty)$ we get*

$$\lim_{n\to\infty} \left\| S^*_{n,p_n,q_n} f - f \right\|_{x^2} = 0.$$

Theorem 3.8 ([10]) *Let $p, q \in (0, 1)$ such that $0 < q < p \leq 1$. Then we have*

$$\left| S^*_{n,p,q} (f; x) - f (x) \right| \leq M\omega_2 \left(f, \sqrt{\delta_n (x)} \right) + \omega \left(f, \frac{\left(\sqrt{p} - \sqrt{q}\right)}{\sqrt{p}} x \right),$$

for every $x \in [0, \infty)$ and $f \in C_B [0, \infty)$, where

$$\delta_n (x) = \frac{3x^2 \left(\sqrt{p} - \sqrt{q}\right)}{\sqrt{p}} + \frac{x}{[n]_{p,q}}.$$

Theorem 3.9 ([10]) *Let $q = q_n \in (0, 1)$, $p = p_n \in (q, 1]$ such that $q_n \to 1$, $p_n \to 1$, as $n \to \infty$. Let also $f \in C^*_m[0, \infty)$ and let $f^*(z) = f(z^2)$, $z \in [0, \infty)$. For all $t > 0$ and $x > 0$, we have*

$$\left| S^*_{n,p_n,q_n} (f, x) - f(x) \right| \leq 2\omega \left(f^*, \sqrt{\frac{1}{[n]_{p_n q_n}} + \frac{2x(\sqrt{p_n} - \sqrt{q_n})}{\sqrt{p_n}}} \right)$$

Theorem 3.10 ([10]) *Let $0 < \alpha \leq 1$ and E be any subset of the interval $[0, \infty)$. Then, if $f \in C_B [0, \infty)$ is locally in $Lip (\alpha)$, i.e., the condition*

$$|f (y) - f (x)| \leq L |y - x|^{\alpha}, \ y \in E \text{ and } x \in [0, \infty) \tag{3.4.6}$$

holds, then, for each $x \in [0, \infty)$, we have

$$\left| S^*_{n,p,q} (f; x) - f (x) \right| \leq L \left\{ \delta_n^{\frac{\alpha}{2}} (x) + 2 (d (x, E))^{\alpha} \right\},$$

where L is a constant depending on α and f; and $d (x, E)$ is the distance between x and E defined by

$$d (x, E) = \inf \{|t - x| : t \in E\}.$$

Theorem 3.11 ([10]) *Let $f \in C_B [0, \infty)$ and $0 < \alpha \leq 1$. Then, for all $x \in [0, \infty)$ we have*

$$\left| S^*_{n,p,q} (f; x) - f (x) \right| \leq \tilde{\omega}_a (f, x) \delta_n^{\frac{\alpha}{2}} (x).$$

Chapter 4
(p, q)-Integral Operators

Two important integral modifications of the discrete operators are due to Kantorovich and Durrmeyer. These type of modifications could be used to approximate integral functions. There are several q analogues of integral modifications. In this chapter we present approximation properties of some of the operators based on (p, q)-integral.

4.1 Kantorovich Type Operators

4.1.1 (p, q)-Bernstein–Kantorovich Operators

Recently Acar–Aral–Mohiuddine [7] proposed a Kantorovich type modification of (p, q)-Bernstein operators using the (p, q)-integral (discussed in [5]) as

$$\overline{B}_n^{p,q}(f, x) = [n+1]_{p,q} \sum_{k=0}^{n} b_{n,k}^{p,q}(1, x) p^{-k} \int_{q[k]_{p,q}/[n+1]_{p,q}}^{[k+1]_{p,q}/[n+1]_{p,q}} f(t) d_{p,q}t \qquad (4.1.1)$$

where

$$b_{n,k}^{p,q}(1, x) = \begin{bmatrix} n \\ k \end{bmatrix}_{p,q} p^{[k(k-1)-n(n-1)]/2} x^k (1 \ominus x)_{p,q}^{n-k}.$$

Acar–Aral–Mohiuddine [7] obtained the following uniform convergence and other following direct estimates for the operators (4.1.1):

V. Gupta et al., *Recent Advances in Constructive Approximation Theory*,
Springer Optimization and Its Applications 138,
https://doi.org/10.1007/978-3-319-92165-5_4

Theorem 4.1 ([7]) *Let $p = p_n$ and $q = q_n$ be the sequences defined in Remark 2 [7]. Then for every $f \in C[0, 1]$, the operators $\overline{B}_n^{p,q}(f, x)$ converge to f uniformly for n sufficiently large.*

Theorem 4.2 ([7]) *Let $0 < q < p \leq 1$ and $x \in [0, 1]$. If $f \in C[0, 1]$, then we have*

$$|\overline{B}_n^{p,q}(f, x) - f(x)| \leq 2\omega(f, \delta_n(p, q; x)).$$

Theorem 4.3 ([7]) *Let $0 < q < p \leq 1$ and $x \in [0, 1]$. If $f \in C[0, 1]$, then there exists a positive constant C such that*

$$|\overline{B}_n^{p,q}(f, x) - f(x)| \leq C\omega_2(f, \sqrt{\delta_n(p, q)}).$$

Theorem 4.4 ([7]) *Let $0 < q < p \leq 1$ and $x \in [0, 1]$. If $f \in Lip_M\alpha$, then the inequality*

$$|\overline{B}_n^{p,q}(f, x) - f(x)| \leq M\delta_n^{\alpha}(p, q; x))$$

holds true.

Theorem 4.5 ([7]) *Let $0 < q < p \leq 1$ and $x \in [0, 1]$. If f has a continuous derivative, then the inequality*

$$|\overline{B}_n^{p,q}(f, x) - f(x)| \leq M\delta_n^{\alpha}(p, q) + 2\delta_n(p, q)\omega(f', \delta_n(p, q))$$

holds true, where M is a positive constant.

Very recently Finta [88] proposed the following (p, q)-Bernstein–Kantorovich operators based on Riemann type q integral (3.2.2) as

$$\overline{B}_{n,p,q}(f, x) = \frac{[n+1]_{p,q}}{p^n} \sum_{k=0}^{n} p^{[k(k-1)-n(n-1)]/2} x^k (1 \ominus x_{p,q})^{n-k}$$
$$q^{-k} \int_{p^{n+1}[k]_{p,q}/[n+1]_{p,q}}^{p^n[k+1]_{p,q}/[n+1]_{p,q}} f(u) d_{q/p}^R u, \tag{4.1.2}$$

where $f \in C[0, 1]$, $x \in [0, 1]$. The main purpose for which Finta [88] considered the Riemann q integral while defining the above (p, q) variant was that in the (p, q) integral, considered by Sadjang [209] over $[0, a]$, some of the nodes lie outside the interval in which the integral is defined. It was observed that the definition of (p, q) integral considered by Sadjang [209] is not optimal.

Finta [88] proved the following estimate:

Theorem 4.6 *If the sequences* (p_n) *and* (q_n) *satisfy* $0 < q_n < p_n \leq 1$ *for* $n = 1, 2, 3, \ldots$ *and* $p_n \to 1, q_n \to 1, p_n^n \to 1$ *as* $n \to \infty$, *then*

$$|\overline{B}_{n,p,q}(f, x) - f(x)| \leq 2\omega(f, \sqrt{\delta_n(x)}),$$

where

$$\delta_n(x) = \left[2(1 - p_n^n)\frac{[n]_{q_n/p_n}}{[n+1]_{q_n/p_n}} + \left(1 - \frac{[n]_{q_n/p_n}}{[n+1]_{q_n/p_n}}\right)^2 - \frac{[n]_{q_n/p_n}}{[n+1]^2_{q_n/p_n}}\right]x^2$$
$$+\frac{3[n]_{q_n/p_n}x}{[n+1]_{q_n/p_n}} + \frac{1}{[n+1]^2_{q_n/p_n}}.$$

4.1.2 (p, q)-Baskakov–Kantorovich Operators

For $x \in [0, \infty)$, the (p, q)-variant of Baskakov–Kantorovich operators (see [118]) is defined as

$$\overline{V}_n^{p,q}(f, x) = [n]_{p,q} \sum_{k=0}^{\infty} b_{n,k}^{p,q}(x) p^{-k} q^k \int_{[k]_{p,q}/q^{k-1}[n]_{p,q}}^{[k+1]_{p,q}/q^k[n]_{p,q}} f(t) d_{p,q}t \quad (4.1.3)$$

where $b_{n,k}^{p,q}(x)$ is as defined in (3.4.3).

The following Lorentz type lemma for (p, q)-Baskakov basis, established in [118], will be used in the sequel.

Lemma 4.1 *For* $n, k \geq 0$, *we have*

$$x(1 + px)D_{p,q}b_{n,k}^{p,q}(x) = \left(\frac{p^{n-1}[k]_{p,q}}{q^{k-1}[n]_{p,q}} - qx\right)\frac{[n]_{p,q}}{qp^{n-1}}b_{n,k}^{p,q}(qx).$$

Proof By simple computation using the definition of (p, q)-derivative, we have

$$D_{p,q}\left(\frac{1}{(1 \oplus x)_{p,q}^{n+k}}\right) = -\frac{p\,[n+k]_{p,q}}{(1 \oplus px)_{p,q}^{n+k+1}}, \; D_{p,q}x^k = [k]_{p,q}x^{k-1}.$$

Applying the product rule

$$D_{p,q}(f(x)g(x)) = f(px)D_{p,q}g(x) + g(qx)D_{p,q}f(x),$$

for (p, q)-derivative, we can write

$$\begin{aligned} & D_{p,q}\left(\frac{x^k}{(1\oplus x)^{n+k}_{p,q}}\right) \\ &= [k]_{p,q}\frac{x^{k-1}}{(1\oplus qx)^{n+k}_{p,q}} - p^{k+1}[n+k]_{p,q}\frac{x^k}{(1\oplus px)^{n+k+1}_{p,q}} \\ &= [k]_{p,q}\frac{x^{k-1}}{(1\oplus qx)^{n+k}_{p,q}} - [n+k]_{p,q}\frac{x^k}{(1+px)p^{n-1}(1\oplus qx)^{n+k}_{p,q}}. \end{aligned}$$

Thus using

$$[n+k]_{p,q} = p^n[k]_{p,q} + q^k[n]_{p,q},$$

we get

$$\begin{aligned} & x(1+px)D_{p,q}\left(\frac{x^k}{(1\oplus x)^{n+k}_{p,q}}\right) \\ &= \left[[k]_{p,q}(1+px)p^{n-1} - [n+k]_{p,q}\,x\right]\frac{x^k}{p^{n-1}(1\oplus qx)^{n+k}_{p,q}} \\ &= \left[[k]_{p,q} - \frac{q^k[n]_{p,q}\,x}{p^{n-1}}\right]\frac{x^k}{(1\oplus qx)^{n+k}_{p,q}} \\ &= \left[\frac{p^{n-1}[k]_{p,q}}{q^{k-1}[n]_{p,q}} - qx\right]\frac{[n]_{p,q}}{qp^{n-1}}\frac{(qx)^k}{(1\oplus qx)^{n+k}_{p,q}}. \end{aligned}$$

Therefore, we have

$$x(1+px)D_{p,q}b^{p,q}_{n,k}(x) = \left(\frac{p^{n-1}[k]_{p,q}}{q^{k-1}[n]_{p,q}} - qx\right)\frac{[n]_{p,q}}{qp^{n-1}}b^{p,q}_{n,k}(qx).$$

■

Remark 4.1 We may note here that for the special case $p = q = 1$ in the above lemma, we may capture at once the Lorentz type relation of the Baskakov operators, viz.

$$x(1+x)\frac{d}{dx}[b_{n,k}(x)] = (k-nx)b_{n,k}(x),$$

where the Baskakov basis is given by

$$b_{n,k}(x) = \binom{n+k-1}{k} \frac{x^k}{(1+x)^{n+k}}.$$

The moments of (p, q)-Baskakov operators, satisfy the following:

Lemma 4.2 ([118]) *If we define*

$$T_{n,m}^{p,q}(x) := V_{n,p,q}(e_m, x) = \sum_{k=0}^{\infty} b_{n,k}^{p,q}(x) \left(\frac{p^{n-1}[k]_{p,q}}{q^{k-1}[n]_{p,q}} \right)^m,$$

where $e_i = t^i, i = 0, 1, 2, \ldots$, *then for* $m \geq 1$, *we have the following recurrence relation:*

$$\begin{aligned} &[n]_{p,q}\, T_{n,m+1}^{p,q}(qx) \\ &= qp^{n-1}x(1+px)D_{p,q}[T_{n,m}^{p,q}(x)] + [n]_{p,q}\, qxT_{n,m}^{p,q}(qx). \end{aligned}$$

In particular, we have

$$V_{n,p,q}(e_0, x) = 1,\; V_{n,p,q}(e_1, x) = x$$

and

$$V_{n,p,q}(e_2, x) = x^2 + \frac{p^{n-1}x}{[n]_{p,q}} \left(1 + \frac{p}{q}x\right).$$

Lemma 4.3 ([118]) *For* $x \in [0, \infty]$, $0 < q < p \leq 1$, *the moments of operators (4.1.3) are given by*

1. $\overline{V}_n^{p,q}(e_0, x) = 1$
2. $\overline{V}_n^{p,q}(e_1, x) = \frac{1}{[2]_{p,q}[n]_{p,q}} + \frac{x}{qp^{n-1}}$
3. $\overline{V}_n^{p,q}(e_2, x) = \frac{[n+1]_{p,q}x^2}{[n]_{p,q}q^3p^{2n-2}} + \frac{x}{p^{n-1}q[n]_{p,q}}\left[\frac{1}{q} + \frac{(2p+q)p}{[3]_{p,q}}\right] + \frac{1}{[3]_{p,q}[n]_{p,q}^2}.$

For $f \in C_B[0, \infty)$ the Steklov mean is defined as

$$f_h(t) = \frac{4}{h^2} \int_0^{\frac{h}{2}} \int_0^{\frac{h}{2}} [2f(t+u+v) - f(t+2(u+v))]\, du dv \quad (4.1.4)$$

By simple computation, it is observed that

(i) $\|f_h - f\|_{C_B} \leq \widetilde{\omega}_2(f, h)$.
(ii) If f is continuous and $f_h', f'' \in C_B$, then

$$\|f_h'\|_{C_B} \leq \frac{5}{h}\widetilde{\omega}(f, h),\; \|f_h''\|_{C_B} \leq \frac{9}{h^2}\widetilde{\omega}_2(f, h),$$

where the first and second order modulus of continuity for $\delta \geq 0$ are respectively defined as

$$\widetilde{\omega}(f,\delta) = \sup_{\substack{x,u,v\geq 0 \\ |u-v|\leq\delta}} |f(x+u) - f(x+v)|$$

and

$$\widetilde{\omega}_2(f,\delta) = \sup_{\substack{x,u,v\geq 0 \\ |u-v|\leq\delta}} |f(x+2u) - 2f(x+u+v) + f(x+2v)|.$$

Theorem 4.7 ([118]) *Let $q \in (0,1)$ and $p \in (q,1]$. The operator $K_n^{p,q}$ maps the space C_B into C_B and*

$$\left\| \overline{V}_n^{p,q}(f) \right\|_{C_B} \leq \|f\|_{C_B}.$$

Theorem 4.8 ([118]) *Let $q \in (0,1)$ and $p \in (q,1]$. If $f \in C_B$, then*

$$\begin{aligned}
&\left|\overline{V}_n^{p,q}(f,x) - f(x)\right| \\
&\leq 5\widetilde{\omega}\left(f, \frac{1}{\sqrt{[n]_{p,q}}}\right)\left(\frac{1}{[2]_{p,q}\sqrt{[n]_{p,q}}} + \left(\frac{1}{qp^{n-1}} - 1\right)x\right) \\
&\quad + \frac{9}{2}\widetilde{\omega}_2\left(f, \frac{1}{\sqrt{[n]_{p,q}}}\right)\left[\left(\frac{[n+1]_{p,q}}{q^3p^{2n-2}} - \frac{2[n]_{p,q}}{qp^{n-1}} + [n]_{p,q}\right)x^2\right. \\
&\quad \left. + \left(\frac{1}{p^{n-1}q}\left[\frac{1}{q} + \frac{(2p+q)p}{[3]_{p,q}}\right] - \frac{2}{[2]_{p,q}}\right)x + \frac{1}{[3]_{p,q}[n]_{p,q}} + 2\right].
\end{aligned}$$

Proof For $x \geq 0$ and $n \in \mathbb{N}$ and using the Steklov function f_h defined by (4.1.4), we can write

$$\begin{aligned}
&\left|\overline{V}_n^{p,q}(f,x) - f(x)\right| \\
&\leq \overline{V}_n^{p,q}(|f - f_h|, x) + \left|\overline{V}_n^{p,q}(f_h - f_h(x), x)\right| + |f_h(x) - f(x)|.
\end{aligned}$$

First by Theorem 4.7 and property (i) of Steklov mean, we have

$$\overline{V}_n^{p,q}(|f - f_h|, x) \leq \left\|\overline{V}_n^{p,q}(f - f_h)\right\|_{C_B} \leq \|f - f_h\|_{C_B} \leq \widetilde{\omega}_2(f,h).$$

Also, by Taylor's expansion, we have

$$\begin{aligned} &\left|\overline{V}_n^{p,q}\left(f_h - f_h\left(x\right), x\right)\right| \\ &\leq \left|f_h'\left(x\right)\right| \overline{V}_n^{p,q}\left(t-x, x\right) + \frac{1}{2}\left\|f''\right\|_{C_B} \overline{V}_n^{p,q}\left(\left(t-x\right)^2, x\right). \end{aligned}$$

By Lemma 4.3, we have

$$\begin{aligned} \left|\overline{V}_n^{p,q}\left(f_h - f_h\left(x\right), x\right)\right| \leq \frac{5}{h}\widetilde{\omega}\left(f,h\right)\left(\frac{1}{[2]_{p,q}[n]_{p,q}} + \frac{x}{qp^{n-1}} - x\right) \\ + \frac{9}{2h^2}\widetilde{\omega}_2\left(f,h\right)\overline{V}_n^{p,q}\left(\left(t-x\right)^2, x\right), \end{aligned}$$

where

$$\begin{aligned} &\overline{V}_n^{p,q}\left(\left(t-x\right)^2, x\right) \\ &= \frac{[n+1]_{p,q}x^2}{[n]_{p,q}q^3p^{2n-2}} + \frac{x}{p^{n-1}q[n]_{p,q}}\left[\frac{1}{q} + \frac{(2p+q)p}{[3]_{p,q}}\right] \\ &\quad + \frac{1}{[3]_{p,q}[n]_{p,q}^2} - 2x\left(\frac{1}{[2]_{p,q}[n]_{p,q}} + \frac{x}{qp^{n-1}}\right) + x^2 \\ &= \left(\frac{[n+1]_{p,q}}{[n]_{p,q}q^3p^{2n-2}} - \frac{2}{qp^{n-1}} + 1\right)x^2 \\ &\quad + \left(\frac{1}{p^{n-1}q[n]_{p,q}}\left[\frac{1}{q} + \frac{(2p+q)p}{[3]_{p,q}}\right] - \frac{2}{[2]_{p,q}[n]_{p,q}}\right)x \\ &\quad + \frac{1}{[3]_{p,q}[n]_{p,q}^2} \end{aligned}$$

for $x \geq 0$, $h > 0$. Setting

$$h = \sqrt{\frac{1}{[n]_{p,q}}},$$

we get the desired result. ■

Theorem 4.9 ([118]) *Let $f \in C_B[0,\infty)$. Then for all $n \in \mathbb{N}$, there exists an absolute constant $C > 0$ such that*

$$|\overline{V}_n^{p,q}(f,x) - f(x)| \leq C\omega_2(f, \delta_n(x)) + \omega(f, \alpha_n(x)),$$

where

$$\delta_n(x) = \left\{ \overline{V}_n^{p,q}((t-x)^2, x) + (\overline{V}_n^{p,q}((t-x), x))^2 \right\}^{1/2}$$

and

$$\alpha_n(x) = \left| \frac{1}{[2]_{p,q}[n]_{p,q}} + \left(\frac{1}{qp^{n-1}} - 1 \right) x \right|.$$

Proof For $x \in [0, \infty)$, we consider the auxiliary operators $\overline{\overline{V}}_n^{p,q}(f, x)$ defined by

$$\overline{\overline{V}}_n^{p,q}(f, x) = \overline{V}_n^{p,q}(f, x) + f(x) - f\left(\frac{1}{[2]_{p,q}[n]_{p,q}} + \frac{x}{qp^{n-1}} \right).$$

It is observed that $\overline{\overline{V}}_n^{p,q}(f, x)$ preserve linear functions. Let $x \in [0, \infty)$ and $g \in W^2$. Applying the Taylor's formula

$$g(t) = g(x) + g'(x)(t-x) + \int_x^t (t-u)g''(u)du,$$

we have

$$\begin{aligned}
&\overline{\overline{V}}_n^{p,q}(g, x) - g(x) \\
&= \overline{\overline{V}}_n^{p,q}\left(\int_x^t (t-u)g''(u)du, x \right) \\
&= \overline{V}_n^{(p,q)}\left(\int_x^t (t-u)g''(u)du, x \right) \\
&\quad - \int_x^{\frac{1}{[2]_{p,q}[n]_{p,q}} + \frac{x}{qp^{n-1}}} \left(\frac{1}{[2]_{p,q}[n]_{p,q}} + \frac{x}{qp^{n-1}} - u \right) g''(u)du \\
&= \overline{V}_n^{(p,q)}\left(\int_x^t (t-u)g''(u)du, x \right) \\
&\quad - \int_x^{\frac{1}{[2]_{p,q}[n]_{p,q}} + \frac{x}{qp^{n-1}}} \left(\frac{1}{[2]_{p,q}[n]_{p,q}} + \frac{x}{qp^{n-1}} - u \right) g''(u)du.
\end{aligned}$$

On the other hand,

$$\left| \int_x^t (t-u)g''(u)du \right| \leq \|g''\| \int_x^t |t-u|du \leq (t-x)^2 \|g''\|,$$

and

$$\left|\int_x^{\frac{1}{[2]_{p,q}[n]_{p,q}}+\frac{x}{qp^{n-1}}}\left(\frac{1}{[2]_{p,q}[n]_{p,q}}+\frac{x}{qp^{n-1}}-u\right)g''(u)du\right|$$

$$\leq\left(\frac{1}{[2]_{p,q}[n]_{p,q}}+\frac{x}{qp^{n-1}}-x\right)^2\|g''\|.$$

Therefore, we have

$$\begin{aligned}
&|\overline{\overline{V}}_n^{p,q}(g,x)-g(x)|\\
&=\left|\overline{V}_n^{p,q}\left(\int_x^t(t-u)g''(u)du,x\right)\right|\\
&+\left|\int_x^{\frac{1}{[2]_{p,q}[n]_{p,q}}+\frac{x}{qp^{n-1}}}\left(\frac{1}{[2]_{p,q}[n]_{p,q}}+\frac{x}{qp^{n-1}}-u\right)g''(u)du\right|\\
&\leq\|g''\|\overline{V}_n^{(p,q)}((t-x)^2,x)+\left(\frac{1}{[2]_{p,q}[n]_{p,q}}+\frac{x}{qp^{n-1}}-x\right)^2\|g''\|\\
&=\delta_n^2(x)\|g''\|.
\end{aligned}$$

Also, we have

$$|\overline{\overline{V}}_n^{p,q}(f,x)|\leq|\overline{V}_n^{p,q}(f,x)|+2\|f\|\leq 3\|f\|.$$

Therefore,

$$\begin{aligned}
&|\overline{V}_n^{p,q}(f,x)-f(x)|\\
&\leq|\overline{\overline{V}}_n^{p,q}(f-g,x)-(f-g)(x)|+\left|f\left(\frac{1}{[2]_{p,q}[n]_{p,q}}+\frac{x}{qp^{n-1}}\right)-f(x)\right|\\
&+|\overline{\overline{V}}_n^{p,q}(g,x)-g(x)|\\
&\leq|\overline{\overline{V}}_n^{p,q}(f-g,x)|+|(f-g)(x)|+\left|f\left(\frac{1}{[2]_{p,q}[n]_{p,q}}+\frac{x}{qp^{n-1}}\right)-f(x)\right|\\
&+|\overline{\overline{V}}_n^{p,q}(g,x)-g(x)|\\
&\leq 4\|f-g\|+\omega\left(f,\left|\frac{1}{[2]_{p,q}[n]_{p,q}}+\left(\frac{1}{qp^{n-1}}-1\right)x\right|\right)+\delta_n^2(x)\|g''\|.
\end{aligned}$$

Finally taking the infimum on the right-hand side over all $g\in W^2$, we get

$$|\overline{V}_n^{p,q}(f,x)-f(x)|\leq 4K_2(f,\delta_n^2(x))+\omega(f,\alpha_n(x)).$$

By the property of K-functional, we have

$$|\overline{V}_n^{p,q}(f, x) - f(x)| \leq C\omega_2(f, \delta_n(x)) + \omega(f, \alpha_n(x)).$$

This completes the proof of the theorem. ∎

Finally, Gupta in [118] discussed the following weighted approximation theorem.

Theorem 4.10 ([118]) *Let $p = p_n$ and $q = q_n$ satisfy $0 < q_n < p_n \leq 1$ and for n sufficiently large $p_n \to 1$, $q_n \to 1$, $q_n^n \to 1$ and $p_n^n \to 1$. For each $f \in C_{x^2}^*[0, \infty)$, we have*

$$\lim_{n\to\infty} \left\| \overline{V}_n^{p_n,q_n}(f) - f \right\|_{x^2} = 0.$$

Proof Following [118] and references therein, in order to complete the proof of the theorem, it is sufficient to verify the following three conditions

$$\lim_{n\to\infty} \left\| \overline{V}_n^{p_n,q_n}(e_\nu, x) - x^\nu \right\|_{x^2} = 0, \quad \nu = 0, 1, 2. \tag{4.1.5}$$

Since $\overline{V}_n^{p_n,q_n}(e_0, x) = 1$ the first condition of (4.1.5) is fulfilled for $\nu = 0$. We can write

$$\begin{aligned}&\left\| \overline{V}_n^{p_n,q_n}(e_1, x) - x \right\|_{x^2}\\ &\leq \left(\frac{1}{[2]_{p_n,q_n}[n]_{p_n,q_n}} + \frac{(1 - q_n p_n^{n-1})x}{q_n p_n^{n-1}} \right) \sup_{x\in[0,\infty)} \frac{1}{1 + x^2}.\end{aligned}$$

and

$$\begin{aligned}&\left\| \overline{V}_n^{p_n,q_n}(e_2, x) - x^2 \right\|_{x^2}\\ &\leq \left(\frac{[n+1]_{p_n,q_n}x^2}{[n]_{p_n,q_n}q_n^3 p_n^{2n-2}} + \frac{x}{p_n^{n-1}q_n[n]_{p_n,q_n}} \left[\frac{1}{q_n} + \frac{(2p_n + q_n)p_n}{[3]_{p_n,q_n}} \right]\right.\\ &\left.+ \frac{1}{[3]_{p_n,q_n}[n]_{p_n,q_n}^2} - x^2 \right) \sup_{x\in[0,\infty)} \frac{1}{1 + x^2}\end{aligned}$$

which implies that

$$\lim_{n\to\infty} \left\| \overline{V}_n^{p_n,q_n}(e_\nu, x) - x^\nu \right\|_{x^2} = 0, \nu = 1, 2.$$

Thus the proof is complete. ∎

Remark 4.2 For $q \in (0, 1)$ and $p \in (q, 1]$ it is seen in [118] that

$$\lim_{n\to\infty} [n]_{p,q} = 1/(p-q).$$

In order to obtain convergence estimates of (p, q)-Baskakov–Kantorovich operators, we assume $p = (p_n)$, $q = (q_n)$ such that $0 < q_n < p_n \leq 1$ and for n sufficiently large $p_n \to 1$, $q_n \to 1$, $p_n^n \to 1$, $q_n^n \to 1$ and

$$\lim_{n\to\infty} [n]_{p_n,q_n} = \infty.$$

One may consider $p_n^n \to a$ and $q_n^n \to b$ for which $[n]_{p,q} \to \infty$ as $n \to \infty$.

4.1.3 (p, q)-Szász–Mirakyan–Kantorovich Operators

Sharma and Gupta in [213] proposed the (p, q)-Szász–Mirakyan–Kantorovich operators for $n \in \mathbb{N}$ and $f : [0, \infty) \to \mathbb{R}$ as follows:

$$\overline{S}_n^{p,q}(f, x) = [n]_{p,q} \sum_{k=0}^{\infty} s_{n,k}^{p,q}(x) \frac{q^{k-2}}{p^k} \int_{[k]_{p,q}/q^{k-3}[n]_{p,q}}^{[k+1]_{p,q}/q^{k-2}[n]_{p,q}} f(t) d_{p,q}t \quad (4.1.6)$$

where

$$s_{n,k}^{p,q}(x) = \frac{1}{E_{p,q}([n]_{p,q}x)} q^{\frac{k(k-1)}{2}} \frac{[n]_{p,q}^k x^k}{[k]_{p,q}!}.$$

Remark 4.3 For $q \in (0, 1)$ and $p \in (q, 1]$, by simple computations

$$\lim_{n\to\infty} [n]_{p,q} = 1/(p-q)\,.$$

In order to obtain results for order of convergence of the operator, we take $q_n \in (0, 1)$, $p_n \in (q_n, 1]$ such that $\lim_{n\to\infty} p_n = 1$ and $\lim_{n\to\infty} q_n = 1$, so that

$$\lim_{n\to\infty} \frac{1}{[n]_{p_n,q_n}} = 0\,.$$

Such a sequence can always be constructed. For example, we can consider

$$q_n = 1 - 1/n \ \text{ and } \ p_n = 1 - 1/2n\,.$$

Clearly

$$\lim_{n\to\infty} p_n^n = e^{-1/2}, \quad \lim_{n\to\infty} q_n^n = e^{-1}$$

and

$$\lim_{n\to\infty} \frac{1}{[n]_{p_n,q_n}} = 0 .$$

In [213] some direct results have been established for the operators (4.1.6). The notations for the class of functions are as mentioned in previous section:

Theorem 4.11 ([213]) *Let $(p_n)_n$ and $(q_n)_n$ be the sequences defined in Remark 4.3. Then for each $f \in C[0,\infty)$, $\overline{S}_n^{(p_n,q_n)}(f;x)$ converges uniformly to f.*

Theorem 4.12 ([213]) *Let $(p_n)_n$ and $(q_n)_n$ be the sequences defined in Remark 4.3. Let $f \in C_B[0,\infty)$. Then for all $n \in \mathbb{N}$, there exists an absolute constant $C > 0$ such that*

$$|\overline{S}_n^{(p_n,q_n)}(f;x) - f(x)| \le C\omega_2(f,\delta_n(x)) + \omega(f,\alpha_n(x)),$$

where

$$\delta_n(x) = \left\{\Phi_2^{(p_n,q_n)}(x) + (\Phi_1^{(p_n,q_n)}(x))^2\right\}^{\frac{1}{2}},$$

$$\Phi_1^{(p,q)}(x) = (q-1)x + \frac{q^2}{(p+q)[n]_{p,q}},$$

$$\Phi_2^{(p,q)}(x) = (pq-2q+1)x^2 + \left(\frac{2q^4+3pq^3+p^2q^2}{(p^2+pq+q^2)[n]_{p,q}} - \frac{2q^2}{(p+q)[n]_{p,q}}\right)x$$
$$+\frac{q^4}{(p^2+pq+q^2)[n]_{p,q}^2}.$$

and

$$\alpha_n(x) = \left|\frac{q_n^2}{[n]_{p_n,q_n}(p_n+q_n)} + (q_n-1)x\right|.$$

Theorem 4.13 ([213]) *Let $(p_n)_n$ and $(q_n)_n$ be the sequences defined in Remark 4.3. Then for $f \in C_{x^2}[0,\infty)$, defined in Section 3.4, $\omega_{a+1}(f;\delta)$ be the modulus of continuity on the interval $[0,a+1] \subset [0,\infty)$, $a > 0$, and for every $n > 1$,*

$$\|\overline{S}_n^{(p_n,q_n)}(f;x) - f\|_{C[0,a]} \le 6M_f(1+a^2)\lambda_n + 2\omega_{a+1}(f;\sqrt{\lambda_n}).$$

Here,

$$\lambda_n = (1-p_nq_n)a^2 + \frac{1}{[n]_{p_n,q_n}(p_n+q_n)(p_n^2+p_nq_n+q_n^2)}\left(\frac{6a}{p_n+q_n} + \frac{1}{[n]_{p_n,q_n}}\right).$$

Theorem 4.14 ([213]) *Let $0 < q_n < p_n \le 1$, such that $p_n \to 1$, $p_n \to 1$, $p_n^n \to a$ and $q_n^n \to b$ as $n \to \infty$. For any $f \in C_{x^2}^*[0,\infty)$, such that $f', f'' \in C_{x^2}^*[0,\infty)$*

(defined in Section 3.4), we have

$$\lim_{n\to\infty}[n]_{p_n,q_n}|\overline{S}_n^{(p_n,q_n)}(f;x)-f(x)|=(\alpha x+1/2)f'(x)+x(\gamma x+1)f''(x)/2$$

uniformly on $[0, A]$ *for any* $A > 0$. *Here*

$$\alpha=\lim_{n\to\infty}[n]_{p_n,q_n}(q_n-1) \text{ and } \gamma=[n]_{p_n,q_n}\lim_{n\to\infty}(p_nq_n-2q_n+1)\,.$$

4.2 Durrmeyer Type Operators

4.2.1 (p, q)-Bernstein–Durrmeyer Operators

Using the form (3.3.2) of (p, q)-Beta function Gupta and Aral proposed the (p, q)-analogue of Bernstein–Durrmeyer operator for $x \in [0, 1]$ as

$$\widetilde{B}_n^{p,q}(f;x)=[n+1]_{p,q}\sum_{k=0}^{n}p^{-[n^2+3n-k^2-k]/2}b_{n,k}^{p,q}(1,x)$$
$$\int_0^1 b_{n,k}^{p,q}(p,pqt)f(t)\,d_{p,q}t \tag{4.2.1}$$

where

$$b_{n,k}^{p,q}(1,x)=\begin{bmatrix} n\\ k\end{bmatrix}_{p,q}p^{[k(k-1)-n(n-1)]/2}x^k(1\ominus x)_{p,q}^{n-k}.$$

and

$$b_{n,k}^{p,q}(p,pqt)=\begin{bmatrix} n\\ k\end{bmatrix}_{p,q}(pt)^k\,(p\ominus pqt)_{p,q}^{n-k}.$$

Lemma 4.4 ([128]) *For the operators defined by (4.2.1) we have for* $x \in [0, 1]$ *the following moments*

$$\widetilde{B}_n^{p,q}(1;x)=1,\ \widetilde{B}_n^{p,q}(t;x)=\frac{p^n+q[n]_{p,q}x}{[n+2]_{p,q}},$$

$$\widetilde{B}_n^{p,q}(t^2;x)=\frac{p^{2n}[2]_{p,q}}{[n+2]_{p,q}[n+3]_{p,q}}+\frac{(2q^2+qp)p^n[n]_{p,q}x}{[n+2]_{p,q}[n+3]_{p,q}}$$
$$+\frac{q^3[n]_{p,q}[x^2[n]_{p,q}+p^{n-1}x(1-x)]}{[n+2]_{p,q}[n+3]_{p,q}}.$$

Lemma 4.5 ([128]) *Let* $n > 3$ *be a given natural number and let* $q_0 = q_0(n) \in (0, p)$ *be the least number such that*

$$p^{2n+1}q - p^{n+1}q^{n+1} + p^{2n-1}q^3 - p^{n-1}q^{n+3} + p^{2n}q^2 - p^n q^{n+2} - 2p^{2n+3} + 2p^n q^{n+3} > 0$$

for every $q \in (q_0, 1)$. *Then*

$$\widetilde{B}_n^{p,q}((t-x)^2, x) \leq \frac{2}{[n+2]_{p,q}} \left(\varphi^2(x) + \frac{1}{[n+3]_{p,q}} \right),$$

where $\varphi^2(x) = x(1-x), \ x \in [0, 1]$.

Next we consider the class:

$$W^2 = \left\{ g \in C[0, 1] : g'', g'' \in C[0, 1] \right\} \text{ for } \delta > 0 \, .$$

The $K-$functional is defined as

$$K_2(f, \delta) = \inf\left\{ ||f - g|| + \eta ||g''|| : g \in W^2 \right\},$$

where the norm-$\|.\|$ denotes the uniform norm on $C[0, 1]$. Applying the well-known inequality due to DeVore and Lorentz [78], there exists a absolute constant $C > 0$ such that

$$K_2(f, \delta) \leq C\omega_2(f, \sqrt{\delta}), \tag{4.2.2}$$

where ω_2 denotes the usual second order modulus of continuity.

The first main result of [128] is the following local theorem:

Theorem 4.15 ([128]) *Let* $n > 3$ *be a natural number and let* $q_0 = q_0(n) \in (0, p)$ *be defined as in Lemma 4.5. Then there exists an absolute constant* $C > 0$ *such that*

$$|\widetilde{B}_n^{p,q}(f, x) - f(x)| \leq C\omega_2 \left(f, [n+2]_{p,q}^{-1/2} \delta_n(x) \right) + \omega \left(f, \frac{1-x}{[n+2]_{p,q}} \right),$$

where

$$f \in C[0, 1], \ \delta_n^2(x) = \varphi^2(x) + \frac{1}{[n+3]_{p,q}}, \ x \in [0, 1] \text{ and } q \in (q_0, 1).$$

Proof For $f \in C[0, 1]$ we define

$$\widetilde{D}_n^{p,q}(f, x) = \widetilde{B}_n^{p,q}(f, x) + f(x) - f\left(\frac{p^n + q[n]_{p,q}x}{[n+2]_{p,q}} \right).$$

Then, by Lemma 4.4, we immediately get

$$\widetilde{D}_n^{p,q}(1,x) = \widetilde{B}_n^{p,q}(1,x) = 1 \tag{4.2.3}$$

and

$$\widetilde{D}_n^{p,q}(t,x) = \widetilde{B}_n^{p,q}(t,x) + x - \frac{p^n + q[n]_{p,q}x}{[n+2]_{p,q}} = x. \tag{4.2.4}$$

By Taylor's formula

$$g(t) = g(x) + (t-x)g'(x) + \int_x^t (t-u)g''(u)\,du,$$

we get

$$\begin{aligned}
\widetilde{D}_n^{p,q}(g,x) &= g(x) + \widetilde{D}_n^{p,q}\left(\int_x^t (t-u)\,g''(u)\,du, x\right) \\
&= g(x) + \widetilde{B}_n^{p,q}\left(\int_x^t (t-u)g''(u)\,du, x\right) \\
&\quad - \int_x^{\frac{p^n+q[n]_{p,q}x}{[n+2]_{p,q}}} \left(\frac{p^n+q[n]_{p,q}x}{[n+2]_{p,q}} - u\right) g''(u)\,du.
\end{aligned}$$

Thus

$$\begin{aligned}
|\widetilde{D}_n^{p,q}(g,x) - g(x)| &\le \widetilde{B}_n^{p,q}\left(\left|\int_x^t |t-u||g''(u)|\,du\right|, x\right) \\
&\quad + \left|\int_x^{\frac{p^n+q[n]_{p,q}x}{[n+2]_{p,q}}} \left|\frac{p^n+q[n]_{p,q}x}{[n+2]_{p,q}} - u\right| |g''(u)|\,du\right| \\
&\le \widetilde{B}_n^{p,q}((t-x)^2, x)\|g''\| + \left(\frac{p^n+q[n]_{p,q}x}{[n+2]_{p,q}} - x\right)^2 \|g''\|
\end{aligned}$$

Also, we have

$$\widetilde{B}_n^{p,q}((t-x)^2, x) + \left(\frac{p^n+q[n]_{p,q}x}{[n+2]_{p,q}} - x\right)^2 \tag{4.2.5}$$

$$\le \frac{2}{[n+2]_{p,q}}\left(\varphi^2(x) + \frac{1}{[n+3]_{p,q}}\right) + \left(\frac{p^n - ([n+2]_{p,q} - q[n]_{p,q})x}{[n+2]_{p,q}}\right)^2.$$

Obviously

$$1 \leq [n+2]_{p,q} - q[n]_{p,q} \leq 2. \tag{4.2.6}$$

Then, using (4.2.6), we get

$$\begin{aligned}
&\left(\frac{p^n - ([n+2]_{p,q} - q[n]_{p,q})x}{[n+2]_{p,q}}\right)^2 \delta_n^{-2}(x) \\
&\quad = \frac{p^{2n} - 2p^n([n+2]_{p,q} - q[n]_{p,q})x + ([n+2]_{p,q} - q[n]_{p,q})^2 x^2}{[n+2]_{p,q}^2} \\
&\qquad \times \frac{[n]_{p,q}}{[n]_{p,q}x(1-x)+1} \\
&\quad \leq \frac{p^{2n} - 2p^n x + 4x^2}{[n+2]_{p,q}} \cdot \frac{[n]_{p,q}}{[n+2]_{p,q}} \cdot \frac{1}{[n]_{p,q}x(1-x)+1},
\end{aligned}$$

i.e.,

$$\left(\frac{p^n - ([n+2]_{p,q} - q[n]_{p,q})x}{[n+2]_{p,q}}\right)^2 \delta_n^{-2}(x) \leq \frac{3}{[n+2]_{p,q}}, \tag{4.2.7}$$

for $n \in \mathbb{N}$. In conclusion, by (4.2.5) and (4.2.7), for $x \in [0,1]$, we obtain

$$\widetilde{B}_n^{p,q}((t-x)^2, x) + \left(\frac{p^n + q[n]_{p,q}x}{[n+2]_{p,q}} - x\right)^2 \leq \frac{5}{[n+2]_{p,q}} \delta_n^2(x). \tag{4.2.8}$$

Hence, with the conditions $n > 3$ and $x \in [0,1]$, we have

$$|\widetilde{D}_n^{p,q}(g, x) - g(x)| \leq \frac{5}{[n+2]_{p,q}} \, \delta_n^2(x) \, \|g''\|. \tag{4.2.9}$$

Furthermore, for $f \in C[0,1]$ we obtain $||D_n^{p,q}(f,x)|| \leq ||f||$, thus

$$|\widetilde{D}_n^{p,q}(f,x)| \leq |\widetilde{B}_n^{p,q}(f,x)| + |f(x)| + \left| f\left(\frac{p^n + q[n]_{p,q}x}{[n+2]_{p,q}}\right) \right| \leq 3\|f\|. \tag{4.2.10}$$

for all $f \in C[0,1]$.

Now, for $f \in C[0,1]$ and $g \in W^2$, we get

$$\begin{aligned}
&|\widetilde{B}_n^{p,q}(f,x) - f(x)| \\
&= \left| \widetilde{D}_n^{p,q}(f,x) - f(x) + f\left(\frac{p^n + q[n]_{p,q}x}{[n+2]_{p,q}}\right) - f(x) \right| \\
&\leq |\widetilde{D}_n^{p,q}(f-g,x)| + |\widetilde{D}_n^{p,q}(g,x) - g(x)| + |g(x) - f(x)|
\end{aligned}$$

$$+\left| f\left(\frac{p^n+q[n]_{p,q}x}{[n+2]_{p,q}}\right)-f(x)\right|$$

$$\leq 4\,\|f-g\|+\frac{5}{[n+2]_{p,q}}\cdot\delta_n^2(x)\cdot\|g''\|+\omega\left(f,\left|\frac{p^n-([n+2]_{p,q}-q[n]_{p,q})x}{[n+2]_{p,q}}\right|\right)$$

$$\leq 5\left(\|f-g\|+\frac{1}{[n+2]_{p,q}}\cdot\delta_n^2(x)\cdot\|g''\|\right)+\omega\left(f,\frac{1-x}{[n+2]_{p,q}}\right),$$

where we have used (4.2.9) and (4.2.10). Taking the infimum on the right-hand side over all $g\in W^2$, we obtain at once

$$|\widetilde{B}_n^{p,q}(f,x)-f(x)|\leq 5\;K_2\left(f,\frac{1}{[n+2]_{p,q}}\delta_n^2(x)\right)+\omega\left(f,\frac{1-x}{[n+2]_{p,q}}\right).$$

Finally, in view of (4.2.2), we find

$$|\widetilde{B}_n^{p,q}(f,x)-f(x)|\;\leq\;C\;\omega_2\left(f,[n+2]_{p,q}^{-1/2}\delta_n(x)\right)+\omega\left(f,\frac{1-x}{[n+2]_{p,q}}\right).$$

This completes the proof of the theorem. ■

The weighted modulus of continuity of second order for $f\in C[0,1]$ and $\varphi(x)=\sqrt{x(1-x)}$ is defined as:

$$\omega_2^{\varphi}(f,\sqrt{\delta})\;=\;\sup_{0<h\leq\sqrt{\delta}}\;\sup_{x,x\pm h\varphi\in[0,1]}\;|f(x+h\varphi(x))-2f(x)+f(x-h\varphi(x))|\,.$$

The corresponding K-functional is defined by

$$\overline{K}_{2,\varphi}(f,\delta)\;=\;\inf\{\|f-g\|+\delta\|\varphi^2g''\|+\delta^2\|g''\|:g\in W^2(\varphi)\},$$

where

$$W^2(\varphi)=\left\{g\in C[0,1]:g'\in AC_{loc}[0,1],\varphi^2g''\in C[0,1]\right\}$$

and $g'\in AC_{loc}[0,1]$ means that g is differentiable and g' is absolutely continuous on every closed interval $[a,b]\subset[0,1]$. By the property due to Ditzian–Totik (see [79, p. 24, Theorem 1.3.1]), we have

$$\overline{K}_{2,\varphi}(f,\delta)\;\leq\;C\;\omega_2^{\varphi}(f,\sqrt{\delta}) \tag{4.2.11}$$

for some absolute constant $C>0$. Moreover, with ψ the admissible step-weight function on $[0,1]$, the Ditzian–Totik moduli of first order is given by

$$\vec{\omega}_{\psi}(f,\delta) = \sup_{0<h\leq\delta} \sup_{x,x\pm h\psi(x)\in[0,1]} |f(x+h\psi(x)) - f(x)|.$$

Now we state and prove the following global direct result of [128]:

Theorem 4.16 ([128]) *Let $n > 3$ be a natural number and let $q_0 = q_0(n) \in (0, p)$ be defined as in Lemma 4.5. Then there exists an absolute constant C such that for $0 < q < p \leq 1$, we have*

$$\|\widetilde{B}_n^{p,q} f - f\| \leq C\, \omega_2^{\varphi}(f, [n+2]_q^{-1/2}) + \vec{\omega}_{\psi}(f, [n+2]_q^{-1}),$$

where $f \in C[0, 1]$, $q \in (q_0, 1)$ and $\psi(x) = 1 - x$, $x \in [0, 1]$.

Proof Let us consider $\widetilde{D}_n^{p,q}(f, x)$ as defined in Theorem 4.15, where $f \in C[0, 1]$. Also, by Taylor's formula with $g \in W^2(\varphi)$, we have

$$g(t) = g(x) + (t-x)\, g'(x) + \int_x^t (t-u)\, g''(u)du.$$

Applying (4.2.3) and (4.2.4), we obtain

$$\widetilde{D}_n^{p,q}(g,x) = g(x) + \widetilde{B}_n^{p,q}\left(\int_x^t t-u)g''(u)\,du, x\right) - \int_x^{\frac{p^n+q[n]_{p,q}x}{[n+2]_{p,q}}} \left(\frac{p^n+q[n]_{p,q}x}{[n+2]_{p,q}} - u\right) g''(u)\,du.$$

Thus we can write

$$\begin{aligned} &|\widetilde{D}_n^{p,q}(g,x) - g(x)| \\ &\leq \widetilde{B}_n^{p,q}\left(\left|\int_x^t |t-u|\cdot|g''(u)|\,du\right|, x\right) \\ &\quad + \left|\int_x^{\frac{p^n+q[n]_{p,q}x}{[n+2]_{p,q}}} \left|\frac{p^n+q[n]_{p,q}x}{[n+2]_{p,q}} - u\right|\cdot|g''(u)|\,du\right|. \end{aligned} \tag{4.2.12}$$

Also, the function δ_n^2 is concave on $[0, 1]$. We have for $u = t + \tau(x-t)$, $\tau \in [0, 1]$, the following estimate

$$\frac{|t-u|}{\delta_n^2(u)} = \frac{\tau|x-t|}{\delta_n^2(t+\tau(x-t))} \leq \frac{\tau|x-t|}{\delta_n^2(t)+\tau(\delta_n^2(x)-\delta_n^2(t))} \leq \frac{|t-x|}{\delta_n^2(x)}.$$

Hence, by (4.2.12), we obtain

$$|\widetilde{D}_n^{p,q}(g,x)-g(x)|$$
$$\leq \widetilde{B}_n^{p,q}\left(\left|\int_x^t \frac{|t-u|}{\delta_n^2(u)}du\right|,x\right)\|\delta_n^2 g''\|+\left|\int_x^{\frac{p^n+q[n]_{p,q}x}{[n+2]_{p,q}}} \frac{\left|\frac{p^n+q[n]_{p,q}x}{[n+2]_{p,q}}-u\right|}{\delta_n^2(u)}du\right|\|\delta_n^2 g''\|$$
$$\leq \frac{1}{\delta_n^2(x)}\widetilde{B}_n^{p,q}((t-x)^2,x)\|\delta_n^2 g''\|+\frac{1}{\delta_n^2(x)}\left(\frac{p^n+q[n]_{p,q}x}{[n+2]_{p,q}}-x\right)^2\|\delta_n^2 g''\|.$$

For $x\in[0,1]$, in view of (4.2.8) and

$$\delta_n^2(x)\cdot|g''(x)|=|\varphi^2(x)g''(x)|+\frac{1}{[n+2]_{p,q}}\cdot|g''(x)|\leq\|\varphi^2 g''\|+\frac{1}{[n+2]_{p,q}}\|g''\|,$$

we get

$$|\widetilde{D}_n^{p,q}(g,x)-g(x)|\leq\frac{5}{[n+2]_{p,q}}\cdot\left(\|\varphi^2 g''\|+\frac{1}{[n+2]_{p,q}}\cdot\|g''\|\right) \tag{4.2.13}$$

Obviously using $[n]_{p,q}\leq[n+2]_{p,q}$, (4.2.10), and (4.2.13), we find for $f\in C[0,1]$, that

$$|\widetilde{B}_n^{p,q}(f,x)-f(x)|\leq|\widetilde{D}_n^{p,q}(f-g,x)|$$
$$+|\widetilde{D}_n^{p,q}(g,x)-g(x)|+|g(x)-f(x)|+\left|f\Big(\frac{p^n+q[n]_{p,q}x}{[n+2]_{p,q}}\Big)-f(x)\right|$$
$$\leq 4\|f-g\|+\frac{5}{[n+2]_{p,q}}\|\varphi^2 g''\|+\frac{5}{[n+2]_{p,q}}\|g''\|+\left|f\Big(\frac{p^n+q[n]_{p,q}x}{[n+2]_{p,q}}\Big)-f(x)\right|.$$

Finally, taking the infimum on the right-hand side over all $g\in W^2(\varphi)$, we obtain

$$|\widetilde{B}_n^{p,q}(f,x)-f(x)|\leq 5\overline{K}_{2,\varphi}\left(f,\frac{1}{[n+2]_{p,q}}\right) \tag{4.2.14}$$
$$+\left|f\left(\frac{p^n+q[n]_{p,q}x}{[n+2]_{p,q}}\right)-f(x)\right|$$

Also, we have

$$\left|f\Big(\frac{p^n+q[n]_{p,q}x}{[n+2]_{p,q}}\Big)-f(x)\right|=\left|f\Big(x+\psi(x)\frac{p^n-([n+2]_{p,q}-q[n]_{p,q})x}{[n+2]_{p,q}\psi(x)}\Big)-f(x)\right|$$
$$\leq \sup_{t,t+\psi(t)\frac{p^n-([n+2]_{p,q}-q[n]_{p,q})x}{[n+2]_{p,q}}\in[0,1]}\left|f\Big(t+\psi(t)\frac{p^n-([n+2]_{p,q}-q[n]_{p,q})x}{[n+2]_{p,q}\psi(x)}\Big)-f(t)\right|$$

$$\leq \vec{\omega}_\psi \left(f, \frac{|p^n - ([n+2]_{p,q} - q[n]_{p,q})x|}{[n+2]_{p,q}\psi(x)} \right)$$

$$\leq \vec{\omega}_\psi \left(f, \frac{1-x}{[n+2]_{p,q}\psi(x)} \right) = \vec{\omega}_\psi \left(f, \frac{1}{[n+2]_{p,q}} \right).$$

Hence, by (4.2.14) and (4.2.11), we get

$$\|\widetilde{B}_n^{p,q} f - f\| \leq C\, \omega_2^\varphi(f, [n+2]_{p,q}^{-1/2}) + \omega_\psi(f, [n+2]_{p,q}^{-1}).$$

This completes the proof of the theorem. ■

Remark 4.4 For $q \in (0, 1)$ and $p \in (q, 1]$ it is obvious that

$$\lim_{n\to\infty} [n]_{p,q} = 1/(p-q).$$

Thus the above theorems do not give an approximation result. If we choose $q_n = e^{-1/n}$ and $p_n = e^{-1/(n+1)}$ such that $0 < q_n < p_n \leq 1$, then

$$\lim_{n\to\infty} p_n = \lim_{n\to\infty} q_n = 1 \text{ and } \lim_{n\to\infty} p_n^n = \lim_{n\to\infty} q_n^n = 1/e.$$

Also we have

$$\lim_{n\to\infty} [n]_{p_n,q_n} = \infty.$$

Since $[n+2]_{p,q} = [2]_{p,q}\, p^n + q^2\, [n]_{p,q}$ we can write

$$\lim_{n\to\infty} \frac{1}{[n]_{p_n,q_n}} = \lim_{n\to\infty} \frac{1}{[n+2]_{p_n,q_n}} = 0 \quad \text{and} \quad \lim_{n\to\infty} \frac{[n]_{p_n,q_n}}{[n+2]_{p_n,q_n}} = 1.$$

The (p, q)-analogue of the genuine Bernstein–Durrmeyer operator proposed by Gupta in [119] for $x \in [0, 1]$ is defined as

$$\widetilde{G}_n^{p,q}(f, x) = [n-1]_{p,q} \sum_{k=1}^{n-1} p^{-[n^2-k^2-n+k-2]/2} b_{n,k}^{p,q}(1, x) \int_0^1 b_{n-2,k-1}^{p,q}(p, pqt) f(pt)\, d_{p,q}t$$
$$+ b_{n,0}^{p,q}(1, x) f(0) + b_{n,n}^{p,q}(1, x) f(1) \tag{4.2.15}$$

where

$$b_{n,k}^{p,q}(1, x) = \begin{bmatrix} n \\ k \end{bmatrix}_{p,q} p^{[k(k-1)-n(n-1)]/2} x^k (1 \ominus x)_{p,q}^{n-k}$$

$$b_{n,k}^{p,q}(p, pqt) = \begin{bmatrix} n \\ k \end{bmatrix}_{p,q} (pt)^k\, (p \ominus pqt)_{p,q}^{n-k}.$$

For the genuine Bernstein–Durrmeyer operators (4.2.15) Gupta in [119] proved the following two direct estimates:

Theorem 4.17 ([119]) *Let $f \in C[0, 1]$. Then there exists an absolute constant $C > 0$ such that*

$$|\widetilde{G}_n^{p,q}(f, x) - f(x)| \le C\,\omega_2\left(f, \sqrt{\frac{x(1-x)}{[n+1]_{p,q}}}\right).$$

Theorem 4.18 ([119]) *There exists an absolute constant $C > 0$ such that*

$$\|\widetilde{G}_n^{p,q} f - f\| \le C\,\omega_2^{\varphi}(f, [n+1]_q^{-1/2}),$$

where $\varphi = \sqrt{x(1-x)}$, $f \in C[0, 1]$.

4.2.2 Limit (p, q)-Bernstein–Durrmeyer Operators

Recently Finta and Gupta in [90] considered the following slightly modified representation of (p, q)-Bernstein–Durrmeyer operators as

$$\begin{aligned}(\tilde{D}_n^{p,q} f)(x) &\equiv \tilde{D}_n^{p,q}(f; x)\\ &= [n+1]_{p,q} \sum_{k=0}^{n} p^{-(n^2+3n-k^2-k)/2} b_{n,k}^{p,q}(1, x)\\ &\quad \int_0^1 \widetilde{b}_{n,k}^{p,q}(p, pqt) f(pt)\, d_{p,q}t,\end{aligned} \tag{4.2.16}$$

where $f \in C[0, 1]$, $x \in [0, 1]$,

$$b_{n,k}^{p,q}(1, x) = \begin{bmatrix} n \\ k \end{bmatrix}_{p,q} p^{(k(k-1)-n(n-1))/2} x^k (1 \ominus x)_{p,q}^{n-k}$$

and

$$\widetilde{b}_{n,k}^{p,q}(p, pqt) = \begin{bmatrix} n \\ k \end{bmatrix}_{p,q} (pt)^k (p \ominus pqt)_{p,q}^{n-k}.$$

The auxiliary results proved in [90] are the following lemmas:

Lemma 4.6 *With the notation*

$$\lambda_{n,k}^{p,q}(f) = [n+1]_{p,q}\, p^{-(n^2+3n-k^2-k)/2} \int_0^1 \widetilde{b}_{n,k}^{p,q}(p, pqt) f(pt)\, d_{p,q}t,$$

where $k = 0, 1, \ldots, n$ *and* $f \in C[0,1]$, *we have for* $x \in [0,1]$ *that*

$$\begin{aligned}
&\tilde{D}_n^{p,q}(f;x) - \tilde{D}_{n+1}^{p,q}(f;x) \\
&= b_{n+1,0}^{p,q}(1,x)\{\lambda_{n,0}^{p,q}(f) - \lambda_{n+1,0}^{p,q}(f)\} + \sum_{k=1}^{n} b_{n+1,k}^{p,q}(1,x)\left\{\lambda_{n,k}^{p,q}(f)\frac{[n+1-k]_{p,q}}{[n+1]_{p,q}}p^k\right. \\
&\quad \left. +\lambda_{n,k-1}^{p,q}(f)\frac{[k]_{p,q}}{[n+1]_{p,q}}q^{n+1-k} - \lambda_{n+1,k}^{p,q}(f)\right\} + b_{n+1,n+1}^{p,q}(1,x)\{\lambda_{n,n}^{p,q}(f) - \lambda_{n+1,n+1}^{p,q}(f)\}.
\end{aligned}$$

Lemma 4.7 *For*

$$\widetilde{b}_{n,k}^{p,q}(p,pqt) = \begin{bmatrix} n \\ k \end{bmatrix}_{p,q} (pt)^k (p \ominus pqt)_{p,q}^{n-k}, \quad k = 0, 1, \ldots, n,$$

we have

$$\begin{aligned}
&[n+1]_{p,q}\, p^{-(n^2+3n-k^2-k)/2} \int_0^1 \widetilde{b}_{n,k}^{p,q}(p,pqt)\, d_{p,q}t = 1, \\
&[n+1]_{p,q}\, p^{-(n^2+3n-k^2-k)/2} \int_0^1 \widetilde{b}_{n,k}^{p,q}(p,pqt)t\, d_{p,q}t = p^{n-k}\frac{[k+1]_{p,q}}{[n+2]_{p,q}}, \\
&[n+1]_{p,q}\, p^{-(n^2+3n-k^2-k)/2} \int_0^1 \widetilde{b}_{n,k}^{p,q}(p,pqt)t^2\, d_{p,q}t = p^{2(n-k)}\frac{[k+1]_{p,q}[k+2]_{p,q}}{[n+2]_{p,q}[n+3]_{p,q}}.
\end{aligned}$$

Lemma 4.8 *For* $x \in [0,1]$, *it holds*

$$\begin{aligned}
\tilde{D}_n^{p,q}(1;x) &= 1, \quad \tilde{D}_n^{p,q}(t;x) = \frac{p^{n+1} + pq[n]_{p,q}x}{[n+2]_{p,q}}, \\
\tilde{D}_n^{p,q}(t^2;x) &= \frac{p^{2n+2}[2]_{p,q}}{[n+2]_{p,q}[n+3]_{p,q}} + \frac{(2q^2+qp)p^{n+2}[n]_{p,q}x}{[n+2]_{p,q}[n+3]_{p,q}} \\
&\quad + \frac{q^3[n]_{p,q}[p^2[n]_{p,q}x^2 + p^{n+1}x(1-x)]}{[n+2]_{p,q}[n+3]_{p,q}}.
\end{aligned}$$

Remark 4.5 If $p = p(n)$ and $q = q(n)$ such that $0 < q(n) < p(n) \leq 1$ and $q(n) \to 1$ as $n \to \infty$, then, by Korovkin's theorem, $\tilde{D}_n^{p,q}(f;x)$ converges uniformly to $f(x)$ for $x \in [0,1]$, as $n \to \infty$. Indeed, the estimates

$$\begin{aligned}
|\tilde{D}_n^{p,q}(t;x) - x| &\leq \frac{1}{[n+2]_{q/p}} + \left|\frac{q}{p}\frac{[n]_{q/p}}{[n+2]_{q/p}} - 1\right|, \\
|\tilde{D}_n^{p,q}(t^2;x) - x^2| &\leq \frac{[2]_{q/p}}{[n+2]_{q/p}[n+3]_{q/p}} + \left(2\left(\frac{q}{p}\right)^3 + \frac{q}{p}\right)\frac{[n]_{q/p}}{[n+2]_{q/p}}\frac{1}{[n+3]_{q/p}} \\
&\quad + \left(\frac{q}{p}\right)^3\frac{[n]_{q/p}}{[n+2]_{q/p}}\frac{1}{4[n+3]_{q/p}} + \left|\left(\frac{q}{p}\right)^3\frac{[n]_{q/p}}{[n+2]_{q/p}}\frac{[n]_{q/p}}{[n+3]_{q/p}} - 1\right|,
\end{aligned}$$

and the facts that

$$[n]_{q_n/p_n} \to \infty \text{ and } \frac{[n]_{q_n/p_n}}{[n+2]_{q_n/p_n}} \to 1 \text{ as } n \to \infty,$$

imply our statement.

More precisely, Finta and Gupta [90] applied the following result (see [87, p. 393, Theorem 2.1] and [87, p. 394, Corollary 2.1]):

Theorem 4.19 *Let Λ be a set of parameters and for $\lambda \in \Lambda$ let $(L_n^\lambda)_{n\geq 1}$ be a sequence of positive linear operators on $C[0,1]$. If there exist the positive sequences $(\alpha_n)_{n\geq 1}$ and $(\beta_n)_{n\geq 1}$ such that*

a) $\alpha_n \to 0$ as $n \to \infty$,
b) there exists $C_1 > 0$ with

$$\beta_n + \beta_{n+1} + \ldots + \beta_{n+m-1} \leq C_1 \alpha_n$$

for all $n, m \geq 1$,
c) there exists $C_2 > 0$ with

$$\|L_n^\lambda g - L_{n+1}^\lambda g\| \leq C_2 \beta_n \|g'\|$$

for all $n \geq 1$ and $g \in C^1[0,1]$,

then there exists $C_3 = C_3(\|L_1^\lambda e_0\|) > 0$ and a positive linear operator $L_\infty^\lambda : C[0,1] \to C[0,1]$ such that

$$\|L_n^\lambda f - L_\infty^\lambda f\| \leq C_3 \omega(f, \alpha_n)$$

for all $f \in C[0,1]$ and $n = 1, 2, \ldots$

We mention that $\|\cdot\|$ denotes the uniform norm on $C[0,1]$, $e_0(x) = 1$ for $x \in [0,1]$, and the sequences $(\alpha_n)_{n\geq 1}$ and $(\beta_n)_{n\geq 1}$ may depend on λ.

In the following theorem Finta and Gupta [90] proved the existence of the limit (p,q)-Bernstein–Durrmeyer operator.

Theorem 4.20 *Let $\tilde{D}_n^{p,q} f$ be defined by (4.2.16), where p and q are fixed. Then there exist an absolute constant $C > 0$ and a positive linear operator $\tilde{D}_\infty^{p,q} : C[0,1] \to C[0,1]$ such that*

$$\|\tilde{D}_n^{p,q} f - \tilde{D}_\infty^{p,q} f\| \leq C\omega\left(f, \left(\frac{q}{p}\right)^{n/2}\right)$$

for all $f \in C[0,1]$ and $n = 1, 2, \ldots$

Proof We have

$$[n+1]_{p,q} = p^k[n+1-k]_{p,q} + q^{n+1-k}[k]_{p,q} \text{ for } k = 0, 1, \ldots, n+1.$$

Using the notation of Lemma 4.6, we obtain for $f \in C[0, 1]$, that

$$\begin{aligned}
\lambda_{n,k}^{p,q}(f) & \frac{[n+1-k]_{p,q}}{[n+1]_{p,q}} p^k + \lambda_{n,k-1}^{p,q}(f) \frac{[k]_{p,q}}{[n+1]_{p,q}} q^{n+1-k} - \lambda_{n+1,k}^{p,q}(f) \\
&= \frac{[n+1-k]_{p,q}}{[n+1]_{p,q}} p^k \{\lambda_{n,k}^{p,q}(f) - \lambda_{n+1,k}^{p,q}(f)\} \\
&+ \frac{[k]_{p,q}}{[n+1]_{p,q}} q^{n+1-k} \{\lambda_{n,k-1}^{p,q}(f) - \lambda_{n+1,k}^{p,q}(f)\}. \quad (4.2.17)
\end{aligned}$$

Let $g \in C^1[0, 1]$ and $x_k = p^{n+2-k} \frac{[k+1]_{p,q}}{[n+3]_{p,q}}$ for $k = 0, 1, \ldots, n$. Obviously, by using the fact $[n]_{p,q} = p^{n-1}[n]_{q/p}$, we have

$$x_k = p^{n+2-k} \frac{p^k[k+1]_{q/p}}{p^{n+2}[n+3]_{q/p}} = \frac{[k+1]_{q/p}}{[n+3]_{q/p}} \in [0, 1],$$

where $k = 0, 1, \ldots, n$.

Furthermore

$$g(pt) = g(x_k) + \int_{x_k}^{pt} g'(u)\, du, \text{ where } t \in [0, 1] \text{ is arbitrary.}$$

Hence, by the definition of $\lambda_{n,k}^{p,q}(g)$ and Lemma 4.7, we obtain

$$\begin{aligned}
&\lambda_{n,k}^{p,q}(g) - \lambda_{n+1,k}^{p,q}(g) \\
&= [n+1]_{p,q} p^{-(n^2+3n-k^2-k)/2} \int_0^1 \widetilde{b}_{n,k}^{p,q}(p, pqt) \left[g(x_k) + \int_{x_k}^{pt} g'(u)\, du \right] d_{p,q}t \\
&-[n+2]_{p,q} p^{-((n+1)^2+3(n+1)-k^2-k)/2} \\
&\int_0^1 \widetilde{b}_{n+1,k}^{p,q}(p, pqt) \left[g(x_k) + \int_{x_k}^{pt} g'(u)\, du \right] d_{p,q}t \\
&= [n+2]_{p,q} p^{-((n+1)^2+3(n+1)-k^2-k)/2} \int_0^1 \widetilde{b}_{n+1,k}^{p,q}(p, pqt) \left(\int_{x_k}^{pt} g'(u)\, du \right) \\
&\left\{ \frac{[n+1]_{p,q}}{[n+2]_{p,q}} p^{n+2} \frac{\widetilde{b}_{n,k}^{p,q}(p, pqt)}{\widetilde{b}_{n+1,k}^{p,q}(p, pqt)} - 1 \right\} d_{p,q}t \quad (4.2.18)
\end{aligned}$$

for $k = 0, 1, \ldots, n$. On the other hand, using

$$[n+2]_{p,q} = p^{k+1}[n+1-k]_{p,q} + q^{n+1-k}[k+1]_{p,q},$$

we have

$$\begin{aligned}
&\frac{[n+1]_{p,q}}{[n+2]_{p,q}} p^{n+2} \frac{b_{n,k}^{p,q}(p, pqt)}{b_{n+1,k}^{p,q}(p, pqt)} - 1 \\
&= \frac{[n+1]_{p,q}}{[n+2]_{p,q}} p^{n+2} \frac{\begin{bmatrix} n \\ k \end{bmatrix}_{p,q} (pt)^k (p \ominus pqt)_{p,q}^{n-k}}{\begin{bmatrix} n+1 \\ k \end{bmatrix}_{p,q} (pt)^k (p \ominus pqt)_{p,q}^{n+1-k}} - 1 \\
&= \frac{[n+1]_{p,q}}{[n+2]_{p,q}} p^{n+2} \frac{[n+1-k]_{p,q}}{[n+1]_{p,q}} \frac{1}{p^{n+1-k} - pq^{n+1-k}t} - 1 \\
&= \frac{[n+1-k]_{p,q}}{[n+2]_{p,q}} \frac{p^{n+2}}{p^{n+1-k} - pq^{n+1-k}t} - 1 \\
&= q^{n+1-k} \left(\frac{[n+1-k]_{p,q}}{[n+2]_{p,q}} \frac{p^{k+2}t}{p^{n+1-k} - pq^{n+1-k}t} - \frac{[k+1]_{p,q}}{[n+2]_{p,q}} \right).
\end{aligned}$$

The equality

$$[n+2]_{p,q} = p^{k+1}[n+1-k]_{p,q} + q^{n+1-k}[k+1]_{p,q}$$

implies that

$$\frac{[n+1-k]_{p,q}}{[n+2]_{p,q}} \leq p^{-(k+1)}. \tag{4.2.19}$$

Analogously, the equality

$$[n+2]_{p,q} = q^{k+1}[n+1-k]_{p,q} + p^{n+1-k}[k+1]_{p,q}$$

implies that

$$\frac{[k+1]_{p,q}}{[n+2]_{p,q}} \leq p^{-(n+1-k)}. \tag{4.2.20}$$

Finally, the function

$$t \to p^{k+2}t/(p^{n+1-k} - pq^{n+1-k}t)$$

is increasing on [0, 1], therefore

$$\frac{p^{k+2}t}{p^{n+1-k}-pq^{n+1-k}t} \le \frac{p^{k+2}}{p^{n+1-k}-pq^{n+1-k}} = \frac{p^{k+2}}{p^{n+1-k}\left(1-p(\frac{q}{p})^{n+1-k}\right)}$$

$$\le \frac{p^{k+2}}{p^{n+1-k}\left(1-p\frac{q}{p}\right)} = \frac{p^{k+2}}{p^{n+1-k}(1-q)}. \quad (4.2.21)$$

Combining (4.2.18)–(4.2.21), and applying Lemma 4.7 and Hölder's inequality, we find

$$|\lambda_{n,k}^{p,q}(g) - \lambda_{n+1,k}^{p,q}(g)|$$

$$\le \|g'\|[n+2]_{p,q}p^{-[(n+1)^2+3(n+1)-k^2-k]/2}$$

$$\times \int_0^1 \widetilde{b}_{n+1,k}^{p,q}(p,pqt)|pt-x_k|q^{n+1-k}\left(p^{-(k+1)}\frac{p^{k+2}}{p^{n+1-k}(1-q)} + p^{-(n+1-k)}\right) d_{p,q}t$$

$$\le \frac{1+p-q}{1-q}\|g'\|\left(\frac{q}{p}\right)^{n+1-k}$$

$$\left\{[n+2]_{p,q}p^{-((n+1)^2+3(n+1)-k^2-k)/2}\int_0^1 \widetilde{b}_{n+1,k}^{p,q}(p,pqt)(pt-x_k)^2\, d_{p,q}t\right\}^{1/2}. \quad (4.2.22)$$

Using Lemma 4.7, we get

$$[n+2]_{p,q}p^{-((n+1)^2+3(n+1)-k^2-k)/2}\int_0^1 \widetilde{b}_{n+1,k}^{p,q}(p,pqt)(pt-x_k)^2\, d_{p,q}t$$

$$= p^{2(n+2-k)}\frac{[k+1]_{p,q}[k+2]_{p,q}}{[n+3]_{p,q}[n+4]_{p,q}} - 2p^{n+2-k}\frac{[k+1]_{p,q}}{[n+3]_{p,q}}p^{n+2-k}\frac{[k+1]_{p,q}}{[n+3]_{p,q}}$$

$$+\left(p^{n+2-k}\frac{[k+1]_{p,q}}{[n+3]_{p,q}}\right)^2$$

$$= p^{2(n+2-k)}\frac{[k+1]_{p,q}}{[n+3]_{p,q}}\left(\frac{[k+2]_{p,q}}{[n+4]_{p,q}} - \frac{[k+1]_{p,q}}{[n+3]_{p,q}}\right)$$

$$= p^{2(n+2-k)}\frac{[k+1]_{p,q}}{[n+3]_{p,q}}(pq)^{k+1}\frac{[n+2-k]_{p,q}}{[n+3]_{p,q}[n+4]_{p,q}}.$$

Analogously to (4.2.20) and (4.2.19), we find that

$$\frac{[k+1]_{p,q}}{[n+3]_{p,q}} \le p^{-(n+2-k)} \quad \text{and} \quad \frac{[n+2-k]_{p,q}}{[n+3]_{p,q}} \le p^{-(k+1)}. \quad (4.2.23)$$

Further, using $[n]_{p,q} = p^{n-1}[n]_{q/p}$, we get

$$[n+4]_{p,q} = p^{n+3}[n+4]_{q/p} \geq p^{n+3}.$$

Hence, by (4.2.22)–(4.2.23), we have for $k = 0, 1, \ldots, n$, that

$$\begin{aligned}
|\lambda_{n,k}^{p,q}(g) - \lambda_{n+1,k}^{p,q}(g)| &\leq \frac{1+p-q}{1-q}\|g'\| \\
&\quad \left(\frac{q}{p}\right)^{n+1-k} \left\{p^{2(n+2-k)} p^{-(n+2-k)} (pq)^{k+1} p^{-(k+1)} p^{-(n+3)}\right\}^{1/2} \\
&= \frac{1+p-q}{1-q}\|g'\| \left(\frac{q}{p}\right)^{n+1-k} \left(\frac{q}{p}\right)^{(k+1)/2} \\
&= \frac{1+p-q}{1-q}\|g'\| \left(\frac{q}{p}\right)^{(2n-k+3)/2} \\
&\leq \frac{1+p-q}{1-q}\|g'\| \left(\frac{q}{p}\right)^{n/2} \left(\frac{q}{p}\right)^{3/2}. \qquad (4.2.24)
\end{aligned}$$

Analogously to (4.2.18), we obtain for $g \in C^1[0, 1]$ and

$$y_k = p^{n+2-k} \frac{[k]_{p,q}}{[n+2]_{p,q}} \in [0, 1], \text{ where } k = 1, 2, \ldots, n+1,$$

that

$$\begin{aligned}
&\lambda_{n,k-1}^{p,q}(g) - \lambda_{n+1,k}^{p,q}(g) \\
&= [n+1]_{p,q} p^{-(n^2+3n-(k-1)^2-(k-1))/2} \int_0^1 \widetilde{b}_{n,k-1}^{p,q}(p, pqt) \left(\int_{y_k}^{pt} g'(u)\, du\right) \\
&\quad \times \left\{1 - \frac{[n+2]_{p,q}}{[n+1]_{p,q}} p^{-(n+2-k)} \frac{\widetilde{b}_{n+1,k}^{p,q}(p, pqt)}{\widetilde{b}_{n,k-1}^{p,q}(p, pqt)}\right\} d_{p,q}t. \qquad (4.2.25)
\end{aligned}$$

But

$$\widetilde{b}_{n+1,k}^{p,q}(p, pqt) = \frac{[n+1]_{p,q}}{[k]_{p,q}} pt \widetilde{b}_{n,k-1}^{p,q}(p, pqt)$$

for $k = 1, 2, \ldots, n+1$, hence, in view of (4.2.25), Lemma 4.7, and Hölder's inequality, we find that

$$
\begin{aligned}
&|\lambda^{p,q}_{n,k-1}(g) - \lambda^{p,q}_{n+1,k}(g)| \\
&\leq \|g'\| [n+1]_{p,q} p^{-(n^2+3n-(k-1)^2-(k-1))/2} \int_0^1 \widetilde{b}^{p,q}_{n,k-1}(p, pqt)|pt - y_k| \\
&\quad \times \left(1 + \frac{[n+2]_{p,q}}{[n+1]_{p,q}} p^{-(n+2-k)} \frac{[n+1]_{p,q}}{[k]_{p,q}} pt\right) d_{p,q}t \\
&\leq \|g'\| \left(1 + p^{-(n+1-k)} \frac{[n+2]_{p,q}}{[k]_{p,q}}\right) \\
&\quad \times \left\{[n+1]_{p,q} p^{-(n^2+3n-(k-1)^2-(k-1))/2} \int_0^1 \widetilde{b}^{p,q}_{n,k-1}(p, pqt)(pt - y_k)^2 \, d_{p,q}t\right\}^{1/2} \\
&= \|g'\| \left(1 + p^{-(n+1-k)} \frac{[n+2]_{p,q}}{[k]_{p,q}}\right) \\
&\quad \times \left\{[n+1]_{p,q} p^{-(n^2+3n-(k-1)^2-(k-1))/2} \int_0^1 \widetilde{b}^{p,q}_{n,k-1}(p, pqt)(pt - y_k)^2 \, d_{p,q}t\right\}^{1/2} \\
&= \|g'\| \left(1 + p^{-(n+1-k)} \frac{[n+2]_{p,q}}{[k]_{p,q}}\right) \left\{p^{2(n+2-k)} \frac{[k]_{p,q}[k+1]_{p,q}}{[n+2]_{p,q}[n+3]_{p,q}}\right. \\
&\quad \left. -2p^{n+2-k} \frac{[k]_{p,q}}{[n+2]_{p,q}} p^{n+2-k} \frac{[k]_{p,q}}{[n+2]_{p,q}} + \left(p^{n+2-k} \frac{[k]_{p,q}}{[n+2]_{p,q}}\right)^2\right\} \\
&= \|g'\| \left(1 + p^{-(n+1-k)} \frac{[n+2]_{p,q}}{[k]_{p,q}}\right) p^{n+2-k} \left(\frac{[k]_{p,q}}{[n+2]_{p,q}} \left(\frac{[k+1]_{p,q}}{[n+3]_{p,q}} - \frac{[k]_{p,q}}{[n+2]_{p,q}}\right)\right)^{1/2}
\end{aligned}
$$

Thus, we have

$$
\begin{aligned}
|\lambda^{p,q}_{n,k-1}(g) - \lambda^{p,q}_{n+1,k}(g)| = \|g'\| &\left(p^{n+2-k} \frac{[k]_{p,q}}{[n+2]_{p,q}} + p\right) \\
&\left(\frac{[n+2]_{p,q}}{[k]_{p,q}} (pq)^k \frac{[n+2-k]_{p,q}}{[n+2]_{p,q}[n+3]_{p,q}}\right)^{1/2}
\end{aligned} \tag{4.2.26}
$$

On the other hand, by $[n]_{p,q} = p^{n-1}[n]_{q/p}$, we have

$$
[n+2]_{p,q} = p^{n+1}[n+2]_{q/p} \geq p^{n+1}[k]_{q/p} = p^{n+2-k}[k]_{p,q} \text{ for } k = 1, 2, \ldots, n+1.
$$

Furthermore

$$
[n+3]_{p,q} = p^{k+1}[n+2-k]_{p,q} + q^{n+2-k}[k+1]_{p,q}, \; k = 1, 2, \ldots, n+1.
$$

Thus

$$\frac{[n+2-k]_{p,q}}{[n+3]_{p,q}} \leq p^{-(k+1)}.$$

Moreover

$$[k]_{p,q} = p^{k-1}[k]_{q/p} \geq p^{k-1}[1]_{q/p} = p^{k-1} \text{ for } k = 1, 2, \ldots, n+1.$$

Hence, by (4.2.26), we have

$$\begin{aligned} |\lambda_{n,k-1}^{p,q}(g) - \lambda_{n+1,k}^{p,q}(g)| &\leq (1+p)\|g'\| \left((pq)^k p^{-(k-1)} p^{-(k+1)}\right)^{1/2} \\ &= (1+p)\|g'\| \left(\frac{q}{p}\right)^{k/2}. \end{aligned} \tag{4.2.27}$$

Since

$$\frac{[n+1-k]_{p,q}}{[n+1]_{p,q}} = p^{-k}\frac{[n+1-k]_{q/p}}{[n+1]_{q/p}} \leq p^{-k}$$

for $k = 0, 1, \ldots, n$, and

$$\frac{[k]_{p,q}}{[n+1]_{p,q}} = p^{-(n+1-k)}\frac{[k]_{q/p}}{[n+1]_{q/p}} \leq p^{-(n+1-k)}$$

for $k = 1, 2, \ldots, n+1$, we obtain, in view of (4.2.17), (4.2.24), and (4.2.27), that

$$\begin{aligned} &\left| \lambda_{n,k}^{p,q}(g)\frac{[n+1-k]_{p,q}}{[n+1]_{p,q}} p^k + \lambda_{n,k-1}^{p,q}(g)\frac{[k]_{p,q}}{[n+1]_{p,q}} q^{n+1-k} - \lambda_{n+1,k}^{p,q}(g) \right| \\ &\leq \frac{[n+1-k]_{p,q}}{[n+1]_{p,q}} p^k |\lambda_{n,k}^{p,q}(g) - \lambda_{n+1,k}^{p,q}(g)| + \frac{[k]_{p,q}}{[n+1]_{p,q}} q^{n+1-k} |\lambda_{n,k-1}^{p,q}(g) - \lambda_{n+1,k}^{p,q}(g)| \\ &\leq \frac{1+p-q}{1-q}\|g'\| \left(\frac{q}{p}\right)^{n/2} \left(\frac{q}{p}\right)^{3/2} + (1+p)\|g'\| \left(\frac{q}{p}\right)^{n+1-k} \left(\frac{q}{p}\right)^{k/2} \\ &\leq \frac{1+p-q}{1-q} \left(\frac{q}{p}\right)^{n/2} \left(\frac{q}{p}\right)^{3/2} + (1+p)\|g'\| \left(\frac{q}{p}\right)^{(2n-k+2)/2} \\ &\leq \|g'\| \left(\frac{q}{p}\right)^{n/2} \left\{ \frac{1+p-q}{1-q} \left(\frac{q}{p}\right)^{3/2} + \sqrt{\frac{q}{p}}(1+p) \right\}. \end{aligned}$$

This means that we may choose $\beta_n = (q/p)^{n/2}$, $n \geq 1$ (see Theorem 4.19). Then for all $n, m \geq 1$, we have

$$\beta_n + \beta_{n+1} + \ldots + \beta_{n+m-1} = \left(\frac{q}{p}\right)^{n/2} + \left(\frac{q}{p}\right)^{(n+1)/2} + \ldots + \left(\frac{q}{p}\right)^{(n+m-1)/2}$$
$$= \left(\frac{q}{p}\right)^{n/2} \frac{1-\left(\frac{q}{p}\right)^{m/2}}{1-\left(\frac{q}{p}\right)^{1/2}} < \frac{\sqrt{p}}{\sqrt{p}-\sqrt{q}} \left(\frac{q}{p}\right)^{n/2}.$$

Thus we may choose $\alpha_n = (q/p)^{n/2}$, $n \geq 1$. Applying Theorem 4.19, we get the statement of our theorem. ■

In the next theorem we shall estimate the error $|\tilde{D}_{\infty}^{p,q}(f;x) - f(x)|$ with the aid of the modulus of continuity.

Theorem 4.21 *For the limit (p,q)-Bernstein–Durrmeyer operator $\tilde{D}_{\infty}^{p,q}$, we have*

$$|\tilde{D}_{\infty}^{p,q}(f;x) - f(x)| \leq 2\omega\left(f, \sqrt{\delta_{p,q}(x)}\right)$$

for all $f \in C[0,1]$ and $x \in [0,1]$, where

$$\delta_{p,q}(x) = \frac{1}{p^4}(p-q)\{2p^2 + (3p+1)x + (p^3-1)x^2\}.$$

Proof For p, q fixed, in view of Lemma 4.8, we have

$$\tilde{D}_{\infty}^{p,q}(1,x) = 1. \tag{4.2.28}$$

Furthermore, by Lemma 4.8 and by the identity $[n]_{p,q} = p^{n-1}[n]_{q/p}$, we have

$$\tilde{D}_n^{p,q}(t;x) = \frac{p^{n+1} + p^n q[n]_{q/p}x}{p^{n+1}[n+2]_{q/p}} = \frac{p + q[n]_{q/p}x}{p[n+2]_{q/p}} \to \frac{p + \frac{q}{1-\frac{q}{p}}x}{p\frac{1}{1-\frac{q}{p}}} = \frac{p-q+qx}{p} =: \tilde{D}_{\infty}^{p,q}(t;x)$$

as $n \to \infty$, and analogously, we have

$$\tilde{D}_n^{p,q}(t^2;x) = \frac{p^{2n+2}[2]_{p,q}}{[n+2]_{p,q}[n+3]_{p,q}} + \frac{(2q^2+qp)p^{n+2}[n]_{p,q}x}{[n+2]_{p,q}[n+3]_{p,q}}$$
$$+ \frac{q^3[n]_{p,q}[p^2[n]_{p,q}x^2 + p^{n+1}x(1-x)]}{[n+2]_{p,q}[n+3]_{p,q}}$$
$$\to \frac{(p+q)(p-q)^2}{p^3} + \frac{q(p+2q)(p-q)x}{p^3} + \frac{q^3}{p^3}\left\{x^2 + \frac{p-q}{p}x(1-x)\right\}$$
$$=: \tilde{D}_{\infty}^{p,q}(t^2;x)$$

as $n \to \infty$. Then

$$\begin{aligned}
&\tilde{D}_{\infty}^{p,q}((t-x)^2;x)\\
&=\frac{1}{p^3}(p+q)(p-q)^2+\frac{q^3}{p^4}(p-q)x(1-x)+\frac{1}{p^3}(pq+2q^2-2p^2)(p-q)x\\
&\quad+\frac{1}{p^3}(p^2-pq-q^2)(p-q)x^2\\
&\le\frac{2}{p^2}(p-q)+\frac{1}{p^4}(p-q)x(1-x)+\frac{3}{p^3}(p-q)x+\frac{1}{p}(p-q)x^2\\
&=\delta_{p,q}(x).
\end{aligned} \tag{4.2.29}$$

For the modulus of continuity, we have

$$\omega(f,\lambda\delta)\le(1+\lambda)\omega(f,\delta),\ \lambda\ge 0.$$

Then

$$|f(t)-f(x)|\le\omega(f,|t-x|)\le(1+\delta^{-1}|t-x|)\omega(f,\delta)$$

for $t, x \in [0, 1]$. Hence, by (4.2.28), Hölder's inequality and (4.2.29), we obtain

$$\begin{aligned}
&|\tilde{D}_{\infty}^{p,q}(f;x)-f(x)|\\
&\le\tilde{D}_{\infty}^{p,q}(|f(t)-f(x)|;x)\le\omega(f,\delta)\left(1+\delta^{-1}\tilde{D}_{\infty}^{p,q}(|t-x|;x)\right)\\
&\le\omega(f,\delta)\left\{1+\delta^{-1}(\tilde{D}_{\infty}^{p,q}((t-x)^2;x))^{1/2}\right\}\le\omega(f,\delta)\left\{1+\delta^{-1}\sqrt{\delta_{p,q}(x)}\right\}.
\end{aligned}$$

Choosing $\delta=\sqrt{\delta_{p,q}(x)}$, we get the assertion of the theorem. ■

Remark 4.6 If $p = p(q)$ and $q \to 1$, then Theorem 4.21 implies that $\tilde{D}_{\infty}^{p,q}(f;x)$ converges uniformly to $f(x)$ for $x \in [0, 1]$.

4.2.3 *(p, q)-Baskakov–Durrmeyer Operators*

Using (p, q)-Beta function of second kind, Aral and Gupta [36] proposed for $x \in [0, \infty)$ the (p, q) analogue of Baskakov–Durrmeyer operators as

$$\widetilde{V}_n^{p,q}(f;x)=[n-1]_{p,q}\sum_{k=0}^{\infty}b_{n,k}^{p,q}(x)q^{[k(k+1)-2]/2}p^{(k+1)(k+2)/2}$$
$$\int_0^{\infty}\begin{bmatrix}n+k-1\\k\end{bmatrix}_{p,q}\frac{t^k}{(1\oplus pt)_{p,q}^{k+n}}f(p^kt)d_{p,q}t \tag{4.2.30}$$

where $b_{n,k}^{p,q}(x)$ is as defined in (3.4.3).

Lemma 4.9 ([36]) *For $x \in [0, \infty)$, the operators (4.2.30) satisfy*

1. $\widetilde{V}_n^{p,q}(1; x) = 1,$
2. $\widetilde{V}_n^{p,q}(t; x) = \frac{1}{qp^2[n-2]_{p,q}} + \frac{[2]_{p,q}}{p^2q^2[n-2]_{p,q}}x + \frac{1}{p^n}x,$
3. $\widetilde{V}_n^{p,q}(t^2; x) = \frac{[2]_{p,q}}{p^5q^3[n-2]_{p,q}[n-3]_{p,q}} + \frac{[2]^2_{p,q}[n]_{p,q}x}{p^{n+3}[n-2]_{p,q}[n-3]_{p,q}}$
$+ \left(\frac{1}{p^{2n}} + \frac{[2]_{p,q}}{p^{n+2}q^2[n-2]_{p,q}} + \frac{[3]_{p,q}}{p^{n+3}q^3[n-3]_{p,q}} + \frac{p^nq[2]_{p,q}[3]_{p,q}+[n]_{p,q}p^5}{p^{n+5}q^6[n-2]_{p,q}[n-3]_{p,q}}\right)x^2.$

Theorem 4.22 ([36]) *Let $p = p_n$ and $q = q_n$ satisfy $0 < q_n < p_n \le 1$ and for n sufficiently large $p_n \to 1$, $q_n \to 1$, $q_n^n \to 1$ and $p_n^n \to 1$. For each $f \in C^*_{x^2}[0, \infty)$ (defined in Section 3.4), we have*

$$\lim_{n\to\infty} \left\| \widetilde{V}_n^{p_n,q_n}(f) - f \right\|_{x^2} = 0.$$

Theorem 4.23 ([36]) *Let $p = p_n$ and $q = q_n$ satisfy $0 < q_n < p_n \le 1$ and for n sufficiently large $p_n \to 1$, $q_n \to 1$, $q_n^n \to 1$ and $p_n^n \to 1$. For each $f \in C^*_{x^2}[0, \infty)$, we have*

$$\lim_{n\to\infty} \sup_{x\in[0,\infty)} \frac{\left|\widetilde{V}_n^{p_n,q_n}(f, x) - f(x)\right|}{\left(1+x^2\right)^{1+\alpha}} = 0.$$

Also, using the linear approximating method viz. Steklov mean as defined in Section 3.4, the following direct estimate was proved.

Theorem 4.24 ([36]) *Let $q \in (0, 1)$ and $p \in (q, 1]$. If $f \in C_B[0, \infty)$, then*

$$\begin{aligned}
&\left|\widetilde{V}_n^{p,q}(f; x) - f(x)\right| \\
&\le 5\omega_1\left(f; \frac{1}{\sqrt{[n-2]_{p,q}}}\right)\left[\frac{1}{qp^2\sqrt{[n-2]_{p,q}}} + \frac{[2]_{p,q}x}{p^2q^2\sqrt{[n-2]_{p,q}}}\right. \\
&\quad \left. + \left(\frac{1}{p^n} - 1\right)\sqrt{[n-2]_{p,q}}x\right] \\
&\quad + \frac{9}{2}\omega_2\left(f; \frac{1}{\sqrt{[n-2]_{p,q}}}\right)\left[\left(\frac{p^{2n} - 2p^n + 1}{p^{2n}}\right)[n-2]_{p,q}x^2\right. \\
&\quad + \left(\frac{[2]_{p,q}}{p^{n+2}q^2} + \frac{[3]_{p,q}[n-2]_{p,q}}{p^{n+3}q^3[n-3]_{p,q}} + \frac{p^nq[2]_{p,q}[3]_{p,q} + [n]_{p,q}p^5}{p^{n+5}q^6[n-3]_{p,q}} - \frac{2[2]_{p,q}}{p^2q^2}\right)x^2 \\
&\quad \left. + \left(\frac{[2]^2_{p,q}[n]_{p,q}}{p^{n+3}[n-3]_{p,q}} - \frac{2}{qp^2}\right)x + \frac{[2]_{p,q}}{p^5q^3[n-3]_{p,q}}\right].
\end{aligned}$$

The operators discussed in [115] provide a better approximation than the usual Baskakov–Durrmeyer operators. Motivated by this, for $n \in \mathbb{N}$, $x \in [0, \infty)$, the (p, q)-Baskakov-Beta operators introduced in [167] are defined by:

$$\widetilde{VB}_n^{p,q}(f, x) = \sum_{k=0}^{\infty} \frac{b_{n,k}^{p,q}(x)}{B_{p,q}(k+1, n)} \int_0^{\infty} \frac{t^k}{(1 \oplus pt)_{p,q}^{n+k+1}} f(q^2 p^{n+k} t) \, d_{p,q}t, \quad (4.2.31)$$

where

$$b_{n,k}^{p,q}(x) = \begin{bmatrix} n+k-1 \\ k \end{bmatrix}_{p,q} p^{k+n(n-1)/2} \, q^{k(k-1)/2} \, \frac{x^k}{(1 \oplus x)_{p,q}^{n+k}}.$$

Theorem 4.25 ([167]) *Let $f \in C_B[0, \infty)$, then for every $x \in [0, \infty)$ and $n > 2$, the following inequality holds:*

$$|\widetilde{VB}_n^{p,q}(f, x) - f(x)| \leqslant \omega \left(f, \frac{|x \left([n]_{p,q} - [n-1]_{p,q}\right) + p^{n-2} q|}{[n-1]_{p,q}} \right)$$
$$+ C \, \omega_2 \left(f, \sqrt{\mu_{n,2}^{p,q}(x) + \left(\frac{x \left([n]_{p,q} - [n-1]_{p,q}\right) + p^{n-2} q}{[n-1]_{p,q}} \right)^2} \right),$$

where C is some positive constant.

In [167], the following direct result was established using Steklov mean as defined in Section 3.4:

Theorem 4.26 ([167]) *Let $f \in C_B[0, \infty)$, then*

$$|\widetilde{VB}n^{p,q}(f, x) - f(x)| \leqslant 5 \, \tilde{\omega} \left(f, \frac{1}{\sqrt{[n-1]_{p,q}}} \right) \mu_{n,1}^{p,q}(x)$$
$$+ \frac{9}{2} \, \tilde{\omega}_2 \left(f, \frac{1}{\sqrt{[n-1]_{p,q}}} \right) \mu_{n,2}^{p,q}(x),$$

where

$$\mu_{n,1}^{p,q}(x) = \frac{x \left([n]_{p,q} - [n-1]_{p,q}\right) + p^{n-2} q}{[n-1]_{p,q}}$$

and

$$\mu_{n,2}^{p,q}(x) = \frac{x^2\left\{[n]_{p,q}\left([n]_{p,q}+\frac{p^n}{q}\right)+q\,[n-1]_{p,q}\,[n-2]_{p,q}-2\,q\,[n]_{p,q}\,[n-2]_{p,q}\right\}}{q\,[n-1]_{p,q}\,[n-2]_{p,q}}$$
$$+\frac{x\left\{[n]_{p,q}\left(p^{n-3}\,q^2+2\,p^{n-2}\,q+p^{n-1}\right)-2\,p^{n-2}\,q^2\,[n-2]_{p,q}\right\}}{q\,[n-1]_{p,q}\,[n-2]_{p,q}}$$
$$+\frac{[2]_{p,q}\,p^{2n-5}\,q}{[n-1]_{p,q}\,[n-2]_{p,q}}.$$

4.2.4 (p, q)-Szász–Durrmeyer Operators

As an application of the (p, q)-Gamma function, Aral and Gupta in [37] introduced (p, q) variant of Szász–Durrmeyer operator for $x \in [0, \infty)$ as

$$\widetilde{S}_{n,p,q}(f;x) = [n]_{p,q}\sum_{k=0}^{\infty} s_{n,k}^{p,q}(x)\int_0^{\infty} p^{k(k-1)/2}\frac{\left([n]_{p,q}\,t\right)^k}{[k]_{p,q}!}$$
$$E_{p,q}\left(-q\,[n]_{p,q}\,t\right) f\left(q^{1-k}p^k t\right) d_{p,q}t, \tag{4.2.32}$$

where

$$s_{n,k}^{p,q}(x) = \frac{1}{E_{p,q}\left([n]_{p,q}\,x\right)}\frac{q^{k(k-1)/2}}{[k]_{p,q}!}\left([n]_{p,q}\,x\right)^k.$$

It may be remarked here that for $p = q = 1$ these operators will reduce to the Szász Durrmeyer operators.

Lemma 4.10 *For $x \in [0, \infty)$, we have*

1. $\widetilde{S}_{n,p,q}(1;x) = 1$
2. $\widetilde{S}_{n,p,q}(t;x) = \frac{q}{[n]_{p,q}} + px$
3. $\widetilde{S}_{n,p,q}(t^2;x) = \frac{p^3}{q}x^2 + \frac{[2]_{p,q}^2 x}{[n]_{p,q}} + \frac{[2]_{p,q}q^2}{p[n]_{p,q}^2}$.

Remark 4.7 We may write

$$\widetilde{S}_{n,p,q}((t-x),x) = \frac{q}{[n]_{p,q}} + (p-1)x$$

$$\widetilde{S}_{n,p,q}((t-x)^2,x) = \frac{(p^3-2pq+q)x^2}{q} + \frac{([2]_{p,q}^2-2q)x}{[n]_{p,q}} + \frac{[2]_{p,q}q^2}{p[n]_{p,q}^2}. \tag{4.2.33}$$

Theorem 4.27 ([37]) *Let $q \in (0,1)$ and $p \in (q,1]$. The operator $\widetilde{S}_{n,p,q}$ maps the space C_B into C_B. It holds*

$$\left\|\widetilde{S}_{n,p,q}(f)\right\|_{C_B} \leq \|f\|_{C_B}.$$

Theorem 4.28 ([37]) *Let $q \in (0,1)$ and $p \in (q,1]$. If $f \in C_B[0,\infty)$, then*

$$\begin{aligned}
&\left|\widetilde{S}_{n,p,q}(f,x) - f(x)\right| \\
&\leq 5\omega\left(f, \frac{1}{\sqrt{[n]_{p,q}}}\right)\left(\frac{q}{\sqrt{[n]_{p,q}}} + \sqrt{[n]_{p,q}}(p-1)x\right) \\
&\quad + \frac{9}{2}\omega_2\left(f, \frac{1}{\sqrt{[n]_{p,q}}}\right)\left[2 + \frac{(p^3-2pq+q)[n]_{p,q}x^2}{q} + ([2]_{p,q}^2 - 2q)x + \frac{[2]_{p,q}q^2}{p[n]_{p,q}}\right].
\end{aligned}$$

Theorem 4.29 ([37]) *Let $p = p_n$ and $q = q_n$ satisfy $0 < q_n < p_n \leq 1$ and for n sufficiently large $p_n \to 1$, $q_n \to 1$, $q_n^n \to 1$ and $p_n^n \to 1$. For each $f \in C_{x^2}^*[0,\infty)$ (defined in Section 3.4), we have*

$$\lim_{n\to\infty} \left\|\widetilde{S}_{n,p_n,q_n}(f) - f\right\|_{x^2} = 0.$$

Theorem 4.30 ([37]) *Let $p = p_n$ and $q = q_n$ satisfy $0 < q_n < p_n \leq 1$ and for n sufficiently large $p_n \to 1$, $q_n \to 1$, $q_n^n \to 1$ and $p_n^n \to 1$ For each $f \in C_{x^2}[0,\infty)$ (see Section 3.4) and $\alpha > 0$, we have*

$$\lim_{n\to\infty} \sup_{x\in[0,\infty)} \frac{\left|\widetilde{S}_{n,p_n,q_n}(f,x) - f(x)\right|}{\left(1+x^2\right)^{1+\alpha}} = 0.$$

Proof For any fixed $x_0 > 0$,

$$\begin{aligned}
&\sup_{x\in[0,\infty)} \frac{\left|\widetilde{S}_{n,p_n,q_n}(f,x) - f(x)\right|}{\left(1+x^2\right)^{1+\alpha}} \\
&= \sup_{x\leq x_0} \frac{\left|\widetilde{S}_{n,p_n,q_n}(f,x) - f(x)\right|}{\left(1+x^2\right)^{1+\alpha}} + \sup_{x\geq x_0} \frac{\left|\widetilde{S}_{n,p_n,q_n}(f,x) - f(x)\right|}{\left(1+x^2\right)^{1+\alpha}} \\
&\leq \left\|\widetilde{S}_{n,p_n,q_n}(f) - f\right\|_{C[0,a]} + \|f\|_{x^2} \sup_{x\geq x_0} \frac{\left|\widetilde{S}_{n,p_n,q_n}\left(1+t^2,x\right)\right|}{\left(1+x^2\right)^{1+\alpha}} \\
&\quad + \sup_{x\geq x_0} \frac{|f(x)|}{\left(1+x^2\right)^{1+\alpha}}.
\end{aligned}$$

■

By Lemma 4.10 and the well-known Korovkin theorem the first term of the above inequality tends to zero for sufficiently large n. By Lemma 4.10 for any fixed $x_0 > 0$ it is easily seen that

$$\sup_{x \geq x_0} \frac{\left|\widetilde{S}_{n,p_n,q_n}\left(1+t^2, x\right)\right|}{\left(1+x^2\right)^{1+\alpha}}$$

tends to zero as $n \to \infty$. We can choose $x_0 > 0$ so large that the last part of above inequality can be made small enough. This completes the proof of the theorem.

A function $f \in C[0, \infty)$ is said to satisfy Lipschitz condition $Lip\ \ \alpha$ on D, $\alpha \in (0, 1]$, $D \subset [0, \infty)$ if

$$|f(t) - f(x)| \leq M_f |t-x|^\alpha, \ t \in [0, \infty) \text{ and } x \in D,$$

where M_f is a constant depending only α and f.

Theorem 4.31 ([37]) *Let $f \in Lip\ \ \alpha$ on D, $D \subset [0, \infty)$ and $\alpha \in (0, 1]$. We have*

$$\left|\widetilde{S}_{n,p,q}(f,x) - f(x)\right| \leq \left(\frac{(p^3-2pq+q)x^2}{q} + \frac{([2]_{p,q}^2 - 2q)x}{[n]_{p,q}} + \frac{[2]_{p,q}q^2}{p[n]_{p,q}^2}\right)^{\alpha/2} + 2d^\alpha(x; D)$$

where $d(x; D)$ represents the distance between x and D.

Proof For $x_0 \in \overline{D}$, the closure of the set D in $[0, \infty)$, we have

$$|f(t) - f(x)| \leq |f(t) - f(x_0)| + |f(x_0) - f(x)|, \quad x \in [0, \infty).$$

Thus, we get

$$\begin{aligned} \left|\widetilde{S}_{n,p,q}(f,x) - f(x)\right| &\leq \widetilde{S}_{n,p,q}(|f(t) - f(x_0)|, x) + |f(x_0) - f(x)| \\ &\leq M_f \widetilde{S}_{n,p,q}\left(|t-x_0|^\alpha, x\right) + M_f |x_0 - x|^\alpha. \quad (4.2.34) \end{aligned}$$

Then, with Hölder's inequality with

$$p := \frac{2}{\alpha} \text{ and } \frac{1}{r} := 1 - \frac{1}{p},$$

we have

$$\widetilde{S}_{n,p,q}\left(|t-x|^\alpha, x\right) \leq \left(\widetilde{S}_{n,p,q}\left(|t-x|^2, x\right)\right)^{\frac{\alpha}{2}} \left(\widetilde{S}_{n,p,q}(1,x)\right)^{1-\frac{\alpha}{2}}. \quad (4.2.35)$$

Also $\widetilde{S}_{n,p,q}$ is monotone

$$\widetilde{S}_{n,p,q}\left(|t-x_0|^\alpha, x\right) \leq \left(\widetilde{S}_{n,p,q}\left(|t-x|^\alpha, x\right)\right)^{\frac{\alpha}{2}} + |x_0 - x|^\alpha.$$

Using (4.2.34), (4.2.35) and (4.2.33), we have the desired result. ■

Now, we present local direct estimate for (p, q)-Szász–Durrmeyer operators using the Lipschitz-type maximal function of order α:

$$\widetilde{\omega}_{\alpha}(f, x) = \sup_{t \neq x,\ t \in [0, \infty)} \frac{|f(t) - f(x)|}{|t - x|^{\alpha}}, \ x \in [0, \infty) \text{ and } \alpha \in (0, 1]. \tag{4.2.36}$$

Theorem 4.32 ([37]) *Let $f \in Lip\alpha$ on D and $f \in C_B[0, \infty)$. Then for all $x \in [0, \infty)$, we have*

$$\left|\widetilde{S}_{n,p,q}(f, x) - f(x)\right| \leq \widetilde{\omega}_{\alpha}(f, x)\left(\frac{(p^3 - 2pq + q)x^2}{q} + \frac{([2]_{p,q}^2 - 2q)x}{[n]_{p,q}} + \frac{[2]_{p,q}q^2}{p[n]_{p,q}^2}\right)^{\frac{\alpha}{2}}$$

Proof From (4.2.36) we obtain

$$|f(t) - f(x)| \leq \widetilde{\omega}_{\alpha}(f, x)\,|t - x|^{\alpha}$$

and

$$\begin{aligned}\left|\widetilde{S}_{n,p,q}(f, x) - f(x)\right| &\leq \widetilde{S}_{n,p,q}(|f(t) - f(x)|, x) \\ &\leq \widetilde{\omega}_{\alpha}(f, x)\,\widetilde{S}_{n,p,q}\left(|t - x|^{\alpha}, x\right).\end{aligned}$$

Applying Hölder's inequality with

$$p := \frac{2}{\alpha} \text{ and } \frac{1}{r} := 1 - \frac{1}{p},$$

we have

$$\left|\widetilde{S}_{n,p,q}(f, x) - f(x)\right| \leq \widetilde{\omega}_{\alpha}(f, x)\,\widetilde{S}_{n,p,q}\left((t - x)^2, x\right)^{\frac{\alpha}{2}}.$$

Using (4.2.33), our assertion follows. ■

4.2.5 (p, q)-Variant of Szász-Beta Operators

In 2006, Gupta and Noor [122] proposed Szász-Beta operators and obtained some direct results in simultaneous approximation. Four years later Gupta and Aral [38] introduced the q variant of the Szász-Beta operators. The (p, q) analogue of Szász-Beta operator is defined by Aral and Gupta in [124] as

$$D_n^{p,q}(f,x)=\sum_{k=1}^{\infty}\frac{s_{n,k}^{p,q}(x)}{B_{p,q}(k,n+1)}\int_0^{\infty}\frac{t^{k-1}}{(1\oplus pt)_{p,q}^{k+n+1}}f(p^{k+1}qt)d_{p,q}t+\frac{f(0)}{E_{p,q}\left([n]_{p,q}x\right)} \tag{4.2.37}$$

where $s_{n,k}^{p,q}(x)$ is as defined in (3.4.4).

Lemma 4.11 ([124]) *For $x\in[0,\infty)$, we obtain*

$$\begin{aligned}
D_n^{p,q}(1,x)&=1,\\
D_n^{p,q}(t,x)&=x,\\
D_n^{p,q}(t^2,x)&=\frac{[2]_{p,q}qx}{p[n-1]_{p,q}}+\frac{p\,[n]_{p,q}\,x^2}{[n-1]_{p,q}},\\
D_n^{p,q}(t^3,x)&=\frac{p^3[n]_{p,q}^2}{q^6\,[n-1]_{p,q}\,[n-2]_{p,q}}x^3\\
&\quad+\left(\frac{(p[2]_{p,q}+p^2)\,[n]_{p,q}}{q^6p^2\,[n-1]_{p,q}\,[n-2]_{p,q}}+\frac{(p^2q+2pq^2)\,[n]_{p,q}}{q^6\,[n-1]_{p,q}\,[n-2]_{p,q}}\right)x^2\\
&\quad+\left(\frac{[2]_{p,q}}{q^5p^3\,[n-1]_{p,q}\,[n-2]_{p,q}}+\frac{(p[2]_{p,q}+p^2)}{q^5p^3\,[n-1]_{p,q}\,[n-2]_{p,q}}\right)x,\\
D_n^{p,q}(t^4,x)&=\frac{[n]_{p,q}^3}{q^{10}\,[n-1]_{p,q}\,[n-2]_{p,q}\,[n-3]_{p,q}}\left(\frac{p^6}{q^2}x^4+\frac{p^3}{q[n]_{p,q}}\left(p^2+2q+3q^2\right)x^3\right.\\
&\quad\left.+\frac{pq(p^2+3pq+3q^2)}{[n]_{p,q}^2}x^2+\frac{q^4}{[n]_{p,q}^3}x\right)\\
&\quad+\frac{\left((p[2]_{p,q}+p^2)+1\right)[n]_{p,q}^2}{p^3q^8\,[n-1]_{p,q}\,[n-2]_{p,q}\,[n-3]_{p,q}}\left(p^3x^3+\frac{(p^2q+2pq^2)}{[n]_{p,q}}x^2+\frac{q^3}{[n]_{p,q}^2}x\right)\\
&\quad+\frac{(p^3+(p[2]_{p,q}+p^2))\,[n]_{p,q}}{p^6q^5\,[n-1]_{p,q}\,[n-2]_{p,q}\,[n-3]_{p,q}}\left(px^2+\frac{qx}{[n]_{p,q}}\right)\\
&\quad+\frac{[2]_{p,q}[3]_{p,q}}{p^6q^3\,[n-1]_{p,q}\,[n-2]_{p,q}\,[n-3]_{p,q}}x.
\end{aligned}$$

For the weighted approximation Aral and Gupta [38] established the following theorem using the moments of Lemma 4.11:

Theorem 4.33 ([124]) *Let $p=p_n$ and $q=q_n$ satisfy $0<q_n<p_n\le 1$ and for n sufficiently large $p_n\to 1$, $q_n\to 1$, $q_n^n\to 1$ and $p_n^n\to 1$. For each $f\in C_{x^2}^*[0,\infty)$ (see Section 3.4), we have*

$$\lim_{n\to\infty}\left\|D_n^{p_n,q_n}(f,x)-f\right\|_{x^2}=0,$$

where

$$\|f\|_{x^2} = \sup_{x\in[0,\,\infty)} \frac{|f(x)|}{1+x^2}.$$

In order to analyze the error estimation on some weighted spaces, Aral and Gupta [38] considered the functions, satisfying the growth condition $|f(t)| \le M(1+t)^m$, for some $M > 0$ and $m > 0$, with the weight

$$\rho(x) = (1+x)^{-m}, \; x \in I = [0, \infty).$$

The polynomial weighted space associated with this weight is defined by

$$C_\rho(I) = \left\{ f \in C(I) : \|f\|_\rho < \infty \right\}$$

where

$$\|f\|_\rho = \sup_{x\in I} \rho(x)\,|f(x)| \tag{4.2.38}$$

Also, for $a \in \mathbb{N}_0$, $b > 0$, $c \ge 0$, suppose

$$\varphi(x) = \sqrt{(1+ax)(bx+c)}.$$

For $\lambda \in [0,1]$, $f \in C_\rho(I)$, the K-functional is given by

$$K_{1,\varphi^\lambda}(f,t)_\rho = \inf\left\{ \|f-g\|_\rho + t\left\|\varphi^\lambda g'\right\|_\rho,\; g \in W^\infty_{1,\lambda}(\varphi) \right\},$$

where $W^\infty_{1,\lambda}(\varphi)$ consists of all functions $g \in C_\rho(I)$ such that $\left\|\varphi^\lambda g'\right\|_\rho < \infty$.

Also this K-functional is associated with the modulus of smoothness in the following norm

$$C_1\omega^2_{\varphi^\lambda}(f,t)_\rho \le K_{1,\varphi^\lambda}(f,t)_\rho \le C_2\omega^2_{\varphi^\lambda}(f,t)_\rho,$$

where for $f \in C_\rho(I)$:

$$\omega^2_{\varphi^\lambda}(f,t)_\rho = \sup_{h\in(0,t]}\sup_{x\in I(\varphi,h)} \left|\rho(x)\,\Delta_{h\varphi(x)} f(x)\right| \tag{4.2.39}$$

and

$$I(\varphi,h) = \{x > 0 : h\varphi(x) \le x\}.$$

The key for the new estimate of [38] is based on the following general result due to [64]:

Theorem 4.34 ([64]) *Fix $a \in \mathbb{N}$ and set $\varphi(x) = \sqrt{x(1+ax)}$.*

Let $L_n : C_\rho(I) \to C(I)$ be a sequence of positive linear operators satisfying the following conditions:

i) $L_n(e_0) = e_0$.

ii) There exists a constant C_1 and a sequence $\{\alpha_n\}$ such that

$$L_n\left((t-x)^2, x\right) \leq C_1 \alpha_n \varphi^2(x).$$

iii) There exists a constant $C_2 = C_2(m)$ such that for each $n \in \mathbb{N}$,

$$L_n\left((1+t)^m, x\right) \leq C_2 (1+x)^m, \ x \geq 0.$$

iv) There exists a constant $C_3 = C_3(m)$ such that for every $m \in \mathbb{N}$,

$$\rho(x) L_n\left(\frac{(t-x)^2}{\rho(t)}, x\right) \leq C_3 \alpha_n \varphi^2(x), \ x \geq 0.$$

Then there exists a constant C such that for any $f \in C_\rho(I)$, $x \geq 0$, $n \in \mathbb{N}$ such that $\alpha_n \leq 1/2\sqrt{2+a}$, it holds

$$\|f - L_n f\|_\rho \leq C \omega_\varphi^2\left(f, \sqrt{\alpha_n}\right)_\rho,$$

where $\omega_\varphi^2(f,t)_\rho$ is the modulus in (4.2.39), with $\lambda = 1$.

Lemma 4.12 ([124]) *For $x \in [0,\infty)$, we have*

$$D_n^{p,q}((t-x)^2, x) \leq \frac{2x(x+1)}{p[n-1]_{p,q}}$$

Lemma 4.13 ([124]) *Let $n \in \mathbb{N}$ and $m = 2$. Then $D_n^{p,q}$ defined by (4.2.37) is an operator from $C_\rho(\mathbb{R})$ into $C_\rho(\mathbb{R})$. Moreover for any $f \in C_\rho(\mathbb{R})$ we have*

$$\left\|D_n^{p,q} f\right\|_\rho \leq \left\{4 + \frac{3}{p[n-1]_{p,q}}\right\} \|f\|_\rho$$

and

$$\rho(x) D_n^{p,q}\left(\frac{(t-x)^2}{\rho(t)}, x\right) \leq C_3 \frac{2x(x+1)}{p[n-1]_{p,q}}, \ x \geq 0.$$

Theorem 4.35 ([124]) *Set $\rho(x) = (1+x)^2$. For any $f \in C_\rho[0,\infty)$, $x \geq 0$, $n \in \mathbb{N}$, one has*

$$\rho(x)\left|f(x) - D_n^{p,q}(f,x)\right|$$

$$\leq C\omega^2\left(f, \sqrt{x(x+1)\left(-3+6p^2+\frac{p^{12}}{q^{12}}-4\frac{p^6}{q^9}+\mathcal{O}\left(\frac{1}{p[n-1]_{p,q}}\right)\right)}\right)_\rho.$$

For $n > 2\sqrt{3}$ one has

$$\left\|f(x) - D_n^{p,q}(f,x)\right\|_\rho$$

$$\leq \omega_\varphi^2\left(f, \left(-3+6p^2+\frac{p^{12}}{q^{12}}-4\frac{p^6}{q^9}+\mathcal{O}\left(\frac{1}{p[n-1]_{p,q}}\right)\right)\right)_\rho,$$

where $\varphi(x) = \sqrt{x(1+x)}$ and $\omega_\varphi^2(f,t)_\rho$ is modulus defined in (4.2.39) with $\lambda = 1$.

Proof It is observed that all the conditions in Theorem 4.34 are verified in Lemma 4.12 and Lemma 4.13 for $m = 2$. From Lemma 4.12, we can see that

$$\varphi(x) = \sqrt{x(x+1)} \text{ and } \alpha_n = 1/\sqrt{p[n-1]_{p,q}}.$$

∎

Consider two functions $f, g \in C_\rho[0,\infty)$ and define the positive bilinear functional

$$D_n^{p,q}(f,g,x) = D_n^{p,q}(fg,x) - D_n^{p,q}(f,x)\,D_n^{p,q}(g,x).$$

To measure the rate of convergence of this positive bilinear functional on weighted spaces we use the modulus of smoothness defined by (4.2.39) and Theorem 4.35.

Theorem 4.36 ([124]) *For any $f \in C_\rho[0,\infty)$, $x \geq 0$, $n \in \mathbb{N}$, one has*

$$\left\|D_n^{p,q}(f,g)\right\|_{\rho^2} \leq \sqrt{C(f)}.\sqrt{C(g)},$$

where $C(f)$ is given by

$$\omega_\varphi^2\left(f^2, \left(-3+6p^2+\frac{p^{12}}{q^{12}}-4\frac{p^6}{q^9}+\mathcal{O}\left(\frac{1}{p[n-1]_{p,q}}\right)\right)^{1/2}\right)$$

$$+\left\{5+\frac{3}{p[n-1]_{p,q}}\right\}\|f\|_\rho\,\omega_\varphi^2\left(f, \left(-3+6p^2+\frac{p^{12}}{q^{12}}-4\frac{p^6}{q^9}+\mathcal{O}\left(\frac{1}{p[n-1]_{p,q}}\right)\right)^{1/2}\right).$$

Proof Using Cauchy–Schwarz inequality, we have

$$|D_n(f, g, x)| \leq \sqrt{D_n(f, f, x)}\sqrt{D_n(g, g, x)}.$$

Also, we have

$$D_n(f, f, x) = D_n^{p,q}\left(f^2, x\right) - f^2(x) + \left[f(x) - D_n^{p,q}(f, x)\right]\left[f(x) + D_n^{p,q}(f, x)\right].$$

By (4.2.38) and Lemma 4.13, we can write

$$|D_n(f, f, x)| \leq D_n^{p,q}\left(f^2, x\right) - f^2(x) + \left\{5 + \frac{3}{p[n-1]_{p,q}}\right\} \|f\|_\rho \left|f(x) - D_n^{p,q}(f, x)\right|.$$

It implies that

$$\frac{|D_n(f, f, x)|}{\rho^2(x)} \leq \frac{\left|D_n^{p,q}(f^2, x) - f^2(x)\right|}{\rho(x)} + \left\{5 + \frac{3}{p[n-1]_{p,q}}\right\} \|f\|_\rho \frac{\left|f(x) - D_n^{p,q}(f, x)\right|}{\rho(x)}$$

Using Theorem 4.35, we have

$$\begin{aligned}
\|D_n(f, f)\|_{\rho^2} &\leq \left\|D_n^{p,q}\left(f^2\right) - f^2\right\|_\rho + \left\{5 + \frac{3}{p[n-1]_{p,q}}\right\} \|f\|_\rho \left\|D_n^{p,q}(f) - f\right\|_\rho \\
&\leq \omega_\varphi^2\left(f^2, \left(-3 + 6p^2 + \frac{p^{12}}{q^{12}} - 4\frac{p^6}{q^9} + \mathcal{O}\left(\frac{1}{p\,[n-1]_{p,q}}\right)\right)^{1/2}\right) \\
&\quad + \left\{5 + \frac{3}{p[n-1]_{p,q}}\right\} \|f\|_\rho \\
&\quad \omega_\varphi^2\left(f, \left(-3 + 6p^2 + \frac{p^{12}}{q^{12}} - 4\frac{p^6}{q^9} + \mathcal{O}\left(\frac{1}{p\,[n-1]_{p,q}}\right)\right)^{1/2}\right).
\end{aligned}$$

This completes the proof of the theorem. ■

Finally the following asymptotic formula was also discussed by Aral and Gupta:

Theorem 4.37 ([124]) *Let $f \in C\left(\mathbb{R}^+\right)$. If $x \in \mathbb{R}^+$, f is two times differentiable in x and f'' is continuous in x, $p = p_n$ and $q = q_n$ satisfy $0 < q_n < p_n \leq 1$ and for n sufficiently large $p_n \to 1$, $q_n \to 1$, $q_n^n \to 1$ and $p_n^n \to 1$. Then the following holds true*

$$\lim_{n\to\infty} [n]_{p_n,q_n}\left[D_n^{p_n,q_n}(f, x) - f(x)\right] = x(1 + \alpha x) f''(x)$$

Using (p,q)-Beta function of second kind, Gupta [120] introduced for $x \in [0,\infty)$ the (p,q) variant of Szász–Mirakyan–Baskakov operators as

$$\widehat{D}_n^{p,q}(f,x) = [n-1]_{p,q} \sum_{k=0}^{\infty} s_{n,k}^{p,q}(x) q^{[k(k+1)-2]/2} p^{(k+1)(k+2)/2}$$
$$\int_0^{\infty} b_{n,k}^{p,q}(t) f(p^k t) d_{p,q}t \tag{4.2.40}$$

where

$$s_{n,k}^{p,q}(x) = \frac{1}{E_{p,q}([n]_{p,q}x)} \frac{q^{k(k-1)/2}[n]_{p,q}^k x^k}{[k]_{p,q}!}, \; b_{n,k}^{p,q}(t) = \begin{bmatrix} n+k-1 \\ k \end{bmatrix}_{p,q} \frac{t^k}{(1 \oplus pt)_{p,q}^{k+n}}.$$

Along with weighted approximation the following direct estimate in terms of modulus of continuity has been established and proved in [120]:

Theorem 4.38 ([120]) *Let $q \in (0,1)$ and $p \in (q,1]$. If $f \in C_B[0,\infty)$, then*

$$\begin{aligned} &\left|\widehat{D}_n^{p,q}(f,x) - f(x)\right| \\ &\le 5\omega\left(f, \frac{1}{\sqrt{[n-2]_{p,q}}}\right)\left(\frac{1}{qp^2\sqrt{[n-2]_{p,q}}} + \frac{([n]_{p,q} - pq^2[n-2]_{p,q})x}{pq^2\sqrt{[n-2]_{p,q}}}\right) \\ &+\frac{9}{2}\omega_2\left(f, \frac{1}{\sqrt{[n-2]_{p,q}}}\right)\left[\left(\frac{[n]_{p,q}^2 - 2q^4[n]_{p,q}[n-3]_{p,q} + pq^6[n-2]_{p,q}[n-3]_{p,q}}{pq^6[n-3]_{p,q}}\right)x^2\right. \\ &\left.+\left(\frac{[q([2]_{p,q}+p)+p^2][n]_{p,q} - 2p^2q^4[n-3]_{p,q}}{p^4q^5[n-3]_{p,q}}\right)x + \frac{[2]_{p,q}}{p^5q^3[n-3]_{p,q}} + 2\right]. \end{aligned}$$

Chapter 5
Univariate Grüss- and Ostrowski-Type Inequalities for Positive Linear Operators

This chapter is dedicated to Grüss-type inequalities on the space of continuous functions defined on a compact metric space and applications of these inequalities in the cases of known operators. Using the least concave majorant of the modulus of continuity we will consider a Grüss inequality for the functional $L(f) = H(f; x)$, where $H : C[a, b] \to C[a, b]$ is a positive linear operator and $x \in [a, b]$ is fixed. These results are motivated by a theorem which can be found in the paper [31] by Andrica and Badea. It is the aim of this chapter to look again at Grüss' inequality from a somewhat different point of view. We are interested on how non-multiplicative can a linear functional be. Moreover, we derive inequalities of the Grüss type using Cauchy's mean value theorem. This study was motivated by Pachpatte's result obtained in [190]. Also, a Grüss inequality on a compact metric space for more than two functions is considered in this chapter. Another renowned classical inequality was introduced by Ostrowski [188] and provides an upper bound for the approximation of the average value by a single value of the function in question. In this chapter some results concerning Ostrowski inequality using the least concave majorant of the modulus of continuity and the second order modulus of smoothness are presented.

5.1 Grüss-Type Inequalities for a Positive Linear Functional

During the last years, Grüss-type inequalities have attracted much attention, because of their applications in several disciplines such as mathematical statistics, econometrics, and actuarial mathematics. We mention here the papers by Landau [154], Karamata [149], and Ostrowski [189]. We note that a whole chapter in a book by Mitrinović et al. [177] is devoted to the inequality we discuss here (see also [81]).

V. Gupta et al., *Recent Advances in Constructive Approximation Theory*,
Springer Optimization and Its Applications 138,
https://doi.org/10.1007/978-3-319-92165-5_5

The functional given by

$$T(f,g) := \frac{1}{b-a}\int_a^b f(t)g(t)dt - \frac{1}{b-a}\int_a^b f(t)dt \cdot \frac{1}{b-a}\int_a^b g(t)dt, \tag{5.1.1}$$

where $f, g : [a,b] \to \mathbb{R}$ are integrable functions, is well known in the literature as the Chebyshev functional (see [71]). The original form of Grüss' inequality estimates the difference between the integral of a product of two functions and the product of integrals of the two functions and was published by Grüss in 1935 (see [138]):

Theorem 5.1 *Let f and g be two functions defined and integrable on $[a,b]$. If $m \leq f(x) \leq M$ and $p \leq g(x) \leq P$ for all $x \in [a,b]$, then we have*

$$|T(f,g)| \leq \frac{1}{4}(M-m)(P-p). \tag{5.1.2}$$

The constant $1/4$ is the best possible.

In 1882, Chebyshev [71] obtained the following inequality.

Theorem 5.2 *If $f, g \in C^1[a,b]$, then*

$$|T(f,g)| \leq \frac{1}{12}\|f'\|_\infty \|g'\|_\infty (b-a)^2 \tag{5.1.3}$$

holds, where $\|f'\|_\infty := \sup_{t\in[a,b]} |f'(t)|$. The constant $\frac{1}{12}$ cannot be improved in the general case.

In 1970, Ostrowski [189] proved the following theorem, which is a combination of the Chebyshev and the Grüss results (5.1.3) and (5.1.2).

Theorem 5.3 *If f is a Lebesgue integrable function on $[a,b]$ satisfying $m \leq f(x) \leq M$, $x \in [a,b]$ and $g : [a,b] \to \mathbb{R}$ is absolutely continuous with $g' \in L_\infty[a,b]$, then the inequality*

$$|T(f,g)| \leq \frac{1}{8}(b-a)(M-m)\|g'\|_\infty \tag{5.1.4}$$

holds. The constant $1/8$ is sharp.

Andrica and Badea [31] extend the Grüss-type inequalities for a positive linear functional as follows:

Theorem 5.4 *Let $I = [a,b]$ be a compact interval of the real axis, $B(I)$ be the space of real-valued and bounded functions defined on I, and $L : B(I) \to \mathbb{R}$ be a* *positive linear functional satisfying $L(e_0) = 1$ where $e_0 : I \ni x \mapsto 1$. Assuming that for $f, g \in B(I)$, i.e., $m \le f(x) \le M$, $p \le g(x) \le P$ for all $x \in I$, we have*

$$|L(fg) - L(f)L(g)| \le \frac{1}{4}(M - m)(P - p). \tag{5.1.5}$$

In [16] the Grüss-type inequality (5.1.5) is extended on a compact metric space for more than two functions.

Theorem 5.5 ([16]) *Let $L : C(X) \to \mathbb{R}$ be a positive, linear functional, $L(1) = 1$, defined on the metric space $C(X)$. The inequality*

$$|L(f_1 f_2 \cdots f_n) - L(f_1)L(f_2) \cdots L(f_n)| \le \frac{1}{4} \sum_{i,j=1,i<j}^{n} \theta(f_i)\theta(f_j) \prod_{k=1,k\neq i,j}^{n} \|f_k\| \tag{5.1.6}$$

holds.

Remark 5.1 ([16]) If $f_3, \ldots, f_n$ are constant functions, relation (5.1.6) reduces to

$$|L(f_1 f_2) - L(f_1)L(f_2)| \le \frac{1}{4}(M_1 - m_1)(M_2 - m_2),$$

where

$$M_i = \max_X f_i, \quad m_i = \min_X f_i, \quad i \in \{1, 2\}.$$

5.2 Grüss-Type Inequalities for Some Positive Linear Operators

A very important tool that we will use is the least concave majorant of the modulus of continuity $\tilde{\omega}(f; \cdot)$. This is given by

$$\tilde{\omega}(f;t) := \begin{cases} \sup\limits_{0\le x\le t\le y\le b-a,\, x\neq y} \dfrac{(t-x)\omega(f;y) + (y-t)\omega(f;x)}{y-x}, & \text{if } 0 \le t \le b-a, \\ \omega(f;b-a), \quad \text{if } t > b-a, \end{cases} \tag{5.2.1}$$

where

$$\omega(f;t) := \sup\{|f(x+h) - f(x)| : x, x+h \in [a,b], 0 \le h \le t\}, \quad f \in C[a,b]$$

is the first moduli of smoothness and it was given in the Ph.D. thesis of Jackson [145].

The following relationship between the different moduli holds:

$$\omega(f;\cdot) \le \tilde{\omega}(f;\cdot) \le 2\omega(f;\cdot).$$

Since some of the error estimates considered in this book are given in terms of the moduli of higher order we give the definition of ω_k, $k \in \mathbb{N}$, as given in 1981 by Schumaker [212]:

Definition 5.1 For $k \in \mathbb{N}$, $t \in \mathbb{R}_+$ and $f \in C[a,b]$ the modulus of smoothness of order k is defined by

$$\omega_k(f;t) := \sup\left\{|\Delta_h^k f(x)| : 0 \le h \le t,\ x, x+kh \in [a,b]\right\}, \tag{5.2.2}$$

where

$$\Delta_h^k f(x) = \sum_{i=0}^{k} (-1)^i \binom{k}{i} f(x+(k-i)h) = \sum_{j=0}^{k} (-1)^{k-j} \binom{k}{j} f(x+jh).$$

Another important tool to measure the smoothness of a function is the so-called Peetre's K-functional, that was introduced by Peetre [195] in 1968.

Definition 5.2 Let $f \in C[a,b]$, $\delta \ge 0$ and $s \in \mathbb{N}$, $s \ge 1$. We denote

$$\begin{aligned} K_s(f;t) &:= K(f;t;C[a,b],C^s[a,b]) \\ &:= \inf\left\{\|f-g\| + t\|g^{(s)}\|,\ g \in C^s[a,b]\right\} \end{aligned} \tag{5.2.3}$$

to be Peetre's K-functional of order s.

The following lemma known as Brudnyĭ's representation theorem establishes the connection between $K_1(f;t)$ and the least concave majorant defined in (5.2.1).

Lemma 5.1 *Every function $f \in C[a,b]$ satisfies the equality*

$$K(f;t;C[a,b],C^1[a,b]) = \frac{1}{2}\tilde{\omega}(f;2t),\ t \ge 0.$$

For more details concerning this lemma one can consult Păltănea's paper [193], Mitjagin and Semenov's paper [176], the book of Rockafellar [206], and the monograph of DeVore and Lorentz [78].

Let $H_n : C[a,b] \to C[a,b]$ be positive linear operators which reproduce constant functions. For a given $x \in [a,b]$ we consider the functional $L(f) = H_n(f;x)$. Set

$$D(f,g) := H_n(fg;x) - H_n(f;x)H_n(g;x).$$

Using the least concave majorant of the modulus of continuity in [16] the following result was proved, which suggests how non-multiplicative the functional $L(f) = H_n(f; x)$ is, for a given $x \in [a, b]$.

Theorem 5.6 ([16]) *If $f, g \in C[a, b]$ and $x \in [a, b]$ is fixed, then the inequality*

$$|D(f, g)| \leq \frac{1}{4}\tilde{\omega}\left(f; 2\sqrt{2H_n\left((e_1 - x)^2; x\right)}\right) \cdot \tilde{\omega}\left(g; 2\sqrt{2H_n\left((e_1 - x)^2; x\right)}\right)$$

holds.

Proof Since $H_n(f; x)$, with $x \in [a, b]$ fixed, is a positive linear functional and can be represented as

$$H_n(f; x) = \int_a^b f(t)d\mu(t)\,,$$

where μ is a probability measure on $[a, b]$, i.e.,

$$\int_a^b d\mu(t) = 1\,.$$

it follows that

$$D(f, f) = H_n(f^2; x) - H_n(f; x)^2 = \int_a^b f^2(t)d\mu(t) - \left(\int_a^b f(s)d\mu(s)\right)^2$$

$$= \int_a^b \left(\int_a^b (f(t) - f(s))d\mu(s)\right)^2 d\mu(t) \leq \|f'\|^2 \int_a^b \left(\int_a^b (t-s)^2 d\mu(s)\right) d\mu(t)$$

$$= 2\|f'\|^2\left[H_n(e_2; x) - H_n(e_1; x)^2\right] \leq 2\|f'\|^2 H_n\left((e_1 - x)^2; x\right),\ \ f \in C^1[a, b].$$

Let $f, g \in C[a, b]$ be fixed, and $r, s \in C^1[a, b]$ be arbitrary. Using the above relation we obtain the following estimates:

$$|D(r, s)| \leq \sqrt{D(r, r)D(s, s)} \leq 2\|r'\|\|s'\|H_n\left((e_1 - x)^2; x\right),$$

$$|D(f, s)| \leq \sqrt{D(f, f)D(s, s)} \leq \|f\|\sqrt{2}\|s'\|\sqrt{H_n\left((e_1 - x)^2; x\right)},$$

$$|D(r, g)| \leq \sqrt{D(g, g)D(r, r)} \leq \|g\|\sqrt{2}\|r'\|\sqrt{H_n\left((e_1 - x)^2; x\right)}.$$

Therefore,

$$|D(f,g)| = |D(f-r+r, g-s+s)|$$

$$\leq |D(f-r, g-s)| + |D(f-r, s)| + |D(r, g-s)| + |D(r,s)|$$

$$\leq \left\{\|f-r\|+\|r'\|\sqrt{2H_n\left((e_1-x)^2;x\right)}\right\}\cdot\left\{\|g-s\|+\|s'\|\sqrt{2H_n\left((e_1-x)^2;x\right)}\right\}.$$

Passing to the infimum over r and $s \in C^1[a,b]$, respectively, we get

$$|D(f,g)| \leq K\left(\sqrt{2H_n\left((e_1-x)^2;x\right)}, f; C^0, C^1\right) K\left(\sqrt{2H_n\left((e_1-x)^2;x\right)}, g; C^0, C^1\right)$$

$$= \frac{1}{4}\tilde{\omega}\left(f; 2\sqrt{2\cdot H_n\left((e_1-x)^2;x\right)}\right)\cdot\tilde{\omega}\left(g; 2\sqrt{2\cdot H_n\left((e_1-x)^2;x\right)}\right),$$

which concludes the proof. ■

Remark 5.2 ([16]) If we choose $H_n = B_n$, the Bernstein operator, then this gives

$$|B_n(fg;x) - B_n(f;x)\cdot B_n(g;x)|$$

$$\leq \frac{1}{4}\tilde{\omega}\left(f; 2\sqrt{2B_n\left((e_1-x)^2;x\right)}\right)\cdot\tilde{\omega}\left(g; 2\sqrt{2B_n\left((e_1-x)^2;x\right)}\right)$$

$$= \frac{1}{4}\tilde{\omega}\left(f; 2\sqrt{\frac{2x(1-x)}{n}}\right)\tilde{\omega}\left(g; 2\sqrt{\frac{2x(1-x)}{n}}\right)$$

$$\leq \tilde{\omega}\left(f; \frac{1}{\sqrt{2n}}\right)\tilde{\omega}\left(g; \frac{1}{\sqrt{2n}}\right), \quad f, g \in C[0,1].$$

Rusu [207] improved the result from Theorem 5.6 by removing the constant $\sqrt{2}$ in the arguments of the least concave majorants as follows.

Theorem 5.7 ([207]) *If $f, g \in C[a,b]$ and $x \in [a,b]$ is fixed, then the inequality*

$$|D(f,g)| \leq \frac{1}{4}\tilde{\omega}\left(f; 2\sqrt{H_n\left((e_1-x)^2;x\right)}\right)\cdot\tilde{\omega}\left(g; 2\sqrt{H_n\left((e_1-x)^2;x\right)}\right) \tag{5.2.4}$$

holds.

In [112], Gonska et al. replaced the second moments $H((e_1-x)^2;x)$ by the smaller quantity $H(e_2;x) - H(e_1;x)^2$, proving that the above approach is not the ideal choice.

Theorem 5.8 (See [112, Theorem 3.1]) *If $L : C[a,b] \to \mathbb{R}$ is a positive linear functional with $L(e_0) = 1$, then for $f, g \in C[a,b]$ we have*

$$|D(f,g)| \le \frac{1}{4}\tilde{\omega}\left(f; 2\sqrt{D(e_1,e_2)}\right)\tilde{\omega}\left(g; 2\sqrt{D(e_1,e_2)}\right),$$

where

$$D(f,g) := L(f\cdot g) - L(f)L(g).$$

Moreover,

$$D\left(\frac{e_1-a}{b-a}, \frac{e_1-a}{b-a}\right) \le \frac{1}{4},$$

with equality holding if and only if

$$L = \frac{1}{2}(\epsilon_a + \epsilon_b), \text{ where } \epsilon_x(f) = f(x),\ x \in \{a,b\}.$$

Corollary 5.1 (See [112, Corollary 5.1]) *If $H_n : C[a,b] \to C[a,b]$ is a positive linear operator which reproduces constant functions, then for $f, g \in C[a,b]$ and $x \in [a,b]$ fixed we have the inequalities:*

$$\begin{aligned}
|D(f,g)| &= |H_n(fg;x) - H_n(f;x)H_n(g;x)| \\
&\le \frac{1}{4}\tilde{\omega}\left(f; 2\sqrt{H_n(e_2;x) - H_n(e_1;x)^2}\right)\tilde{\omega}\left(g; 2\sqrt{H_n(e_2;x) - H_n(e_1;x)^2}\right) \\
&\le \frac{1}{4}\tilde{\omega}\left(f; 2\sqrt{H_n\left((e_1-x)^2;x\right)}\right)\cdot\tilde{\omega}\left(g; 2\sqrt{H_n\left((e_1-x)^2;x\right)}\right).
\end{aligned}$$

In [105], Gonska and Tachev used second order moduli of smoothness instead of the least concave majorant of the first order modulus of continuity and showed in the case of the classical Bernstein operators that in certain cases this leads to better results than those obtained earlier.

5.2.1 *The Classical Hermite–Fejér Interpolation Operator*

The classical Hermite–Fejér interpolation operator is a positive linear operator and can be written as

$$L_n(f;x) = \sum_{k=1}^{n} f(x_k)(1 - x \cdot x_k) \cdot \left(\frac{T_n(x)}{n(x - x_k)} \right)^2, \tag{5.2.5}$$

where $f \in C[-1,1]$ and $x_k = \cos \dfrac{2k-1}{2n}\pi,\ 1 \le k \le n$ are the zeros of $T_n(x) = \cos(n \cdot \arccos)$, the n-th Chebyshev polynomial of the first kind.

Remark 5.3 ([16]) If we choose $H_n = L_n$ in Theorem 5.6, then the classical Hermite–Fejér interpolation operator satisfies

$$\begin{aligned} &|L_n(fg;x) - L_n(f;x)L_n(g;x)| \\ &\le \frac{1}{4}\tilde{\omega}\left(f; \frac{2\sqrt{2}}{\sqrt{n}}|T_n(x)|\right) \cdot \tilde{\omega}\left(g; \frac{2\sqrt{2}}{\sqrt{n}}|T_n(x)|\right). \end{aligned} \tag{5.2.6}$$

The inequality (5.2.6) was improved using Theorem 5.8 by Gonska et al. [112] as follows

Remark 5.4 ([112]) The classical Hermite–Fejér interpolation operator satisfies

$$\begin{aligned} &|L_n(fg;x) - L_n(f;x)L_n(g;x)| \\ &\le \frac{1}{4}\tilde{\omega}\left(f; \frac{2}{\sqrt{n}}|T_n(x)|\sqrt{1 - \frac{1}{n}T_{n-1}^2(x)}\right) \cdot \tilde{\omega}\left(g; \frac{2\sqrt{2}}{\sqrt{n}}|T_n(x)|\sqrt{1 - \frac{1}{n}T_{n-1}^2(x)}\right). \end{aligned}$$

5.2.2 *The Convolution-Type Operator*

For every function $f \in C[-1,1]$, and any natural number n, the convolution-type operator $G_{m(n)}$ is defined by

$$G_{m(n)}(f,t) := \pi^{-1} \int_{-\pi}^{\pi} f\left(\cos(\arccos t + v)\right) K_{m(n)}(v)dv,$$

where the kernel $K_{m(n)}$ is a positive and even trigonometric polynomial of degree $m(n)$ satisfying

$$\int_{-\pi}^{\pi} K_{m(n)}(v)dv = \pi,$$

meaning that, $G_{m(n)}(1,t) = 1$ for $t \in [-1,1]$.

For each $f \in C[-1,1]$ the integral $G_{m(n)}(f,\cdot)$ is an algebraic polynomial of degree $m(n)$. Moreover,

$$K_{m(n)}(v) = \frac{1}{2} + \sum_{k=1}^{m(n)} \rho_{k,m(n)} \cdot \cos kv,\ \ v \in [-\pi,\pi].$$

In order to give a Grüss-type inequality for the convolution-type operator $G_{m(n)}$, we need a result that goes back to Lehnhoff [155]:

Lemma 5.2 ([155]) *For $x \in [-1, 1]$ we have*

$$G_{m(n)}\left((e_1 - x)^2, x\right) = x^2\left\{\frac{3}{2} - 2\rho_{1,m(n)} + \frac{1}{2}\rho_{2,m(n)}\right\} + (1 - x^2)\left\{\frac{1}{2} - \frac{1}{2}\rho_{2,m(n)}\right\}.$$

The first moment of the convolution-type operator (see [68]) is given by

$$G_{m(n)}(e_1 - x; x) = x\left[\rho_{1,m(n)} - 1\right].$$

If $K_{m(n)}$ is the Fejér–Korovkin kernel with $m(n) = n - 1$, then it is known that (see [170])

$$\rho_{1,n-1} = \cos\frac{\pi}{n+1}, \quad \rho_{2,n-1} = \frac{n}{n+1}\cos\frac{2\pi}{n+1} + \frac{1}{n+1}. \tag{5.2.7}$$

Using the relations (5.2.7) we get

$$G_{n-1}\left((e_1 - x)^2; x)\right) \le 4\left(\frac{\pi}{n+1}\right)^2.$$

If we consider in Theorem 5.7 the convolution-type operators with the Fejér–Korovkin kernel we have the following result:

Theorem 5.9 ([207]) *If we consider the convolution-type operator with the Fejér–Korovkin kernel, we have*

$$\begin{aligned}|D(f; g)| &= |G_{n-1}(fg; x) - G_{n-1}(f; x)G_{n-1}(g; x)|\\ &\le \frac{1}{4}\tilde{\omega}\left(f; \frac{4\pi}{n+1}\right)\tilde{\omega}\left(g; \frac{4\pi}{n+1}\right).\end{aligned}$$

Let $m(n) = n \in \mathbb{N}_0$ and we consider the de La Vallée Poussin kernel by

$$V_n(\nu) = \frac{(n!)^2}{(2n)!}\left(2\cos\left(\frac{\nu}{2}\right)\right)^{2n}$$

with

$$\rho_{1,n} = \frac{n}{n+1}, \quad \rho_{2,n} = \frac{(n-1)n}{(n+1)(n+2)}.$$

Using Theorem 5.7, Rusu obtained the following result:

Theorem 5.10 ([207]) *If we consider the convolution-type operator with the de La Vallée Poussin kernel, we have*

$$|D(f;g)| = |G_n(fg;x) - G_n(f;x)G_n(g;x)|$$
$$\leq \frac{1}{4}\tilde{\omega}\left(f; \frac{2\sqrt{2}}{\sqrt{n+1}}\right) \cdot \tilde{\omega}\left(g; \frac{2\sqrt{2}}{\sqrt{n+1}}\right).$$

Using Theorem 5.8, the following result presenting a Grüss-type inequality for the convolution operator with de La Vallée Poussin kernel, we have

Theorem 5.11 ([112]) *If we consider the convolution-type operator with the de La Vallée Poussin kernel, we have*

$$|D(f;g)| = |G_n(fg;x) - G_n(f;x)G_n(g;x)|$$
$$\leq \frac{1}{4}\tilde{\omega}\left(f; \frac{2\sqrt{2-\frac{x^2}{n+1}}}{\sqrt{n+1}}\right) \cdot \tilde{\omega}\left(g; \frac{2\sqrt{2-\frac{x^2}{n+1}}}{\sqrt{n+1}}\right).$$

This result is a slight improvement of the one from Theorem 5.10.

Finally, we consider the operator of degree $m(n) = 2n-2$, $n \in \mathbb{N}$ and the Jackson kernel

$$J_{2n-2}(v) = \frac{3}{2n(2n^2+1)}\left(\frac{\sin\left(n\frac{v}{2}\right)}{\sin\left(\frac{v}{2}\right)}\right)^4$$

with

$$\rho_{1,2n-2} = \frac{2n^2-2}{2n^2+1},\ \rho_{2,2n-2} = \frac{2n^3-11n+9}{n(2n^2+1)}.$$

Theorem 5.12 ([207]) *If we consider the convolution-type operator with the Jackson kernel we have*

$$|D(f,g)| = |G_{2n-2}(fg;x) - G_{2n-2}(f;x)G_{2n-2}(g;x)|$$
$$\leq \frac{1}{4}\tilde{\omega}\left(f; \frac{2\sqrt{3}}{n}\right)\tilde{\omega}\left(g; \frac{2\sqrt{3}}{n}\right).$$

5.2.3 King Operators

King [150] defined the following operators.

Definition 5.3 Let $(r_n(x))$ be a sequence of continuous functions with $0 \leq r_n(x) \leq 1$. Let $K_n : C[0,1] \to C[0,1]$ be given by:

$$K_n(f;x) = \sum_{k=0}^{n} \binom{n}{k} (r_n(x))^k (1 - r_n(x))^{n-k} f\left(\frac{k}{n}\right)$$
$$= \sum_{k=0}^{n} v_{n,k}(x) f\left(\frac{k}{n}\right),$$

for $f \in C[0, 1], 0 \leq x \leq 1$.

The second moment of King operators is given by

$$K_n((e_1 - x)^2; x) = \frac{1}{n} r_n(x)(1 - r_n(x)) + (r_n(x) - x)^2.$$

King [150] proved that for special choices of $r_n(x) = r_n^*(x)$, the following result holds.

Theorem 5.13 (See Theorem 1.3. in [101]) *Let $(K_n^*)_{n \in \mathbb{N}}$ be the sequence of operators defined before with*

$$r_n^*(x) := \begin{cases} x^2, & \text{for } n = 1, \\ -\frac{1}{2(n-1)} + \sqrt{\frac{n}{n-1} x^2 + \frac{1}{4(n-1)^2}}, & \text{for } n = 2, 3, \ldots \end{cases}$$

Then we get $K_n^*(e_2; x) = x^2$, for $n \in \mathbb{N}$, $x \in [0, 1]$ and $K_n^*(e_1; x) \neq e_1(x)$. K_n^* is not a polynomial operator.

The second moment of the operators K_n^* is given by

$$K_n^*((e_1 - x)^2; x) = 2x(x - r_n^*(x)).$$

By letting $H_n = K_n$ in Theorem 5.7, Rusu [208] obtained the following Grüss-type inequality for King operators.

Theorem 5.14 ([208]) *If we consider the King operators we have*

$$|D(f, g)| = |K_n(fg; x) - K_n(f; x)K_n(g; x)|$$
$$\leq \frac{1}{4} \cdot \tilde{\omega}\left(f; 2\sqrt{K_n((e_1 - x)^2; x)}\right) \cdot \tilde{\omega}\left(g; 2\sqrt{K_n((e_1 - x)^2; x)}\right).$$

The following result concerning the special King-type operators V_n^* was given in [113]:

Theorem 5.15 ([113]) *If we consider the operator K_n^* that reproduces not only constant functions but also e_2, we obtain the inequality*

$$\begin{aligned}|D(f,g)| &= \left|K_n^*(fg;x) - K_n^*(f;x)K_n^*(g;x)\right| \\ &\leq \frac{1}{4}\cdot\tilde{\omega}\left(f;2\sqrt{K_n^*((e_1-x)^2;x)}\right)\cdot\tilde{\omega}\left(g;2\sqrt{K_n^*((e_1-x)^2;x)}\right) \\ &= \frac{1}{4}\cdot\tilde{\omega}\left(f;2\sqrt{2x(x-K_n^*(e_1;x))}\right)\cdot\tilde{\omega}\left(g;2\sqrt{2x(x-K_n^*(e_1;x))}\right) \\ &\leq \frac{1}{4}\cdot\tilde{\omega}\left(f;2\sqrt{\frac{x(1-x)}{n}}\right)\cdot\tilde{\omega}\left(g;2\sqrt{\frac{x(1-x)}{n}}\right).\end{aligned}$$

5.2.4 A Piecewise Linear Interpolation Operator S_{Δ_n}

The operator $S_{\Delta_n} : C[0,1] \to C[0,1]$, defined by

$$S_{\Delta_n}(f;x) = \frac{1}{n}\sum_{k=0}^{n}\left[\frac{k-1}{n},\frac{k}{n},\frac{k+1}{n};|\alpha-x|\right]_\alpha f\left(\frac{k}{n}\right),$$

where $[a,b,c;f] = [a,b,c;f(\alpha)]_\alpha$ denotes the divided difference of a function $f : D \to \mathbb{R}$ on (distinct knots) $\{a,b,c\} \subset D$, $D \subset \mathbb{R}$, interpolates the function at the points $0, \frac{1}{n}, \ldots, \frac{k}{n}, \ldots, \frac{n-1}{n}, 1$ (see [111]).

Also, the operator S_{Δ_n} can be given as follows:

$$S_{\Delta_n}f(x) := \sum_{k=0}^{n} f\left(\frac{k}{n}\right)u_{n,k}(x),$$

for $f \in C[0,1]$ and $x \in [0,1]$, where $u_{n,k} \in C[0,1]$ are piecewise linear and continuous functions, such that

$$u_{n,k}\left(\frac{l}{n}\right) = \delta_{kl},\ k,l = 0,\ldots,n.$$

For $x \in \left[\frac{k-1}{n},\frac{k}{n}\right]$, the second moment of the operator S_{Δ_n} is given by

$$S_{\Delta_n}((e_1-x)^2;x) = \left(x - \frac{k-1}{n}\right)\left(\frac{k}{n}-x\right),$$

which is maximal when $x = \frac{2k-1}{2n}$. Therefore,

$$S_{\Delta_n}((e_1-x)^2;x) \leq \frac{1}{4n^2}.$$

By taking $H_n = V_n$ in Theorem 5.7, the Grüss-type inequality for S_{Δ_n} is given in the following.

Theorem 5.16 ([113]) *If $f, g \in C[0,1]$ and $x \in [0,1]$ is fixed, then the inequality*

$$\begin{aligned}
|D(f,g)| &= \left|S_{\Delta_n}(fg;x) - S_{\Delta_n}(f;x)S_{\Delta_n}(g;x)\right| \\
&\leq \frac{1}{4}\tilde{\omega}\left(f; 2\cdot\sqrt{S_{\Delta_n}((e_1-x)^2;x)}\right)\cdot\tilde{\omega}\left(g; 2\cdot\sqrt{S_{\Delta_n}((e_1-x)^2;x)}\right) \\
&\leq \frac{1}{4}\tilde{\omega}\left(f;\frac{1}{n}\right)\cdot\tilde{\omega}\left(g;\frac{1}{n}\right)
\end{aligned}$$

holds.

5.3 Estimates via Cauchy's Mean Value Theorem

Let $L : C[a,b] \to \mathbb{R}$ be a linear positive functional. We denote by

$$T(f,g) = L(fg) - L(f)\cdot L(g), \quad f, g \in C[a,b].$$

In this section we will study non-multiplicativity for the functional L using Cauchy's mean value theorem.

Theorem 5.17 ([16]) *If $L : C[a,b] \to \mathbb{R}$ is a linear positive functional, with $L(1) = 1$, then*

i) there is $(\eta,\theta) \in [a,b]\times[a,b]$ such that

$$T(f,g) = \frac{f'(\eta)}{h'(\eta)}\cdot\frac{g'(\theta)}{h'(\theta)}\cdot T(h,h).$$

ii) $|T(f,g)| \leq \left\|\frac{f'}{h'}\right\|\cdot\left\|\frac{g'}{h'}\right\|\cdot|T(h,h)|$, where $f, g, h \in C^1[a,b]$ and $h'(t) \neq 0$ for each $t \in [a,b]$.

Remark 5.5 ([16]) If in Theorem 5.17 we consider $h(x)=x$, $x\in[a,b]$, and $L(f)=\frac{1}{b-a}\int_a^b f(x)dx$, then

i) there is $(\eta,\theta) \in [a,b]\times[a,b]$ such that

$$\frac{1}{b-a}\int_a^b f(x)g(x)dx - \frac{1}{(b-a)^2}\int_a^b f(x)dx\cdot\int_a^b g(x)dx = \frac{(b-a)^2}{12}f'(\eta)\cdot g'(\theta).$$

This identity was found by Ostrowski [189] in 1970.

ii) $$\left|\frac{1}{b-a}\int_a^b f(x)g(x)dx - \frac{1}{(b-a)^2}\int_a^b f(x)dx \int_a^b g(x)dx\right|$$
$$\leq \frac{(b-a)^2}{12} \sup_{x\in[a,b]} |f'(x)| \sup_{x\in[a,b]} |g'(x)|.$$

This inequality was proved by Chebyshev [71] in 1882.

Theorem 5.18 ([16]) *If $L : C[a,b] \to \mathbb{R}$ is a linear positive functional, with $L(1) = 1$, then the following inequality is verified:*

$$|T(f,h) + T(g,h)| \leq |T(h,h)| \cdot \left(\left\|\frac{f'}{h'}\right\| + \left\|\frac{g'}{h'}\right\|\right),$$

where $f, g, h \in C^1[a,b]$ and $h'(t) \neq 0$ for each $t \in [a,b]$.

In the paper [189] Ostrowski defined the concept of synchronous functions. The functions $f, g : [a,b] \to \mathbb{R}$ are called synchronous, if we have, for any couple of points x, y from $[a,b]$, $f(x) \geq f(y)$ if and only if $g(x) \geq g(y)$.

In the case that f, g are synchronous, we get $T(f,g) \geq 0$.

Theorem 5.19 ([16]) *If $L : C[a,b] \to \mathbb{R}$ is a linear positive functional, with $L(1) = 1$, then the following inequality is satisfied:*

$$|T(f,g)| \leq \frac{1}{2}\left[\left\|\frac{f'}{h'}\right\| |T(g,h)| + \left\|\frac{g'}{h'}\right\| |T(f,h)|\right], \tag{5.3.1}$$

where $f, g, h \in C^1[a,b]$, $h'(t) \neq 0$ for each $t \in [a,b]$ and the functions f, g, respectively g, h are synchronous.

5.4 Grüss-Type Inequalities on Compact Metric Spaces

In this section we consider some results concerning Grüss-type inequalities in C(X), the set of continuous functions defined on a compact metric space X. Before giving these results, we will recall the definition for the (metric) modulus of continuity and its least concave majorant (see [98]). Let (X, d) be a compact metric space and $d(X) < \infty$ be the diameter of the compact space X.

Definition 5.4 Let $f \in C(X)$. If, for $t \in [0, \infty)$, the quantity

$$\omega_d(f;t) := \sup\{|f(x) - f(y)|\,;\, x, y \in X,\ d(x,y) \leq t\}$$

is the (metric) modulus of continuity, then its least concave majorant is given by

$$\widetilde{\omega}_d(f;t) = \begin{cases} \sup\limits_{0\le x\le t\le y\le d(X), x\ne y} \frac{(t-x)\omega_d(f;y)+(y-t)\omega_d(f;x)}{y-x} & \text{for } 0 \le t \le d(X)\,, \\ \omega_d(f;d(X)) & \text{if } t > d(X)\,. \end{cases}$$

Denote

$$Lip_r = \left\{ g \in C(X) \,\middle|\, |g|_{Lip_r} := \sup_{d(x,y)>0} \frac{|g(x)-g(y)|}{d^r(x,y)} < \infty \right\},\ 0 < r \le 1.$$

Lip_r is a dense subspace of $C(X)$ equipped with the supremum norm $\|\cdot\|$ and $|\cdot|_{Lip_r}$ is a seminorm on Lip_r.

The K-functional with respect to $(Lip_r, |\cdot|_{Lip_r})$ is given by

$$K(t;f;C(X),Lip_r) := \inf_{g\in Lip_r} \{\|f-g\| + t\cdot |g|_{Lip_r}\},$$

for $f \in C(X)$ and $t \ge 0$.

In the next result we give the relationship between the K-functional and the least concave majorant of the (metric) modulus of continuity (see [176]).

Lemma 5.3 *Every continuous function f on X satisfies*

$$K\left(\frac{t}{2};f;C(X),Lip_1\right) = \frac{1}{2}\cdot\widetilde{\omega}_d(f;t),\ 0\le t\le d(X).$$

The result concerning the non-multiplicativity of positive linear operators reproducing linear function was remarkably generalized by Rusu [207] replacing $([a,b],|\cdot|)$ by a compact metric space (X,d), $H_n((e_1-x)^2;x)$ by $H_n(d^2(\cdot,x);x)$, and $K(\cdot,f;C[a,b],C^1[a,b])$ by $K(\cdot,f;C(X),Lip_1)$.

Theorem 5.20 (See [207, Theorem 3.1]) *If $f,g \in C(X)$, where (X,d) is a compact metric space, and $x\in X$, then the inequality*

$$|D(f,g)| \le \frac{1}{4}\widetilde{\omega}_d\left(f;4\sqrt{H_n(d^2(\cdot,x);x)}\right)\cdot\widetilde{\omega}_d\left(g;4\sqrt{H_n(d^2(\cdot,x);x)}\right) \quad (5.4.1)$$

holds, where $H_n(d^2(\cdot,x);x)$ is the second moment of the operator H_n.

Let $A,B : C(X)\to\mathbb{R}$, satisfying $A(e_0)=B(e_0)=1$. Denote by

$$D_{A,B}(f,g) = A(fg)+B(fg)-A(f)B(g)-B(f)A(g).$$

In [95] the quantity $D_{A,B}(f,g)$ was bounded using the least concave majorants of the moduli of continuity for the functions that define this quantity.

Theorem 5.21 (See [95, Theorem 3.1]) *Let (X, d) be a compact metric space with diameter $d(X) > 0$ and let $A, B : C(X) \to \mathbb{R}$ be linear positive functionals reproducing constant functions ($A(e_0) = B(e_0) = 1$). If $f \in Lip_{r_1}$ and $g \in Lip_{r_2}$ with $r_1, r_2 \in (0, 1]$, then the following inequality holds*

$$D_{A,B}(f, g) \leq |f|_{Lip_{r_1}} |g|_{Lip_{r_2}} A_y B_x \left(d^{r_1+r_2}(x, y)\right),$$

where A_y denotes the fact that its input is viewed as a function of y (in the exact same fashion, we have defined B_x).

The next result gives the bounds of $D_{A,B}(f, g)$ using the least concave majorant $\tilde{\omega}_d$.

Theorem 5.22 (See [95, Theorem 3.3]) *Let $f, g \in C(X)$ be two continuous functions on the compact metric space (X, d). If A, B are two positive linear functionals, $A, B : C(X) \to \mathbb{R}$ reproducing constants ($A(e_0) = B(e_0) = 1$, then the following inequality*

$$|D_{A,B}(f, g)| \leq \tilde{\omega}_d \left(f; \sqrt{A_x B_y (d^2(x, y))}\right) \tilde{\omega}_d \left(g; \sqrt{A_x B_y (d^2(x, y))}\right)$$

holds.

A probabilistic interpretation of the above result is given in [95]. Let (U, V) and $(\tilde{U}, \tilde{V})$ be two-dimensional continuous random vectors having the joint pdfs $\rho_{U,V} : [a, b] \times [a, b] \to \mathbb{R}_+$ and $\rho_{\tilde{U},\tilde{V}} : [a, b] \times [a, b] \to \mathbb{R}_+$, respectively. Let $X = [a, b] \times [a, b]$ and $A, B : C(X) \to \mathbb{R}$ be the linear positive functionals defined by

$$A(h) = \int_a^b \int_a^b h(u, v) \rho_{U,V}(u, v) du dv,$$

$$B(h) = \int_a^b \int_a^b h(u, v) \rho_{\tilde{U},\tilde{V}}(u, v) du dv.$$

Using Theorem 5.22, the following probabilistic inequality was obtained in [95]:

$$\begin{aligned} &|COV[f(U), g(V)] + COV[f(\tilde{U}), g(\tilde{V})] \\ &+ (E[g(V)] - E[g(\tilde{V})])(E[f(U)] - E[f(\tilde{U})])| \leq \tilde{\omega}(f; \sqrt{\tau}) \tilde{\omega}(g; \sqrt{\tau}), \end{aligned} \tag{5.4.2}$$

where $COV[U, V]$ denotes the covariance of the random variables U and V, $E[W]$ denotes the expectation of the random variable W, $\sigma_W^2 := VAR[W]$ denotes the variance (dispersion) of the random variable W, and $\tau := \tau_{U,V,\tilde{U},\tilde{V}}$ is given by

$$\tau_{U,V,\tilde{U},\tilde{V}} = \sigma_U^2 + \sigma_V^2 + \sigma_{\tilde{U}}^2 + \sigma_{\tilde{V}}^2 + (E[U] - E[\tilde{U}])^2 + (E[V] - E[\tilde{V}])^2. \tag{5.4.3}$$

Remark 5.6 (See [95, Remark 4.1])

i) For a single random vector (U, V), i.e., $\rho_{U,V} \equiv \rho_{\tilde{U},\tilde{V}}$, inequality (5.4.2) becomes

$$|COV(f(U), g(V)| \le \frac{1}{2}\tilde{\omega}\left(f; \sqrt{2(\sigma_U^2 + \sigma_V^2)}\right)\tilde{\omega}\left(g; \sqrt{2(\sigma_U^2 + \sigma_V^2)}\right), \tag{5.4.4}$$

which establishes covariance bounds using the least concave majorant of the modulus of continuity.

ii) If f and g are Lipschitz functions with Lipschitz constants L_f and L_g respectively, we can bound $\tilde{\omega}(f; \tau_{U,V,\tilde{U},\tilde{V}})$ and $\tilde{\omega}(g; \tau_{U,V,\tilde{U},\tilde{V}})$ by $L_f\tau_{U,V,\tilde{U},\tilde{V}}$ and $L_f\tau_{U,V,\tilde{U},\tilde{V}}$, respectively, to obtain

$$\begin{aligned}&|COV[f(U), g(V)] + COV[f(\tilde{U}), g(\tilde{V})]\\ &+ (E[g(V)] - E[g(\tilde{V})])(E[f(U)] - E[f(\tilde{U})])| \le L_f L_g \tau_{U,V,\tilde{U},\tilde{V}},\end{aligned}$$

In particular, for a single random vector, inequality (5.4.4) can be relaxed to

$$|COV(f(U), g(V))| \le L_f L_g(\sigma_U^2 + \sigma_V^2).$$

5.5 Grüss Inequalities via Discrete Oscillations

Gonska et al. [113] obtained a new Grüss-type inequality which involves oscillations of functions. This result is better than (5.2.4) in the sense that the oscillations of functions are relative only to certain points, while in (5.2.4) the oscillations, expressed in terms of $\tilde{\omega}$, are relative to the whole interval $[a, b]$.

Let X be an arbitrary set and $B(X)$ the set of all real-valued, bounded functions on X. Take $a_n \in \mathbb{R}$, $n \ge 0$, such that $\sum_{n=0}^{\infty}|a_n| < \infty$ and $\sum_{n=0}^{\infty} a_n = 1$. Furthermore, let $x_n \in X$, $n \ge 0$, be arbitrary mutually distinct points of X. For $f \in B(X)$ set $f_n := f(x_n)$. Let $L : B(X) \to \mathbb{R}$ be a functional defined as $Lf = \sum_{n=0}^{\infty} a_n f_n$.

Gonka et al. [113] obtained the following relation concerning the functional L:

$$\begin{aligned}L(f\cdot g) - L(f)\cdot L(g) &= \sum_{n=0}^{\infty} a_n f_n g_n - \sum_{n=0}^{\infty} a_n f_n \cdot \sum_{m=0}^{\infty} a_m g_m\\ &= \sum_{0\le n<m<\infty} a_n a_m (f_n - f_m)(g_n - g_m).\end{aligned}$$

Theorem 5.23 ([113]) *The Chebyshev–Grüss inequality for the above linear, not necessarily positive, functional L is given by:*

$$|L(fg) - L(f) \cdot L(g)| \leq osc_L(f) \cdot osc_L(g) \cdot \sum_{0 \leq n < m < \infty} |a_n a_m|,$$

where $f, g \in B(X)$ and the oscillations are given by:

$$osc_L(f) := \sup\{|f_n - f_m| : 0 \leq n < m < \infty\},$$
$$osc_L(g) := \sup\{|g_n - g_m| : 0 \leq n < m < \infty\}.$$

Theorem 5.24 ([113]) *In particular, if $a_n \geq 0$, $n \geq 0$, then L is a positive linear functional and we have:*

$$|L(fg) - L(f) \cdot L(g)| \leq \frac{1}{2} \cdot \left(1 - \sum_{n=0}^{\infty} a_n^2\right) \cdot osc_L(f) \cdot osc_L(g),$$

for $f, g \in B(X)$ and the oscillations are given as above.

5.5.1 Applications for Linear Operators

The result from the previous section was applied in the case of known operators (see [113]), for example, Bernstein, King, piecewise linear interpolation operator S_{Δ_n}, Szász–Mirakyan, Baskakov, Bleimann–Butzer–Hahn, Lagrange. Also, the readers should consult the important books on approximation by linear operators [125, 130, 203]. In the following we will recall some of these results.

Theorem 5.25 ([113]) *The Bernstein operator satisfies the following Grüss-type inequality:*

$$|B_n(f \cdot g; x) - B_n(f; x) \cdot B_n(g; x)| \leq \frac{1}{2}\left(1 - \frac{1}{4^n}\binom{2n}{n}\right) \cdot osc_{B_n}(f) \cdot osc_{B_n}(g),$$

for $x \in [0, 1]$.

Theorem 5.26 ([113]) *The Chebyshev–Grüss inequality for King operator K_1^* is the following:*

$$\left|K_1^*(f \cdot g; x) - K_1^*(f; x) \cdot K_1^*(g; x)\right| \leq \frac{1}{4} \cdot osc_{K_1^*}(f) \cdot osc_{K_1^*}(g)$$
$$= \frac{1}{4} \cdot |f_0 - f_1| \cdot |g_0 - g_1|.$$

Theorem 5.27 ([113]) *For $n = 2, 3, \ldots$ the following Grüss-type inequality for King operator holds :*

$$\left|K_n^*(fg)(x) - K_n^*(f;x)\cdot K_n^*(g;x)\right| \leq \frac{n}{2(n+1)}\cdot osc_{K_n^*}(f)\cdot osc_{K_n^*}(g).$$

Theorem 5.28 ([113]) *The Grüss-type inequality for S_{Δ_n} is*

$$\begin{aligned}\left|S_{\Delta_n}(fg;x) - S_{\Delta_n}(f;x)S_{\Delta_n}(g;x)\right| &\leq \frac{1}{2}\left(1-\sum_{i=0}^{n}u_{n,i}^2(x)\right)osc_{S_{\Delta_n}}(f)osc_{S_{\Delta_n}}(g)\\ &\leq \frac{1}{2}\left(1-\frac{1}{2}\right)osc_{S_{\Delta_n}}(f)osc_{S_{\Delta_n}}(g)\\ &= \frac{1}{4}osc_{S_{\Delta_n}}(f)osc_{S_{\Delta_n}}(g),\end{aligned}$$

with

$$\begin{aligned}osc_{S_{\Delta_n}}(f) &:= \max\left\{|f_k - f_l| : 0 \leq k < l \leq n\right\}\\ osc_{S_{\Delta_n}}(g) &:= \max\left\{|g_k - g_l| : 0 \leq k < l \leq n\right\},\end{aligned}$$

where $f_k := f\left(\frac{k}{n}\right)$.

Let

$$\sigma_n(x) := e^{-2nx}\sum_{k=0}^{\infty}\frac{(nx)^{2k}}{(k!)^2}$$

and

$$\inf_{x\geq 0}\sigma_n(x) := \iota \geq 0.$$

Theorem 5.29 ([113]) *For the Szász–Mirakjan operator we have*

$$\begin{aligned}|S_n(f\cdot g;x) - S_n(f;x)\cdot S_n(g;x)| &\leq \frac{1}{2}(1-\sigma_n(x))\cdot osc_{S_n}(f)\cdot osc_{S_n}(g)\\ &\leq \frac{1}{2}(1-\iota)\cdot osc_{S_n}(f)\cdot osc_{S_n}(g),\end{aligned}$$

where $f, g \in C_b[0,\infty)$, $osc_{S_n}(f) = \sup\{|f_k - f_l| : 0 \leq k < l < \infty\}$, with $f_k := f\left(\frac{k}{n}\right)$ and a similar definition applies to g. $C_b[0,\infty)$ is the set of all continuous, real-valued, bounded functions on $[0,\infty)$.

Lemma 5.4 ([113]) *The relation*

$$\inf_{x\geq 0} \sigma_n(x) = \iota = 0.$$

holds.

Corollary 5.2 ([113]) *The Chebyshev–Grüss inequality for the Szász–Mirakyan operator is:*

$$|S_n(fg; x) - S_n(f; x) \cdot S_n(g; x)| \leq \frac{1}{2} \cdot osc_{S_n}(f) \cdot osc_{S_n}(g),$$

where $f, g \in C_b[0, \infty)$, $osc_{S_n}(f) = \sup\{|f_k - f_l| : 0 \leq k < l < \infty\}$ and a similar definition applies to g.

Let

$$\vartheta_n(x) := \frac{1}{(1+x)^{2n}} \sum_{k=0}^{\infty} \binom{n+k-1}{k}^2 \left(\frac{x}{1+x}\right)^{2k}, \text{ for } x \geq 0$$

and

$$\inf_{x\geq 0} \vartheta_n(x) := \epsilon \geq 0.$$

Theorem 5.30 ([113]) *For the Baskakov operator one has*

$$\begin{aligned} |V_n(f \cdot g; x) - V_n(f; x) \cdot V_n(g; x)| &\leq \frac{1}{2}(1 - \vartheta_n(x)) \cdot osc_{V_n}(f) \cdot osc_{V_n}(g) \\ &\leq \frac{1}{2}(1 - \epsilon) \cdot osc_{V_n}(f) \cdot osc_{V_n}(g), \end{aligned}$$

where $f, g \in C_b[0, \infty)$, $osc_{V_n}(f) = \sup\{|f_k - f_l| : 0 \leq k < l < \infty\}$, $f_k := f\left(\frac{k}{n}\right)$ and a similar definition applies to g.

Lemma 5.5 ([113]) *The relation $\inf_{x\geq 0} \vartheta_n(x) = \epsilon = 0$ holds, for all $n \geq 1$.*

An inequality analogous to the one in Corollary 5.2 follows immediately for the Baskakov operator.

The Bleimann–Butzer–Hahn operators $BH_n : C_b[0, \infty) \to C_b[0, \infty)$ are defined by (see [28])

$$BH_n(f)(x) := \sum_{k=0}^{n} f\left(\frac{k}{n-k+1}\right) \cdot bh_{n,k}(x),$$

for every $f \in C_b[0, \infty)$, $x \in [0, \infty)$, $n \in \mathbb{N}$.

The fundamental functions are given by

$$bh_{n,k}(x) := \frac{1}{(1+x)^n}\binom{n}{k}x^k.$$

Theorem 5.31 ([113]) *The Grüss-type inequality in the case of the Bleimann–Butzer–Hahn operator is:*

$$|T(f,g;x)| \le \frac{1}{2}\left(1 - \frac{1}{4^n}\binom{2n}{n}\right) \cdot osc_{BH_n}(f) \cdot osc_{BH_n}(g),$$

with $f, g \in C_b[0,\infty)$, $x \in [0,\infty)$ and

$$osc_{BH_n}(f) := \sup\left\{|f_k - f_l| : 0 \le k < l \le n\right\},$$

for $f_k := f\left(\frac{k}{n-k+1}\right)$ and a similar definition applies to g.

Let $f \in C[-1,1]$ and $-1 \le x_{1,n} < x_{2,n} < \ldots < x_{n,n} \le 1$, for $n = 1, 2, \ldots$. The Lagrange operator $L_n : C[-1,1] \to \Pi_{n-1}$ is defined by

$$L_n(f;x) := \sum_{k=1}^{n} f(x_{k,n})l_{k,n}(x),$$

where

$$l_{k,n}(x) = \frac{w_n(x)}{w_n'(x_{k,n})(x - x_{k,n})}, \quad 1 \le k \le n,$$

and

$$w_n(x) = \prod_{k=1}^{n}(x - x_{k,n}).$$

The Lebesgue function of the interpolation is

$$\Lambda_n(x) := \sum_{k=1}^{n}\left|l_{k,n}(x)\right|.$$

Theorem 5.32 ([113]) *The Grüss inequality with discrete oscillations for the Lagrange operator is given by*

$$|T(f,g;x)| \le osc_{L_n}(f) \cdot osc_{L_n}(g) \cdot \left(\frac{\Lambda_n^2(x)}{2} - \frac{1}{8}\right),$$

for $f, g \in B[-1,1]$ and $-1 \le x \le 1$.

5.5.2 Grüss-Type Inequalities via Discrete Oscillations for More Than Two Functions

In [16] a Grüss-type inequality on a compact metric space for more than two functions was introduced. A similar interesting result is obtained in [13].

In what follows X is an arbitrary set, $B(X)$ is the set of all real-valued bounded functions on X, and $f^1, \ldots, f^p \in B(X)$. Take $a_n \in \mathbb{R}$, $a_n \geq 0$, $n \geq 0$, such that $\sum_{n=0}^{\infty} a_n = 1$. Furthermore, let $x_n \in X$, $n \geq 0$ be arbitrary mutually distinct points of X. For $f^k \in B(X)$ set $f_n^k := f^k(x_n)$, $k = \overline{1, p}$. Consider a positive linear functional $L : B(X) \to \mathbb{R}$, such that

$$L(f) := \sum_{n=0}^{\infty} a_n f_n .$$

Denote by

$$osc_L(f^k) := \sup \left\{ |f_n^k - f_m^k| : 0 \leq n < m < \infty \right\}.$$

Lemma 5.6 ([13]) *Let $B(X)$ be the set of all real-valued and bounded functions on X and $f^i \in B(X)$, $i = \overline{1, p}$. Then the following inequality holds*

$$osc_L(\prod_{k=1}^{p} f^k) \leq \sum_{i=1}^{p} osc_L(f^i) \prod_{j=1, j \neq i}^{p} \sup_{0 \leq n < \infty} \left\{ |f_n^j| \right\}.$$

Proof This inequality can be proved by induction. ■

The next result is a Grüss-type inequality via discrete oscillations for more than two functions.

Theorem 5.33 ([13]) *For a positive linear functional,*

$$L : B(X) \to \mathbb{R},\ L(f) := \sum_{n=0}^{\infty} a_n f_n,\ a_n \in \mathbb{R},\ a_n \geq 0,\ \sum_{n=0}^{\infty} a_n = 1 ,$$

the Grüss-type inequality via discrete oscillations, involving more than two functions is

$$\left| L(f^1 \cdot \ldots \cdot f^p) - L(f^1) \cdot \ldots \cdot L(f^p) \right|$$
$$\leq \frac{1}{2} \left(1 - \sum_{n=0}^{\infty} a_n^2 \right) \cdot \sum_{i,j=1, i<j}^{p} osc_L(f^i) \cdot osc_L(f^j) \cdot \prod_{k=1, k \neq i, j}^{p} \sup_{0 \leq s < \infty} \{ |f_s^k| \}.$$

Proof By induction the following inequality can be proved

$$\left| L(f^1 \cdot \ldots \cdot f^p) - L(f^1) \cdot \ldots \cdot L(f^p) \right| \tag{5.5.1}$$

$$\leq \left(\sum_{0 \leq n < m < \infty} a_n a_m \right) \cdot \sum_{i,j=1, i<j}^{p} osc_L(f^i) \cdot osc_L(f^j) \cdot \prod_{k=1, k \neq i,j}^{p} \sup_{0 \leq s < \infty} \{|f_s^k|\}.$$

Applying in (5.5.1) the following identity

$$\sum_{0 \leq n < m < \infty} a_n a_m = \frac{1}{2} \left(1 - \sum_{n=0}^{\infty} a_n^2 \right),$$

the theorem is proved. ■

Remark 5.7 The above result is better than what was obtained in [16] in the sense that the oscillations of functions are relative only to certain points, while in [16] the oscillations are relative to the whole compact metric space X.

5.6 Ostrowski Inequalities

In 1938 Ostrowski [188] proved his celebrated inequality which we cite below in the form given by Anastassiou in 1995 (see [29]).

Theorem 5.34 *Let f be in $C^1[a, b]$, $x \in [a, b]$. Then*

$$\left| f(x) - \frac{1}{b-a} \int_a^b f(t) dt \right| \leq \frac{(x-a)^2 + (b-x)^2}{2(b-a)} \cdot \|f'\|_\infty. \tag{5.6.1}$$

The approach considered by Ostrowski and several other mathematicians on the topic were carried out assuming differentiability properties of the functions. The next result extends the validity of the Ostrowski inequality to functions in $C(I)$ by using first and second order moduli of smoothness.

Theorem 5.35 ([12]) *Let $L : C[a, b] \to C[a, b]$ be nonzero, linear, and bounded, and such that $L : C^1[a, b] \to C^1[a, b]$ with $||(Lg)'|| \leq c_L \cdot ||g'||$ for all $g \in C^1[a, b]$. Then for all $f \in C[a, b]$ and $x \in [a, b]$ we have*

$$\left\| Lf(x) - \frac{1}{b-a} \int_a^b Lf(t) dt \right\| \leq ||L|| \cdot \tilde{\omega} \left(f; \frac{c_L}{||L||} \cdot \frac{(x-a)^2 + (b-x)^2}{2(b-a)} \right).$$

Proof Let $A_x : C[a,b] \to \mathbb{R}$ be a linear functional, given by

$$A_x(f) := f(x) - \frac{1}{b-a}\int_a^b f(t)dt.$$

This functional is a bounded linear functional with $||A_x|| \le 2$. We have

$$|A_x(Lf)| \le |Lf(x)| + \frac{1}{b-a}\int_a^b |Lf(t)|dt \le 2 \cdot ||L|| \cdot ||f||_\infty.$$

For $g \in C^1[a,b]$ we can write

$$|A_x(Lg)| = \left| Lg(x) - \frac{1}{b-a}\int_a^b Lg(t)dt \right| \le c_L \cdot \frac{(x-a)^2 + (b-x)^2}{2(b-a)} \cdot ||g'||_\infty.$$

Therefore,

$$\begin{aligned} |A_x(Lf)| &= |(A_x \circ L)(f - g + g)| \\ &\le 2 \cdot ||L|| \cdot ||f - g||_\infty + c_L \cdot \frac{(x-a)^2 + (b-x)^2}{2(b-a)} ||g'||_\infty. \end{aligned}$$

Passing to the infimum over $g \in C^1[a,b]$ and using relation (5.2.3) and Lemma 5.1 we have

$$|A_x(Lf)| \le ||L|| \cdot \tilde{\omega}\left(f; \frac{c_L}{||L||} \cdot \frac{(x-a)^2 + (b-x)^2}{2(b-a)}\right).$$

■

If $L = Id$ is the identity on $C[a,b]$, then $||L|| = c_L = 1$, and in this case we get the following corollary. It implies Ostrowski's classical result.

Corollary 5.3 ([12]) *For $f \in C[a,b]$ we have*

$$\left| f(x) - \frac{1}{b-a}\int_a^b f(t)dt \right| \le \tilde{\omega}\left(f; \frac{(x-a)^2 + (b-x)^2}{2(b-a)}\right).$$

Gavrea and Gavrea [96] were the first to observe the possibility of using moduli of smoothness in this context.

Let

$$Lip[0,1] := \left\{ f \in C[0,1] \,|\, |f|_{Lip} := \sup_{x \ne y} \frac{|f(x) - f(y)|}{|x-y|} < \infty \right\}$$

be the space of Lipschitz functions, $M_1^+[0,1]$ be the set of all probability Borel measures on $[0,1]$ and $\lambda \in M_1^+[0,1]$ a given measure.

Denote by

$$w_\lambda(x) := \int_0^1 |t-x|\, \mathrm{d}\lambda(t).$$

Gonska et al. [110] gave a generalization of the Theorem 5.35 for integrals with respect to probability measures λ and applied the new estimates to iterates of certain positive linear operators and to differences of such mappings.

Theorem 5.36 ([110]) *Let $L : C[0,1] \to C[0,1]$ be nonzero, linear, and bounded. Suppose that $L(Lip[0,1]) \subset Lip[0,1]$ and there exists $c_L > 0$ such that*

$$|Lg|_{Lip} \leq c_L\, |g|_{Lip}\,,$$

for all $g \in Lip[0,1]$. Then for all $f \in C[0,1]$, $\lambda \in M_1^+[0,1]$ and $x \in [0,1]$ we have

$$\left| Lf(x) - \int_0^1 Lf(t)\mathrm{d}\lambda(t) \right| \leq \|L\|\, \widetilde{\omega}\left(f; \frac{c_L}{\|L\|} w_\lambda(x)\right).$$

Corollary 5.4 ([110]) *In the setting of Theorem 5.36 suppose that, moreover, L is a positive linear operator reproducing the constant functions. Then*

$$\left| Lf(x) - \int_0^1 Lf(t)\mathrm{d}\lambda(t) \right| \leq \widetilde{\omega}\left(f; c_L w_\lambda(x)\right)$$

holds, for all $f \in C[0,1]$, $\lambda \in M_1^+[0,1]$ and $x \in [0,1]$.

Proposition 5.1 ([110]) *Let $L : C[0,1] \to C[0,1]$ be a positive linear operator with $Le_0 = e_0$. Suppose that $L(Lip[0,1]) \subset Lip[0,1]$ and there exists $c_L > 0$ such that $|Lg|_{Lip} \leq c_L\, |g|_{Lip}$, $g \in Lip[0,1]$. Let $A : C[0,1] \to C[0,1]$ be a positive linear operator with $Ae_0 = e_0$. For each $x \in [0,1]$ consider the measure $\lambda_x \in M_1^+[0,1]$ defined by*

$$\int_0^1 f(t)\mathrm{d}\lambda_x(t) = Af(x), \;\; f \in C[0,1].$$

Then the inequality

$$\begin{aligned} |(A \circ L)f(x) - Lf(x)| &\leq \widetilde{\omega}\left(f; c_L A(|t-x|, x)\right) \\ &\leq \widetilde{\omega}\left(f; c_L (A((t-x)^2, x))^{\frac{1}{2}}\right) \end{aligned}$$

holds, for all $f \in C[0,1]$, $x \in [0,1]$.

Applications of Proposition 5.1 involving differences of linear operators were given in [110]. In the following we will recall some of these applications.

The Beta-type operators $\overline{\mathbb{B}}_n$ were introduced by Lupaş in his thesis [156]. For $n = 1, 2, 3, \ldots$ and $f \in C[0, 1]$ the Beta-type operators are given by

$$\overline{\mathbb{B}}_n(f; x) := \begin{cases} f(0), & x = 0, \\ \dfrac{1}{B(nx, n - nx)} \displaystyle\int_0^1 t^{nx-1}(1-t)^{n-1-nx} f(t)dt, & 0 < x < 1, \\ f(1), & x = 1, \end{cases} \tag{5.6.2}$$

where $B(\cdot, \cdot)$ is the Euler's Beta function.

The genuine Bernstein–Durrmeyer operators are introduced as a composition of Bernstein operators and Beta operators, namely $U_n = B_n \circ \overline{\mathbb{B}}_n$ (see [57, 114]). These are given in explicit form by

$$U_n(f; x) = (1-x)^n f(0) + x^n f(1) \tag{5.6.3}$$

$$+ (n-1) \sum_{k=1}^{n-1} \left(\int_0^1 f(t) p_{n-2,k-1}(t) dt \right) p_{n,k}(x), \; f \in C[0, 1].$$

Proposition 5.2 (See [110, Example 5.2]) *For the genuine Bernstein–Durrmeyer operator and the Beta operator the following property holds:*

$$\left| U_n f(x) - \overline{\mathbb{B}}_n f(x) \right| \le \tilde{\omega} \left(f; \left(\frac{x(1-x)}{n} \right)^{\frac{1}{2}} \right). \tag{5.6.4}$$

Proof Let $L = \overline{\mathbb{B}}_n$, the Beta operator and $A = B_n$, the classical Bernstein operator. Then $A \circ L = U_n$, the genuine Bernstein–Durrmeyer operator. From Proposition 5.1 we get the inequality (5.6.4). ∎

Proposition 5.3 (See [110, Example 5.3]) *Let $L = B_n$ and $A = \overline{\mathbb{B}}_n$. Then $A \circ L = S_n$ is a Stancu operator. We infer that*

$$|S_n f(x) - B_n f(x)| \le \tilde{\omega} \left(f; \left(\frac{x(1-x)}{n+1} \right)^{\frac{1}{2}} \right).$$

Proposition 5.4 (See [110, Example 5.5]) *Let $L = B_{n+1}$ and $A = B_n$. Then $A \circ L = D_n$, an operator which was investigated in [108]. In this case we have*

$$|D_n f(x) - B_{n+1} f(x)| \le \tilde{\omega} \left(f; \left(\frac{x(1-x)}{n} \right)^{\frac{1}{2}} \right).$$

Corollary 5.5 ([110]) *Let $L : C[0,1] \to C[0,1]$ be a positive linear operator with $Le_0 = e_0$, and μ an invariant measure for L. Suppose that $L(Lip[0,1]) \subset Lip[0,1]$ and there exists $c_L > 0$ such that $|Lg|_{Lip} \le c_L\, |g|_{Lip}$, $g \in Lip[0,1]$. Then the inequality*

$$\left| Lf(x) - \int_0^1 f(t)\mathrm{d}\mu(t) \right| \le \widetilde{\omega}\left(f; c_L w_\mu(x)\right)$$

holds, for all $f \in C[0,1]$, $x \in [0,1]$.

Let $L_n : C[0,1] \to C[0,1]$ be a positive linear operator. Then the powers of L_n are defined by

$$L_n^0 := Id,\ L_n^1 := L_n \text{ and } L_n^{m+1} := L_n \circ L_n^m,\ m \in \mathbb{N}.$$

Corollary 5.6 ([110]) *In the setting of Corollary 5.5 we have*

$$\left| L^m f(x) - \int_0^1 f(t)\mathrm{d}\mu(t) \right| \le \widetilde{\omega}\left(f; c_L^m w_\mu(x)\right), \tag{5.6.5}$$

for all $f \in C[0,1]$, $x \in [0,1]$, $m \ge 1$. Moreover, if $c_L < 1$, then

$$\lim_{m\to\infty} L^m f = \left(\int_0^1 f(t)\mathrm{d}\mu(t)\right) e_0, \text{ uniformly on } [0,1],$$

and, consequently, L has exactly one invariant measure $\mu \in M_1^+[0,1]$.

Chapter 6
Bivariate Grüss-Type Inequalities for Positive Linear Operators

6.1 Bivariate Linear Operators

6.1.1 Bivariate Bernstein Operator

Let $I = [0, 1]$ and $X = [0, 1] \times [0, 1]$ endowed with the Euclidean metric

$$d_2((s,t),(x,y)) := \sqrt{(s-x)^2 + (t-y)^2}, \tag{6.1.1}$$

for $(s, t), (x, y) \in X$. The bivariate Bernstein operators, introduced by Butzer in [65], are given by

$$B_{n_1,n_2}(f; x, y) := \sum_{i_1=0}^{n_1} \sum_{i_2=0}^{n_2} f\left(\frac{i_1}{n_1}, \frac{i_2}{n_2}\right) b_{n_1,i_1}(x) b_{n_2,i_2}(y), \ \ f \in \mathbb{R}^X, \ \ x, y \in I, \tag{6.1.2}$$

where

$$b_{n_1,i_1}(x) := \binom{n_1}{i_1} x^{i_1}(1-x)^{n_1-i_1} \text{ and } b_{n_2,i_2}(y) := \binom{n_2}{i_2} y^{i_2}(1-y)^{n_2-i_2}.$$

The second moment of the bivariate Bernstein polynomial in this case is given by

$$\begin{aligned} B_{n_1,n_2}(d_2^2(\cdot,(x,y));(x,y)) &= \sum_{i_1=0}^{n_1} \sum_{i_2=0}^{n_2} d_2^2\left(\left(\frac{i_1}{n_1}, \frac{i_2}{n_2}\right), (x,y)\right) b_{n_1,i_1}(x) b_{n_2,i_2}(y) \\ &= \frac{x(1-x)}{n_1} + \frac{y(1-y)}{n_2} \leq \frac{1}{4}\left(\frac{1}{n_1} + \frac{1}{n_2}\right). \end{aligned}$$

V. Gupta et al., *Recent Advances in Constructive Approximation Theory*,
Springer Optimization and Its Applications 138,
https://doi.org/10.1007/978-3-319-92165-5_6

6.1.2 Bivariate Szász–Mirakyan Operators

Let $X := [0,\infty) \times [0,\infty)$ and $f \in \mathbb{R}^X$. The bivariate Mirakjan–Favard–Szász operators (see [86])) are defined as

$$S_{n_1,n_2}(f;(x,y)) := e^{-n_1 x} \cdot e^{-n_2 y} \sum_{k_1=0}^{\infty} \sum_{k_2=0}^{\infty} \frac{(n_1 x)^{k_1}}{k_1!} \cdot \frac{(n_2 y)^{k_2}}{k_2!} \cdot f\left(\frac{k_1}{n_1}, \frac{k_2}{n_2}\right). \tag{6.1.3}$$

6.1.3 Bivariate Baskakov Operators

Let $X := [0,\infty) \times [0,\infty)$ and $f \in \mathbb{R}^X$. The bivariate Baskakov operators (see [137])) are defined as

$$\begin{aligned} &V_{n_1,n_2}(f;(x,y)) \\ &:= \sum_{k_1=0}^{\infty} \sum_{k_2=0}^{\infty} \binom{n_1+k_1-1}{k_1}\binom{n_2+k_2-1}{k_2} \frac{x^{k_1}}{(1+x)^{n_1+k_1}} \frac{y^{k_2}}{(1+y)^{n_2+k_2}} f\left(\frac{k_1}{n_1}, \frac{k_2}{n_2}\right) \\ &= \sum_{k_1=0}^{\infty} \sum_{k_2=0}^{\infty} a_{n_1,k_1}(x) \cdot a_{n_2,k_2}(y) \cdot f\left(\frac{k_1}{n_1}, \frac{k_2}{n_2}\right), \end{aligned}$$

for $(x,y) \in X, n_1, n_2 \in \mathbb{N}$.

6.1.4 Bivariate King Operators

Let $r_{n_1}(x)$, $r_{n_2}(y)$ be sequences of continuous functions with $0 \leq r_{n_1}(x) \leq 1$ and $0 \leq r_{n_2}(y) \leq 1$. The bivariate King operator

$$K_{n_1,n_2} : C([0,1] \times [0,1]) \to C([0,1] \times [0,1]) \,,\; n_1 \neq n_2$$

can be defined by

$$K_{n_1,n_2}(f;(x,y)) := \sum_{k_1=0}^{n_1} \sum_{k_2=0}^{n_2} v_{n_1,k_1}(x) \cdot v_{n_2,k_2}(y) \cdot f\left(\frac{k_1}{n_1}, \frac{k_2}{n_2}\right),$$

with

$$v_{n_1,k_1}(x) = \binom{n_1}{k_1}(r_{n_1}(x))^{k_1}(1 - r_{n_1}(x))^{n_1-k_1},$$

$$v_{n_2,k_2}(y) = \binom{n_2}{k_2}(r_{n_2}(y))^{k_2}(1 - r_{n_2}(y))^{n_2-k_2},$$

for $f \in C([0, 1] \times [0, 1])$, $0 \leq x, y \leq 1$.

The second moment of the bivariate King operators is given by

$$K_{n_1,n_2}(d_2^2(\cdot, (x, y)); (x, y)) = \sum_{k_1=0}^{n_1}\sum_{k_2=0}^{n_2} d_2^2\left(\left(\frac{k_1}{n_1}, \frac{k_2}{n_2}\right), (x, y)\right) v_{n_1,k_1}(x)v_{n_2,k_2}(y)$$

$$= \frac{r_{n_1}(x)}{n_1}[1 - r_{n_1}(x)] + [r_{n_1}(x) - x]^2 + \frac{r_{n_2}(y)}{n_2}[1 - r_{n_2}(y)] + [r_{n_2}(y) - y]^2,$$

where $0 \leq r_{n_1}(x) \leq 1$ and $0 \leq r_{n_2}(y) \leq 1$ are continuous functions and for the metric (6.1.1).

For the special choices of $r_{n_1}(x) = r_{n_1}^*(x)$ and $r_{n_2}(y) = r_{n_2}^*(y)$, the second moment of the bivariate special King operators V_{n_1,n_2}^* is:

$$V_{n_1,n_2}^*(d_2^2(\cdot, (x, y)); (x, y)) = 2x(x - r_{n_1}^*(x)) + 2y(y - r_{n_2}^*(y)), \tag{6.1.4}$$

where

$$r_{n_1}^*(x) := \begin{cases} r_1^*(x) = x^2 & \text{, for } n_1 = 1 \\ r_{n_1}^*(x) = -\frac{1}{2(n_1-1)} + \sqrt{\frac{n_1}{n_1-1}x^2 + \frac{1}{4(n_1-1)^2}} & \text{, for } n_1 = 2, 3, \ldots, \end{cases}$$

and the same holding for $r_{n_2}^*(y)$.

6.1.5 Bivariate Hermite–Fejér Interpolation Operators

Let $x_k = \cos\frac{2k-1}{2n}\pi$, $1 \leq k \leq n$ be the zeros of $T_n(x) = \cos(n \cdot \arccos x)$, the n-th Chebyshev polynomial of the first kind. The bivariate Hermite–Fejér interpolation operators are given by

$$H_{2n_1-1,2n_2-1}(f; x, y)$$

$$:= \sum_{k_1=1}^{n_1}\sum_{k_2=1}^{n_2} f(x_{k_1}, y_{k_2})(1 - x\,x_{k_1})(1 - y\,y_{k_2})\left(\frac{T_{n_1}(x)}{n_1(x - x_{k_1})}\right)^2\left(\frac{T_{n_2}(y)}{n_2(y - y_{k_2})}\right)^2,$$

for $n_i \geq 1$, $i = 1, 2$, $f \in C([-1, 1]^2)$ and all $(x, y) \in [-1, 1]^2$.

In [98], Gonska showed that for such Hermite–Fejér operators, the following inequality involving the first absolute moment holds:

$$H_{2n-1}(|e_1 - x|\,;x) \le \frac{4}{n} \cdot |T_n(x)| \cdot \{\sqrt{1-x^2} \cdot \ln n + 1\} \le 10\,|T_n(x)| \cdot \frac{\ln n}{n}.$$

If we consider the metric

$$d_1((s,t),(x,y)) = |s-x| + |t-y|,$$

then the first moment of the bivariate operators is given by

$$\begin{aligned}
&H_{2n_1-1,2n_2-1}\left(d_1(\cdot,(x,y));(x,y)\right)\\
&= H_{2n_1-1}(|\cdot_1 - x|\,;x) + H_{2n_2-1}(|\cdot_2 - y|\,;y)\\
&\le \frac{4}{n_1} \cdot \left|T_{n_1}(x)\right| \cdot \{\sqrt{1-x^2} \cdot \ln n_1 + 1\} + \frac{4}{n_2} \cdot \left|T_{n_2}(y)\right| \cdot \{\sqrt{1-y^2} \cdot \ln n_2 + 1\}\\
&= 4 \cdot \left\{ \left(\frac{1}{n_1} \cdot \left|T_{n_1}(x)\right| \cdot \{\sqrt{1-x^2} \cdot \ln n_1 + 1\} \right)\right.\\
&\left. + \left(\frac{1}{n_2} \cdot \left|T_{n_2}(y)\right| \cdot \{\sqrt{1-y^2} \cdot \ln n_2 + 1\} \right) \right\}.
\end{aligned}$$

If we now consider the Euclidean metric (6.1.1), the second moment of the bivariate Hermite–Fejér operator can be evaluated as follows

$$H_{2n_1-1,2n_2-1}(d_2^2(\cdot,(x,y));(x,y)) = \frac{1}{n_1} \cdot T_{n_1}^2(x) + \frac{1}{n_2} \cdot T_{n_2}^2(y).$$

6.1.6 Bivariate Convolution Operators

For every function $f \in C(X)$, $X := [-1,1] \times [-1,1]$, the bivariate convolution operator $G_{m(n_1),m(n_2)}$ is defined as

$$\begin{aligned}
&G_{m(n_1),m(n_2)}(f;(x,y)) =\\
&\frac{1}{\pi}\int_{-\pi}^{\pi}\int_{-\pi}^{\pi} f(\cos(\arccos(x)+v_1), \cos(\arccos(y)+v_2))K_{m(n_1)}(v_1)K_{m(n_2)}(v_2)\mathrm{d}v_1\mathrm{d}v_2,
\end{aligned}$$

where the kernels $K_{m(n_1)}$, $K_{m(n_2)}$ are positive and even trigonometric polynomials of degrees $m(n_1)$ and $m(n_2)$, satisfying

$$\int_{-\pi}^{\pi} K_{m(n_i)}(v_i)\mathrm{d}v_i = \pi,\ i = 1,2,$$

meaning that $G_{m(n_1)}(1, x) = 1$ and $G_{m(n_2)}(1, y) = 1$, for $x, y \in [-1, 1]$. Both $G_{m(n_i)}(f, \cdot)$, $i = 1, 2$ are algebraic polynomials of degree $m(n_i)$, $i = 1, 2$ and the kernel $K_{m(n_i)}$ has the form:

$$K_{m(n_i)}(\nu_i) = \frac{1}{2} + \sum_{k_i=1}^{m(n_i)} \rho_{k_i, m(n_i)} \cdot \cos(k_i \nu_i),\ i = 1, 2,$$

for $\nu_i \in [-\pi, \pi]$.

The second moment of the bivariate convolution operator in the case of Fejér–Korovkin kernels is given by

$$\begin{aligned}
&G_{n_1-1,n_2-1}(d_2^2(\cdot - (x_1, x_2)); (x_1, x_2)) \\
&= G_{n_1-1}(d_2^2(\cdot - x_1); x_1) + G_{n_2-1}(d_2^2(\cdot - x_2); x_2) \\
&\leq 4\pi^2 \left(\frac{1}{(n_1 + 1)^2} + \frac{1}{(n_2 + 1)^2} \right).
\end{aligned}$$

6.1.7 *Bivariate Piecewise Linear Interpolation Operators at Equidistant Knots*

Let $X = [0, 1] \times [0, 1]$ be endowed with the Euclidean metric. The bivariate piecewise linear interpolation operator at equidistant knots

$$S_{\Delta_{n_1}, \Delta_{n_2}} : C(X) \to C(X)$$

at the points

$$0, \frac{1}{n_1} \dots, \frac{k_1}{n_1}, \dots, \frac{n_1 - 1}{n_1}, 1$$

and

$$0, \frac{1}{n_2} \dots, \frac{k_2}{n_2}, \dots, \frac{n_2 - 1}{n_2}, 1 .$$

respectively, can be explicitly described as

$$S_{\Delta_{n_1}, \Delta_{n_2}}(f; x, y) = \tag{6.1.5}$$

$$\frac{1}{n_1 n_2} \sum_{k_1=0}^{n_1} \sum_{k_2=0}^{n_2} \left[\frac{k_1 - 1}{n_1}, \frac{k_1}{n_1}, \frac{k_1 + 1}{n_1}; |\alpha - x| \right]_\alpha \left[\frac{k_2 - 1}{n_2}, \frac{k_2}{n_2}, \frac{k_2 + 1}{n_2}; |\alpha - y| \right]_\alpha f\left(\frac{k_1}{n_1}, \frac{k_2}{n_2} \right).$$

Denote

$$u_{n_1,k_1}(x) = \frac{1}{n_1}\left[\frac{k_1-1}{n_1}, \frac{k_1}{n_1}, \frac{k_1+1}{n_1}; |\alpha - x|\right]_\alpha, u_{n_1,k_1} \in C([0,1]),$$

with a similar definition holding for $u_{n_2,k_2}(y)$.

The bivariate operator $S_{\Delta_{n_1},\Delta_{n_2}}$ can also be defined by

$$S_{\Delta_{n_1},\Delta_{n_2}} f(x,y) := \sum_{k_1=0}^{n_1}\sum_{k_2=0}^{n_2} f\left(\frac{k_1}{n_1}, \frac{k_2}{n_2}\right) \cdot u_{n_1,k_1}(x) \cdot u_{n_2,k_2}(y),$$

for $f \in C(I)$, $x, y \in X$, $u_{n_i,k_i} \in C[0,1]$, $i = 1, 2$,

$$u_{n_i,k_i}\left(\frac{l_i}{n_i}\right) = \delta_{k_i,l_i},\ k_i, l_i = 0, \ldots, n_i,\ i = 1, 2.$$

For $x \in \left[\frac{k_1-1}{n_1}, \frac{k_1}{n_1}\right]$, $y = \left[\frac{k_2-1}{n_2}, \frac{k_2}{n_2}\right]$, we can obtain the second moment of this operator

$$S_{\Delta_1,\Delta_2}\left(d^2(\cdot,(x,y);x,y)\right) = \sum_{i_1=0}^{n_1}\sum_{i_2=0}^{n_2} u_{n_1,i_1}(x)u_{n_2,i_2}(y)d^2\left(\left(\frac{i_1}{n_1}, \frac{i_2}{n_2}\right), (x,y)\right)$$

$$= \sum_{i_1=0}^{n_1}\sum_{i_2=0}^{n_2} u_{n_1,i_1}(x)u_{n_2,i_2}(y)\left\{\left(\frac{i_1}{n_1} - x\right)^2 + \left(\frac{i_2}{n_2} - y\right)^2\right\}$$

$$= \sum_{i_1=0}^{n_1}\sum_{i_2=0}^{n_2} u_{n_1,i_1}(x)u_{n_2,i_2}(y)\left(\frac{i_1}{n_1} - x\right)^2 + \sum_{i_1=0}^{n_1}\sum_{i_2=0}^{n_2} u_{n_1,i_1}(x)u_{n_2,i_2}(y)\left(\frac{i_2}{n_2} - y\right)^2$$

$$= \sum_{i_1=0}^{n_1} u_{n_1,i_1}(x)\left(\frac{i_1}{n_1} - x\right)^2 + \sum_{i_2=0}^{n_2} u_{n_2,i_2}(y)\left(\frac{i_2}{n_2} - y\right)^2$$

$$= \sum_{i_1=0}^{n_1} \frac{n_1}{2}\left(\frac{i_1}{n_1} - x\right)^2\left\{\left|\frac{i_1+1}{n_1} - x\right| - 2\left|\frac{i_1}{n_1} - x\right| + \left|\frac{i_1-1}{n_1} - x\right|\right\}$$

$$+ \sum_{i_2=0}^{n_2} \frac{n_2}{2}\left(\frac{i_2}{n_2} - y\right)^2\left\{\left|\frac{i_2+1}{n_2} - y\right| - 2\left|\frac{i_2}{n_2} - y\right| + \left|\frac{i_2-1}{n_2} - y\right|\right\}$$

$$= \left(x - \frac{k_1-1}{n_1}\right)\left(\frac{k_1}{n_1} - x\right) + \left(y - \frac{k_2-1}{n_2}\right)\left(\frac{k_2}{n_2} - y\right) \le \frac{1}{4n_1^2} + \frac{1}{4n_2^2}.$$

6.1.8 *Bivariate Lagrange Operator*

Let $X := [-1, 1] \times [-1, 1]$ and $(x_{k_1,n_1}, y_{k_2,n_2}) \in I$, $k_1 = \overline{1, n_1}$, $k_2 = \overline{1, n_2}$. Then the bidimensional Lagrange operator L_{n_1,n_2} is given by

$$L_{n_1,n_2}(f; x, y) := \sum_{k_1=1}^{n_1} \sum_{k_2=1}^{n_2} f(x_{k_1,n_1}, y_{k_2,n_2}) l_{k_1,n_1}(x) l_{k_2,n_2}(y), \tag{6.1.6}$$

for $f \in \mathbb{R}^X$ and the Lagrange fundamental functions are given as usual by

$$l_{k_1,n_1}(x) = \frac{\omega_{n_1}(x)}{\omega'_{n_1}(x_{k_1,n_1})(x - x_{k_1,n_1})}, \quad 1 \le k_1 \le n_1,$$

where $\omega_{n_1}(x) = \prod_{k_1=1}^{n_1} (x - x_{k_1,n_1})$. The fundamental functions $l_{k_2,n_2}(y)$ are defined analogously.

The corresponding Lebesgue functions are

$$\Lambda_{n_1}(x) := \sum_{k_1=1}^{n_1} \left| l_{k_1,n_1}(x) \right| \text{ and } \Lambda_{n_2}(y) := \sum_{k_2=1}^{n_2} \left| l_{k_2,n_2}(y) \right|.$$

6.2 Grüss-Type Inequalities in the Bivariate Case

Let $C(X)$ be the Banach lattice of real valued continuous functions defined on the compact metric space (X, d).

Let $H : C(X^2) \to C(X^2)$ be a positive linear operator reproducing constant function and define

$$D(f, g; x, y) = H(fg; x, y) - H(f; x, y) \cdot H(g; x, y).$$

In order to derive an inequality of Chebyshev–Grüss type we recall a general result given by Rusu in [208].

Let d_2 be the Euclidean metric defined in (6.1.1). Using Theorem 5.20, Rusu [208] obtained the following result for the bivariate case:

Theorem 6.1 (See [208, Theorem 3.3.1]) *If $f, g \in C(X^2)$ and $x, y \in X$ fixed, then the inequality*

$$|D(f, g; x, y)| \le \frac{1}{4} \tilde{\omega}_{d_2} \left(f; 4\sqrt{H\left(d_2^2(\cdot, (x, y)); x, y\right)} \right) \tilde{\omega}_{d_2} \left(g; 4\sqrt{H\left(d_2^2(\cdot, (x, y)); x, y\right)} \right)$$

holds, where $H\left(d_2^2(\cdot, (x, y)); x, y\right)$ is the second moment of the bivariate operator H.

This result was applied in the case of known operators (see [208]), for example, Bernstein, King, Hermite–Fejér, convolution operators, piecewise linear interpolation operators. In the following we will recall some of these results.

Theorem 6.2 (See [208, Theorem 3.3.3]) *If we consider $H = B_{n_1,n_2}$, the bivariate Bernstein operator, in Theorem 6.1, we get*

$$\left|B_{n_1,n_2}(f \cdot g; (x, y)) - B_{n_1,n_2}(f; (x, y)) \cdot B_{n_1,n_2}(g; (x, y))\right|$$
$$\leq \frac{1}{4}\widetilde{\omega}_{d_2}\left(f; 4 \cdot \sqrt{\frac{x(1-x)}{n_1} + \frac{y(1-y)}{n_2}}\right) \cdot \widetilde{\omega}_{d_2}\left(g; 4 \cdot \sqrt{\frac{x(1-x)}{n_1} + \frac{y(1-y)}{n_2}}\right), \tag{6.2.1}$$

which implies

$$|D(f, g; (x, y))| \leq \frac{1}{4}\widetilde{\omega}_{d_2}\left(f; 2\sqrt{\frac{1}{n_1} + \frac{1}{n_2}}\right) \cdot \widetilde{\omega}_{d_2}\left(g; 2\sqrt{\frac{1}{n_1} + \frac{1}{n_2}}\right), \tag{6.2.2}$$

for two functions $f, g \in C(X)$, $X = [0, 1] \times [0, 1]$ and $x, y \in [0, 1]$ fixed.

Theorem 6.3 (See [208, Theorem 3.3.4]) *If we set $H = V^*_{n_1,n_2}$ in Theorem 6.1 and consider the second moments of these operators for the case $n_1 = n_2 = 2, 3, \ldots$, $n_1 \neq n_2$, we obtain the following inequality for the bivariate King operators*

$$|T(f, g; (x_1, x_2))|$$
$$\leq \frac{1}{4}\widetilde{\omega}_{d_2}\left(f; 4\sqrt{\sum_{i=1}^{2} 2x_i(x_i - r^*_{n_i}(x_i))}\right) \cdot \widetilde{\omega}_{d_2}\left(g; 4\sqrt{\sum_{i=1}^{2} 2x_i(x_i - r^*_{n_i}(x_i))}\right)$$
$$= \frac{1}{4}\widetilde{\omega}_{d_2}\left(f; 4\sqrt{\sum_{i=1}^{2} 2x_i\left(x_i + \frac{1}{2(n_i-1)} - \sqrt{\frac{n_i}{n_i-1}x_i^2 + \frac{1}{4(n_i-1)^2}}\right)}\right)$$
$$\cdot\, \widetilde{\omega}_{d_2}\left(g; 4\sqrt{\sum_{i=1}^{2} 2x_i\left(x_i + \frac{1}{2(n_i-1)} - \sqrt{\frac{n_i}{n_i-1}x_i^2 + \frac{1}{4(n_i-1)^2}}\right)}\right).$$

Theorem 6.4 (See [208, Theorem 3.3.5]) *If we consider $H = H_{2n_1-1,2n_2-1}$ in Theorem 6.1, we get the following inequality for the Hermite–Fejér operator*

$$\left|H_{2n_1-1,2n_2-1}(f \cdot g; (x, y)) - H_{2n_1-1,2n_2-1}(f; (x, y)) \cdot H_{2n_1-1,2n_2-1}(g; (x, y))\right|$$
$$\leq \frac{1}{4}\widetilde{\omega}_{d_2}\left(f; 4\sqrt{\frac{1}{n_1}T^2_{n_1}(x) + \frac{1}{n_2}T^2_{n_2}(y)}\right) \cdot \widetilde{\omega}_{d_2}\left(g; 4\sqrt{\frac{1}{n_1}T^2_{n_1}(x) + \frac{1}{n_2}T^2_{n_2}(y)}\right)$$

for two functions $f, g \in C(X)$, $X = [-1, 1] \times [-1, 1]$ and $x_1, x_2 \in [-1, 1]$ fixed.

Theorem 6.5 (See [208, Theorem 3.3.15]) *For $f, g \in C(X)$, where $X = [-1, 1]^2$, the Grüss inequality for the bivariate convolution operator with the Fejér–Korovkin kernel is given by*

$$\begin{aligned}
&|T(f, g; (x_1, x_2))| \\
&\leq \frac{1}{4}\widetilde{\omega}_{d_2}\left(f; 4\sqrt{4\pi^2\left(\frac{1}{(n_1+1)^2} + \frac{1}{(n_2+1)^2}\right)}\right) \\
&\times \widetilde{\omega}_{d_2}\left(g; 4\sqrt{4\pi^2\left(\frac{1}{(n_1+1)^2} + \frac{1}{(n_2+1)^2}\right)}\right) \\
&= \frac{1}{4}\widetilde{\omega}_{d_2}\left(f; 8\pi\sqrt{\frac{1}{(n_1+1)^2} + \frac{1}{(n_2+1)^2}}\right)\widetilde{\omega}_{d_2}\left(g; 8\pi\sqrt{\frac{1}{(n_1+1)^2} + \frac{1}{(n_2+1)^2}}\right).
\end{aligned}$$

Theorem 6.6 (See [208, Theorem 3.3.5]) *If we consider $H = S_{\Delta_{n_1},\Delta_{n_2}}$ in Theorem 6.1, we get the following inequality for bivariate piecewise linear interpolation operator at equidistant knots*

$$\begin{aligned}
|T(f, g; x, y)| &\leq \frac{1}{4} \cdot \widetilde{\omega}_{d_2}\left(f; 4 \cdot \sqrt{S_{\Delta_{n_1},\Delta_{n_2}}(d_2^2(\cdot, (x, y)); (x, y))}\right) \\
&\times \widetilde{\omega}_{d_2}\left(g; 4 \cdot \sqrt{S_{\Delta_{n_1},\Delta_{n_2}}(d_2^2(\cdot, (x, y)); (x, y))}\right) \\
&\leq \frac{1}{4}\widetilde{\omega}_{d_2}\left(f; 2 \cdot \sqrt{\frac{1}{n_1^2} + \frac{1}{n_2^2}}\right) \cdot \widetilde{\omega}_{d_2}\left(g; 2 \cdot \sqrt{\frac{1}{n_1^2} + \frac{1}{n_2^2}}\right)
\end{aligned}$$

for two functions $f, g \in C(X)$, $X = [0, 1] \times [0, 1]$ and $x_1, x_2 \in [0, 1]$ fixed.

6.3 The Composite Bivariate Bernstein Operators

In this chapter we consider a sequence of composite bivariate Bernstein operators and the cubature formula associated with them. The upper bounds for the remainder term of cubature formula are described in terms of moduli of continuity of order two. This is motivated by a recent series of articles by Barbosu et al. (see [50, 51, 53, 54]).

In order to formulate the general result of this section, we recall the following facts:

Definition 6.1 For the compact intervals $I, J \subset \mathbb{R}$, $F \in C(I \times J)$, $r \in \mathbb{N}_0$ and $\delta \in \mathbb{R}_+$, the partial moduli of smoothness of order r are defined as follows

$$\omega_r(F;\delta,0):=\sup\left\{\left|\sum_{\nu=0}^{r}(-1)^{r-\nu}\binom{r}{\nu}F(x+\nu h,y)\right|:(x,y),(x+rh,y)\in I\times J,|h|\le\delta\right\}$$

and

$$\omega_r(F;0,\delta):=\sup\left\{\left|\sum_{\nu=0}^{r}(-1)^{r-\nu}\binom{r}{\nu}F(x,y+\nu h)\right|:(x,y),(x,y+rh)\in I\times J,|h|\le\delta\right\}.$$

The total modulus of smoothness of order r is defined by

$$\omega_r(F;\delta_1,\delta_2):=\sup\left\{\left|\sum_{\nu=0}^{r}(-1)^{r-\nu}\binom{r}{\nu}F(x+\nu h_1,y+\nu h_2)\right|:\right.$$

$$\left.(x,y),(x+rh_1,y+rh_2)\in I\times J,|h_1|\le\delta_1,\ |h_2|\le\delta_2\right\}.$$

Definition 6.2 Let I and J be compact intervals of the real axis and let $L : C(I)\to C(I)$ and $M : C(J)\to C(J)$ be discretely defined operators, i.e.,

$$L(g;x)=\sum_{e\in E}g(x_e)A_e(x),\ g\in C(I),x\in I,$$

where E is a finite index set, then $x_e\in I$ are mutually distinct and $A_e\in C(I)$, $e\in E$.

Analogously,

$$M(h;y)=\sum_{f\in F}h(y_f)B_f(y),\ h\in C(J),y\in J.$$

If L is of the form above, then its parametric extension to $C(I\times J)$ is given by

$$_xL(F;x,y)=L(F_y;x)=\sum_{e\in E}F_y(x_e)A_e(x)=\sum_{e\in E}F(x_e,y)A_e(x).$$

Here F_y, $y\in J$, denote the partial functions of F given by

$$F_y(x)=F(x,y),x\in I\ .$$

Similarly,

$$_yM(F;x,y)=\sum_{f\in F}F(x,y_f)B_f(y).$$

The tensor product of L and M (or M and L) is given by

$$\left({}_xL \circ_y M\right)(F; x, y) = \sum_{e \in E} \sum_{f \in F} F(x_e, y_f) A_e(x) B_f(y).$$

A simplified form of [57, Theorem 37] was formulated in [14] as follows

Theorem 6.7 (See [14, Theorem 2.1]) *Let L and M be discretely defined operators as given above such that*

$$|(g - Lg)(x)| \leq \sum_{\rho=0}^{r} \Gamma_{\rho,L}(x)\omega_\rho(g; \Lambda_{\rho,L}(x)), \ g \in C(I), x \in I,$$

and

$$|(h - Mh)(y)| \leq \sum_{\sigma=0}^{s} \Gamma_{\sigma,M}(y)\omega_\sigma(h; \Lambda_{\sigma,M}(y)), \ h \in C(J), y \in J.$$

Here $\omega_\rho, \rho = 0, \ldots, r$, denote the moduli of order ρ, and Γ and Λ are bounded functions. Analogously for M. Then for $(x, y) \in I \times J$ and $F \in C(I \times J)$ the following holds:

$$\left|\left[F - ({}_xL \circ_y M)F\right](x, y)\right| \leq \sum_{\rho=0}^{r} \Gamma_{\rho,L}(x)\omega_\rho(F; \Lambda_{\rho,L}(x), 0)$$
$$+ \|L\| \sum_{\sigma=0}^{s} \Gamma_{\sigma,M}(y)\omega_\sigma(F; 0, \Lambda_{\sigma,M}(y)),$$

where $\|L\|$ denotes the operator norm of L, which is finite due to the representation of L.

Subsequently, the above results are applied in the case of known bivariate Bernstein operators.

Example 6.1 (See [14, Example 3.1]) If we consider $L = B_{n_1}$ and $M = B_{n_2}$ with two classical Bernstein operators mapping $C[0, 1]$ into $C[0, 1]$, then for $F \in C([0, 1] \times [0, 1])$ and $(x, y) \in [0, 1] \times [0, 1]$:

$$\left({}_xB_{n_1} \circ_y B_{n_2}\right)(F; x, y) = \sum_{i_1=0}^{n_1} \sum_{i_2=0}^{n_2} F\left(\frac{i_1}{n_1}, \frac{i_2}{n_2}\right) p_{n_1,i_1}(x) p_{n_2,i_2}(y),$$

where

$$p_{n,i}(x) = \binom{i}{n} x^i (1 - x)^{n-i}, \ x \in [0, 1],$$

and

$$
\begin{aligned}
&\left|\left[F-\left({}_xB_{n_1}\circ_y B_{n_2}\right)F\right](x,y)\right| \\
&\le \frac{3}{2}\left[\omega_2\left(F;\sqrt{\frac{x(1-x)}{n_1}},0\right)+\omega_2\left(F;0,\sqrt{\frac{y(1-y)}{n_2}}\right)\right] \\
&\le \frac{3}{2}\left[\|F^{(2,0)}\|\frac{x(1-x)}{n_1}+\|F^{(0,2)}\|\frac{y(1-y)}{n_2}\right],\ F\in C^{2,2}([0,1]\times[0,1]).
\end{aligned}
$$

In [14] the following estimate was considered (see [50, Theorem 2.3])

$$
\begin{aligned}
\left|\left[F-\left({}_xB_{n_1}\circ_y B_{n_2}\right)F\right](x,y)\right| &\le \frac{1}{2}\frac{x(1-x)}{n_1}\|F^{(2,0)}\|+\frac{1}{2}\frac{y(1-y)}{n_2}\|F^{(0,2)}\| \\
&+\frac{1}{4}\frac{x(1-x)y(1-y)}{n_1n_2}\|F^{(2,2)}\| \\
&\le \frac{1}{8n_1}\|F^{(2,0)}\|+\frac{1}{8n_2}\|F^{(0,2)}\|+\frac{1}{64n_1n_2}\|F^{(2,2)}\|.
\end{aligned}
$$

Integrating the bivariate Bernstein polynomials for $F\in C([0,1]\times[0,1])$ one obtains the following cubature formula

$$
\int_0^1\int_0^1 F(x,y)dxdy=\frac{1}{(n_1+1)(n_2+1)}\sum_{i_1=0}^{n_1}\sum_{i_2=0}^{n_2}F\left(\frac{i_1}{n_1},\frac{i_2}{n_2}\right)+R_{n_1,n_2}[F], \tag{6.3.1}
$$

where the remainder is bounded as follows:

$$
\left|R_{n_1,n_2}[F]\right|\le\frac{1}{12n_1}\|F^{(2,0)}\|+\frac{1}{12n_2}\|F^{(0,2)}\|+\frac{1}{144n_1n_2}\|F^{2,2}\|,
$$

if $F\in C^{2,2}([0,1]\times[0,1])$.

A new upper bound for the approximation error of cubature formula associated with the bivariate Bernstein operators was described in terms of moduli of continuity of order two in [14].

Theorem 6.8 ([14]) *For the remainder term of the cubature formula (6.3.1), $n_1,n_2\in\mathbb{N}$ and $F\in C([0,1]\times[0,1])$ there holds*

$$
\left|R_{n_1,n_2}[F]\right|\le\frac{3}{2}\left[\int_0^1\omega_2\left(F;\sqrt{\frac{x(1-x)}{n_1}},0\right)dx+\int_0^1\omega_2\left(F;0,\sqrt{\frac{y(1-y)}{n_2}}\right)dy\right].
$$

Moreover, if $F\in C^{2,2}([0,1]\times[0,1])$, then the above implies

$$|R_{n_1,n_2}[F]| \leq \frac{1}{4}\left(\frac{1}{n_1}\|F^{(2,0)}\| + \frac{1}{n_2}\|F^{(0,2)}\|\right).$$

Let $m_1, m_2 \in \mathbb{N}$, $1 \leq k \leq m_1$, $1 \leq l \leq m_1$, and

$$(x, y) \in \left[\frac{k-1}{m_1}, \frac{k}{m_1}\right] \times \left[\frac{l-1}{m_2}, \frac{l}{m_2}\right].$$

The bivariate composite Bernstein operators can be constructed as follows

$$\overline{\mathcal{B}}(f; x, y) = m_1^{n_1} \cdot m_2^{n_2} \sum_{i=0}^{n_1}\sum_{j=0}^{n_2} \binom{n_1}{i}\binom{n_2}{j}\left(x - \frac{k-1}{m_1}\right)^i \left(\frac{k}{m_1} - x\right)^{n_1-i}$$
$$\cdot \left(y - \frac{l-1}{m_2}\right)^j \left(\frac{l}{m_2} - y\right)^{n_2-j} f\left(\frac{k-1}{m_1} + \frac{i}{m_1 n_1}, \frac{l-1}{m_2} + \frac{j}{n_2 m_2}\right)$$

and

$$\left|f(x, y) - \overline{\mathcal{B}}(f; x, y)\right| = \frac{\left(x - \frac{k-1}{m}\right)\left(\frac{k}{m_1} - x\right)}{2n_1}\|f^{(2,0)}\| + \frac{\left(y - \frac{l-1}{m_2}\right)\left(\frac{l}{m_2} - y\right)}{2n_2}\|f^{(0,2)}\|$$
$$+ \frac{\left(x - \frac{k-1}{m}\right)\left(\frac{k}{m_1} - x\right)\left(y - \frac{l-1}{m_2}\right)\left(\frac{l}{m_2} - y\right)}{4n_1 n_2}\|f^{(2,2)}\|,$$

where $f \in C^{2,2}([0, 1] \times [0, 1])$.

The order of convergence of the bivariate composite Bernstein operators was considered involving the second modulus of continuity as follows (see [14])

Theorem 6.9 ([14]) *For* $f \in C([0, 1] \times [0, 1])$, $n_1, n_2, m_1, m_2 \in \mathbb{N}$ *and* $(x, y) \in [0, 1] \times [0, 1]$ *it holds*

$$\left|f(x, y) - \overline{\mathcal{B}}(f; x, y\right| \leq \frac{3}{2}\left\{\omega_2\left(f; \sqrt{\frac{\left(x - \frac{k-1}{m_1}\right)\left(\frac{k}{m_1} - x\right)}{n_1}}, 0\right)\right.$$
$$\left. + \omega_2\left(f; 0, \sqrt{\frac{\left(y - \frac{l-1}{m_2}\right)\left(\frac{l}{m_2} - y\right)}{n_2}}\right)\right\},$$

if $(x, y) \in \left[\frac{k-1}{m_1}, \frac{k}{m_1}\right] \times \left[\frac{l-1}{m_2}, \frac{l}{m_2}\right]$, $1 \leq k \leq m_1$, $1 \leq l \leq m_2$.

In what follows we present an inequality for the bivariate composite Bernstein operators, expressed in terms of the least concave majorant of a modulus of continuity. Let $C(X)$ be the Banach lattice of real valued continuous functions defined on the compact metric space (X, d).

Let $H : C(X^2) \to C(X^2)$ be a positive linear operator reproducing constant function and define

$$D(f, g; x, y) = H(fg; x, y) - H(f; x, y) \cdot H(g; x, y).$$

Using Theorem 6.1 the following inequality of Grüss-type is obtained.

Proposition 6.1 ([14]) *For $f, g \in C(X^2)$ and $x, y \in X$ fixed, the following Grüss-type inequality holds*

$$|\overline{\mathcal{B}}(fg; x, y) - \overline{\mathcal{B}}(f; x, y)\overline{\mathcal{B}}(g; x, y)| \leq \frac{1}{4}\tilde{\omega}_{d_2}\left(f; 4\sqrt{\Psi(x, y)}\right)\tilde{\omega}_{d_2}\left(g; 4\sqrt{\Psi(x, y)}\right)$$

$$\leq \frac{1}{4}\tilde{\omega}_{d_2}\left(f; 2\sqrt{\frac{1}{n_1 m_1^2} + \frac{1}{n_2 m_2^2}}\right) \cdot \tilde{\omega}_{d_2}\left(g; 2\sqrt{\frac{1}{n_1 m_1^2} + \frac{1}{n_2 m_2^2}}\right)$$

where

$$\Psi(x, y) = \frac{\left(x - \frac{k-1}{m_1}\right)\left(\frac{k}{m_1} - x\right)}{n_1} + \frac{\left(y - \frac{l-1}{m_2}\right)\left(\frac{l}{m_2} - y\right)}{n_2}$$

and

$$(x, y) \in \left[\frac{k-1}{m_1}, \frac{k}{m_1}\right] \times \left[\frac{l-1}{m_2}, \frac{l}{m_2}\right].$$

6.4 A Cubature Formula Associated with the Bivariate Bernstein Operators

In [53] Bărbosu and Miclăuş introduced the following cubature formula:

$$\int_0^1 \int_0^1 f(x, y)dxdy = \sum_{k=1}^{m_1} \sum_{l=1}^{m_2} \int_{\frac{k-1}{m_1}}^{\frac{k}{m}} \int_{\frac{l-1}{m_2}}^{\frac{l}{m_2}} f(x, y)dxdy$$

$$\approx \sum_{k=1}^{m_1} \sum_{l=1}^{m_2} \int_{\frac{k-1}{m_1}}^{\frac{k}{m}} \int_{\frac{l-1}{m_2}}^{\frac{l}{m_2}} \overline{\mathcal{B}}(f; x, y)dxdy = \int_0^1 \int_0^1 \overline{\mathcal{B}}(f; x, y)dxdy := \overline{\mathcal{I}}(f).$$

It follows

$$\int_{\frac{k-1}{m_1}}^{\frac{k}{m}} \int_{\frac{l-1}{m_2}}^{\frac{l}{m_2}} \overline{\mathcal{B}}(f; x, y)dxdy$$

$$= m_1^{n_1} m_2^{n_2} \sum_{i=0}^{n_1} \sum_{j=0}^{n_2} \binom{n_1}{i}\binom{n_2}{j} \int_{\frac{k-1}{m_1}}^{\frac{k}{n_1}} \left(x - \frac{k-1}{m_1}\right)^i \left(\frac{k}{m_1} - x\right)^{n_1-i} dx$$

$$\cdot \int_{\frac{l-1}{m_2}}^{\frac{l}{m_2}} \left(y - \frac{l-1}{m_2}\right)^j \left(\frac{l}{m_2} - y\right)^{n_2-j} dyf\left(\frac{k-1}{m_1} + \frac{i}{m_1 n_1}, \frac{l-1}{m_2} + \frac{j}{n_2 m_2}\right)$$

$$= \sum_{i=0}^{n_1} \sum_{j=0}^{n_2} A_{n_1,n_2,m_1,m_2} f\left(\frac{k-1}{m_1} + \frac{i}{m_1 n_1}, \frac{l-1}{m_2} + \frac{j}{n_2 m_2}\right),$$

where

$$A_{n_1,n_2,m_1,m_2} = \frac{1}{m_1 m_2 (n_1+1)(n_2+1)}.$$

In the following some upper bounds of the error of cubature formula associated with the bivariate Bernstein operators are given.

Theorem 6.10 ([14]) *For* $f \in C^{2,2}([0,1] \times [0,1])$ *it follows*

$$\left|\int_0^1 \int_0^1 f(x, y)dxdy - \overline{\mathcal{I}}(f)\right|$$
$$\leq \frac{1}{12 n_1 m_1^2} \| f^{(2,0)} \| + \frac{1}{12 n_2 m_2^2} \| f^{(0,2)} \| + \frac{1}{144 n_1 n_2 m_1^2 m_2^2} \| f^{(2,2)} \|.$$

Theorem 6.11 ([14]) *For* $f \in C^{2,2}([0,1] \times [0,1])$ *it follows*

$$\left|\int_0^1 \int_0^1 f(x, y)dxdy - \overline{\mathcal{I}}(f)\right| \leq \frac{1}{4}\left\{\frac{1}{m_1^2 n_1} \| f^{(2,0)} \| + \frac{1}{m_2^2 n_2} \| f^{(0,2)} \| \right\}.$$

Let (X, d) be a compact metric space and $L : C(X) \to \mathbb{R}$ be a positive linear functional reproducing constants. We consider the positive bilinear functional

$$D(f, g) := L(fg) - L(f)L(g).$$

Theorem 6.12 ([14]) *If $f, g \in C(X)$, (X, d) a compact metric space, then the inequality*

$$|D(f, g)| \leq \frac{1}{4}\tilde{\omega}_d\left(f; 2\sqrt{L^2(d^2(\cdot,\cdot))}\right)\tilde{\omega}_d\left(g; 2\sqrt{L^2(d^2(\cdot,\cdot))}\right)$$

holds.

The following result suggests how non-multiplicative the functional

$$\overline{\mathcal{I}}(f) = \int_0^1\int_0^1 \overline{\mathcal{B}}(f; (x, y))dxdy$$

is.

Corollary 6.1 ([14]) *If $f, g \in C([0, 1] \times [0, 1])$, then*

$$\left|\overline{\mathcal{I}}(fg) - \overline{\mathcal{I}}(f)\overline{\mathcal{I}}(g)\right| \leq \frac{1}{4}\tilde{\omega}_{d_2}\left(f; 2\sqrt{\frac{1}{3}\left(1 + \frac{1}{n_1 m_1^2} + \frac{1}{n_2 m_2^2}\right)}\right) \tag{6.4.1}$$

$$\cdot\,\tilde{\omega}_{d_2}\left(g; 2\sqrt{\frac{1}{3}\left(1 + \frac{1}{n_1 m_1^2} + \frac{1}{n_2 m_2^2}\right)}\right).$$

6.5 Grüss-Type Inequalities via Discrete Oscillations

In this section we present some new bivariate Grüss-type inequalities via discrete oscillations and applications of these inequalities to different bivariate linear operators.

Let I be an arbitrary set and $B(X)$ the set of all real-valued, bounded functions on $X = I^2$. Take $a_n, b_n \in \mathbb{R}$, $n \geq 0$, such that

$$\sum_{n=0}^{\infty} |a_n| < \infty, \; \sum_{n=0}^{\infty} a_n = 1 \text{ and } \sum_{n=0}^{\infty} |b_n| < \infty, \; \sum_{n=0}^{\infty} b_n = 1\,,$$

respectively. Furthermore, let $x_n \in I, n \geq 0$ and $y_m \in I$, $m \geq 0$ be arbitrary mutually distinct points. For $f \in B(X)$ set $f_{n,m} := f(x_n, y_m)$. Now consider the functional $L : B(X) \to \mathbb{R}$, $Lf = \sum_{n=0}^{\infty}\sum_{m=0}^{\infty} a_n b_m f_{n,m}$. The functional L is linear and reproduces constant functions.

Theorem 6.13 ([13]) *The Grüss-type inequality for the above linear functional L is given by:*

$$|L(fg) - L(f) \cdot L(g)| \leq \frac{1}{2} \cdot osc_L(f) \cdot osc_L(g) \cdot \sum_{n,m,i,j=0,\ (n,m)\neq(i,j)}^{\infty} \left|a_n b_m a_i b_j\right|,$$

where $f, g \in B(X)$ and we define the oscillations to be:

$$osc_L(f) := \sup\{\left|f_{n,m} - f_{i,j}\right| : n, m, i, j \geq 0\},$$
$$osc_L(g) := \sup\{\left|g_{n,m} - g_{i,j}\right| : n, m, i, j \geq 0\}.$$

Theorem 6.14 ([13]) *In particular, if $a_n \geq 0$, $b_m \geq 0$, $n, m \geq 0$, then L is a positive linear functional and we have:*

$$|L(fg) - L(f) \cdot L(g)| \leq \frac{1}{2} \cdot \left(1 - \sum_{n=0}^{\infty} a_n^2 \cdot \sum_{m=0}^{\infty} b_m^2\right) \cdot osc_L(f) \cdot osc_L(g),$$

for $f, g \in B(X)$ and the oscillations are given as above.

In [13] the above Grüss-type inequalities involving discrete oscillations were applied for known operators, for example Lagrange, Bernstein, Mirakjan–Favard–Szász, and piecewise linear interpolation operators.

Let

$$D_H(f, g; x, y) = H(f \cdot g; x, y) - H(f; x, y) \cdot H(g; x, y),$$

where (x, y) is fixed and H is a linear operator.

Theorem 6.15 ([13]) *The Grüss-type inequality for the bivariate Lagrange operator defined in (6.1.6) is given by*

$$\left|D_{L_{n_1,n_2}}(f, g; x, y)\right| \leq \frac{1}{2} \cdot osc_{L_{n_1,n_2}}(f) \cdot osc_{L_{n_1,n_2}}(g) \cdot \left[\Lambda_{n_1}^2(x)\Lambda_{n_2}^2(y) - \frac{1}{16}\right],$$

where $f, g \in C(X)$. The oscillation for f is defined by

$$\begin{aligned} osc_{L_{n_1,n_2}}(f) := \max\{&\left|f\left(x_{k_1,n_1}, y_{k_2,n_2}\right) - f\left(x_{m_1,n_1}, y_{m_2,n_2}\right)\right| : \\ &1 \leq k_1, m_1 \leq n_1, 1 \leq k_2, m_2 \leq n_2\}. \end{aligned}$$

A similar definition is given for the oscillation of the second function.

Proof The sums of the squared fundamental functions of a Lagrange interpolation satisfy (see [141]):

$$\sum_{k_1=1}^{n_1} l_{k_1,n_1}^2(x) \geq \frac{1}{4}, \text{ for } -1 \leq x \leq 1,$$

and the same holds for the squares of the fundamental functions with respect to y. Therefore,

$$\sum_{k_1,m_1=1}^{n_1} \left(\sum_{k_2,m_2=1,(k_1,k_2)\neq(m_1,m_2)}^{n_2} \left| l_{k_1,n_1}(x) l_{k_2,n_2}(y) l_{m_1,n_1}(x) l_{m_2,n_2}(y) \right| \right)$$

$$= \left(\sum_{i=1}^{n_1} \sum_{j=1}^{n_2} \left| l_{i,n_1}(x) l_{j,n_2}(y) \right| \right)^2 - \sum_{i=1}^{n_1} \sum_{j=1}^{n_2} l_{i,n_1}^2(x) l_{j,n_2}^2(y)$$

$$= \Lambda_{n_1}^2(x)\Lambda_{n_2}^2(y) - \sum_{i=1}^{n_1} \sum_{j=1}^{n_2} l_{i,n_1}^2(x) l_{j,n_2}^2(y) \leq \Lambda_{n_1}^2(x)\Lambda_{n_2}^2(y) - \frac{1}{16},$$

and the theorem is proved. ■

Let $I = [0, 1]$ and $X = [0, 1] \times [0, 1]$ endowed with the Euclidean metric (6.1.1).

Theorem 6.16 ([13]) *The Grüss-type inequality for the bivariate Bernstein operator (6.1.2) is given by*

$$\left| D_{B_{n_1,n_2}}(f, g; x, y) \right| \leq \frac{1}{2} \cdot \frac{n_2 n_1 + n_2 + n_1}{(n_1 + 1)(n_2 + 1)} \cdot osc_{B_{n_1,n_2}}(f) \cdot osc_{B_{n_1,n_2}}(g),$$

where $f, g \in C(X)$. The oscillation for f is defined by

$$osc_{B_{n_1,n_2}}(f) := \max\{\left| f_{k,l} - f_{s,t} \right| : k, s = 0, \ldots, n_1; l, t = 0, \ldots, n_2\},$$

and $f_{k,l} := f\left(\frac{k}{n_1}, \frac{l}{n_2}\right)$; a similar definition applies to g.

Proof Let $\varphi_{n_1}(x) := \sum_{i_1=0}^{n_1} b_{n_1,i_1}^2(x)$, $x \in X$. Then we get immediately

$$\varphi_{n_1}(x) \geq \frac{1}{n_1 + 1}, \quad x \in X, \tag{6.5.1}$$

and the same holds for $\varphi_{n_2}(y)$, $y \in X$. According to Theorem 6.14, for each $x, y \in I$, $f, g \in B(X)$ we have

$$\left|D_{B_{n_1,n_2}}(f,g;x,y)\right|$$
$$\leq \frac{1}{2}\left(1-\sum_{i_1=0}^{n_1} b_{n_1,i_1}^2(x)\cdot\sum_{i_2=0}^{n_2} b_{n_2,i_2}^2(y)\right)\cdot osc_{B_{n_1,n_2}}(f)\cdot osc_{B_{n_1,n_2}}(g)$$
$$\leq \frac{1}{2}\cdot\frac{n_2n_1+n_2+n_1}{(n_1+1)(n_2+1)}\cdot osc_{B_{n_1,n_2}}(f)\cdot osc_{B_{n_1,n_2}}(g).$$

■

Theorem 6.17 ([13]) *The new Grüss-type inequality for the bivariate Bernstein operator is:*

$$\left|D_{B_{n_1,n_2}}(f,g;x,y)\right|$$
$$\leq \frac{1}{2}\left(1-\binom{2n_1}{n_1}\binom{2n_2}{n_2}\frac{1}{4^{n_1}}\cdot\frac{1}{4^{n_2}}\right)\cdot osc_{B_{n_1,n_2}}(f)\cdot osc_{B_{n_1,n_2}}(g),\ x,y\in X.$$

Proof In order to obtain this result one applies the following inequality (see [113]):

$$\varphi_{n_1}(x)\geq\frac{1}{4^{n_1}}\binom{2n_1}{n_1},\ x\in X, \tag{6.5.2}$$

with equality if and only if $x=1/2$. The same result holds for $\varphi_{n_2}(y)$, if $y\in X$. ■

Denote

$$\sigma_{n_1}(x):=e^{-2n_1x}\sum_{k_1=0}^{\infty}\frac{(n_1x)^{2k_1}}{(k_1!)^2}\ \text{and}\ \sigma_{n_2}(y):=e^{-2n_2y}\sum_{k_2=0}^{\infty}\frac{(n_2y)^{2k_2}}{(k_2!)^2}.$$

Theorem 6.18 ([13]) *The Grüss-type inequality via discrete oscillations for the bivariate Mirakjan–Favard–Szász operators defined in (6.1.3) is given by*

$$\left|D_{M_{n_1,n_2}}(f,g;x,y)\right|\leq\frac{1}{2}\cdot(1-\sigma_{n_1}(x)\cdot\sigma_{n_2}(y))\cdot osc_{M_{n_1,n_2}}(f)\cdot osc_{M_{n_1,n_2}}(g),$$

where

$$f,g\in B(X),\ osc_{M_{n_1,n_2}}(f):=sup\{\left|f_{s,l}-f_{r,t}\right|:s,r,l,t\geq 0\},$$

with

$$f_{s,l}:=f\left(\frac{s}{n_1},\frac{l}{n_2}\right).$$

It was proved in [113] that $\inf_{x\geq 0} \sigma_{n_1}(x) = 0$ and $\inf_{y\geq 0} \sigma_{n_2}(y) = 0$ hold. Then the above inequality takes the following form:

Theorem 6.19 ([13]) *The Grüss-type inequality for the Mirakjan–Favard–Szász operators is given by*

$$|T(f, g; x, y)| \leq \frac{1}{2} osc_{M_{n_1,n_2}}(f) \cdot osc_{M_{n_1,n_2}}(g),$$

where the functions f and g and the oscillations are given as above.

Theorem 6.20 ([13]) *The new Grüss-type inequality for $S_{\Delta_{n_1},\Delta_{n_2}}$ is*

$$\begin{aligned}
&\left|D_{S_{\Delta_{n_1},\Delta_{n_2}}}(f, g; x, y)\right| \\
&\leq \frac{1}{2}\left(1 - \sum_{k_1=0}^{n_1} u_{n_1,k_1}^2(x) \cdot \sum_{k_2=0}^{n_2} u_{n_2,k_2}^2(y)\right) \cdot osc_{S_{\Delta_{n_1},\Delta_{n_2}}}(f) \cdot osc_{S_{\Delta_{n_1},\Delta_{n_2}}}(g) \\
&\leq \frac{3}{8} osc_{S_{\Delta_{n_1},\Delta_{n_2}}}(f) \cdot osc_{S_{\Delta_{n_1},\Delta_{n_2}}}(g).
\end{aligned}$$

Proof In order to obtain this result, we need to find the minimum of the sums

$$\tau_{n_1}(x) := \sum_{k_1=0}^{n_1} u_{n_1,k_1}^2(x) \text{ and } \tau_{n_2}(y) := \sum_{k_2=0}^{n_2} u_{n_2,k_2}^2(y)\,.$$

For particular intervals $x \in \left[\frac{k_1 - 1}{n_1}, \frac{k_1}{n_1}\right]$ and $y \in \left[\frac{k_2 - 1}{n_2}, \frac{k_2}{n_2}\right]$, we get that

$$\tau_{n_1}(x) := \sum_{k_1=0}^{n_1} u_{n_1,k_1}^2(x) = (n_1 x - k_1 + 1)^2 + (k_1 - n_1 x)^2, \text{ for } k_1 = 1, \ldots, n_1$$

and the same for $\tau_{n_2}(y)$. The functions $\tau_{n_1}(x)$ and $\tau_{n_2}(y)$ are minimal if and only if $x = \frac{2k_1 - 1}{2n_1}$, $y = \frac{2k_2 - 1}{2n_2}$ and the minimum value for both $\tau_{n_1}(x)$ and $\tau_{n_2}(y)$ is $\frac{1}{2}$. According to Theorem 6.14 this result is proved. ∎

Chapter 7
Estimates for the Differences of Positive Linear Operators

The results presented in this chapter are motivated by the problem proposed by Lupaş in [158]. One of the questions raised by Lupaş was to provide an estimate for

$$B_n \circ \overline{\mathbb{B}}_n - \overline{\mathbb{B}}_n \circ B_n =: U_n - S_n,$$

where B_n are the Bernstein operators and $\overline{\mathbb{B}}_n$ are the Beta operators.

7.1 Differences of Positive Linear Operators Using the Taylor Expansion

Using the Taylor expansion with Peano remainder, Gonska et al. [107] obtained the following more general results regarding Lupaş' problem.

Theorem 7.1 ([107]) *Let $A, B : C[0,1] \to C[0,1]$ be positive operators such that*

$$(A - B)\left((e_1 - x)^i; x\right) = 0 \; for \; i = 0, 1, 2, 3 \; and \; x \in [0,1].$$

Then for $f \in C^3[0,1]$ there holds

$$|(A-B)(f;x)| \leq \frac{1}{6}(A+B)\left(|e_1 - x|^3; x\right)\tilde{\omega}\left(f'''; \frac{1}{4}\frac{(A+B)\left((e_1-x)^4;x\right)}{(A+B)\left(|e_1-x|^3;x\right)}\right).$$

V. Gupta et al., *Recent Advances in Constructive Approximation Theory*,
Springer Optimization and Its Applications 138,
https://doi.org/10.1007/978-3-319-92165-5_7

Theorem 7.2 ([107]) *If A and B are given as in Theorem 7.1, satisfying $Ae_0 = Be_0 = e_0$, then for all $f \in C[0,1]$, $x \in [0,1]$ we have*

$$|(A-B)(f;x)| \leq c_1\omega_4\left(f;\sqrt[4]{\frac{1}{2}(A+B)\left((e_1-x)^4;x\right)}\right).$$

Here c_1 is an absolute constant independent of f, x, A, and B.

Using the above result the following solution to Lupaş' problem was given in [107]:

Proposition 7.1 *If S_n and U_n are given as above, then*

$$|(S_n-U_n)(f;x)| \leq c_1\omega_4\left(f;\sqrt[4]{\frac{3x(1-x)}{n(n+1)}}\right).$$

Here c_1 is an absolute constant independent of n, f, and x.

Gonska et al. have continued their research on the differences of positive linear operators by providing estimates for such differences in [102, 106, 107]. The inequality of Theorem 7.1 was generalized to the case in which

$$(A-B)\left((e_1-x)^i;x\right) = 0 \text{ for } i = 0,\ldots,n,\ n \geq 0$$

as follows:

Theorem 7.3 ([102]) *Let $A, B : C[0,1] \to C[0,1]$ be positive operators such that*

$$(A-B)\left((e_1-x)^i;x\right) = 0 \textit{ for } i = 0,1,\ldots,n \textit{ and } x \in [0,1].$$

Then for $f \in C^n[0,1]$ there holds

$$|(A-B)(f;x)| \leq \frac{1}{n!}(A+B)\left(|e_1-x|^n;x\right)\tilde{\omega}\left(f^{(n)};\frac{1}{n+1}\frac{(A+B)\left((e_1-x)^{n+1};x\right)}{(A+B)\left(|e_1-x|^n;x\right)}\right).$$

In the following we shall recall more general statements using the moduli of smoothness that were introduced by Gonska and Kovacheva [100] and Gonska [99].

Lemma 7.1 ([100]) *If $f \in C^q[0,1]$, then for all $0 < h \leq \frac{1}{2}$ there are functions $g \in C^{q+2}[0,1]$, such that*

i) $\|f^{(q)} - g^{(q)}\| \leq \frac{3}{4}\omega_2(f^{(q)};h)$,

ii) $\|g^{(q+1)}\| \leq \frac{5}{h}\omega_1(f^{(q)};h)$,

iii) $\|g^{(q+2)}\| \leq \frac{3}{2h^2}\omega_2(f^{(q)};h)$.

Lemma 7.2 ([99]) *Let $I = [0,1]$ and $f \in C^r(I)$, $r \in \mathbb{N}_0$. For any $h \in (0,1]$ and $s \in \mathbb{N}$ there exists a function $f_{h,r+s} \in C^{2r+s}(I)$ with*

i) $\|f^{(j)} - f^{(j)}_{h,r+s}\| \le c\omega_{r+s}(f^{(j)}; h)$, *for* $0 \le j \le r$,
ii) $\|f^{(j)}_{h,r+s}\| \le ch^{-j}\omega_j(f; h)$, *for* $0 \le j \le r+s$,
iii) $\|f^{(j)}_{h,r+s}\| \le ch^{-(r+s)}\omega_{r+s}(f^{(j-r-s)}; h)$, *for* $r+s \le j \le 2r+s$.

Here the constant c depends only on r and s.

Using the Lemma 7.2 a generalization of Theorem 7.2 was given as follows:

Theorem 7.4 ([102]) *If A and B are as in Theorem 7.3, satisfying*

$$Ae_0 = Be_0 = e_0\,,$$

then for all $f \in C[0,1]$, $x \in [0,1]$ we have

$$|(A-B)(f;x)| \le c_1\omega_{n+1}\left(f;\ \sqrt[n+1]{\frac{1}{2}(A+B)\left(|e_1 - x|^{n+1}x\right)}\right).$$

Here c_1 is an absolute constant independent of f, x, A, and B.

Using the above results, some inequalities for the differences of known positive linear operators were obtained in [102]. Let B_n be the Bernstein operator, $\overline{\mathbb{B}}_n$ the Beta-type operator defined in (5.6.2), $U_n = B_n \circ \overline{\mathbb{B}}_n$ the genuine Bernstein–Durrmeyer operator and $D_n = B_n \circ B_{n+1}$ the composition of two Bernstein operators. In the following we present the estimates of the differences of these operators obtained in [102].

Proposition 7.2 ([102]) *The Bernstein operators and the Beta-type operators satisfy the following properties:*

$$i)|(B_{n+1}-\overline{\mathbb{B}}_n)(f;x)| \le \frac{x(1-x)}{n+1}\tilde{\omega}\left(f'';\ \sqrt{\frac{(n+1)(6nx(1-x)+7)}{18n^2}}\right),\ f \in C^2[0,1]$$

$$\le \frac{x(1-x)}{3n\sqrt{n+1}}\sqrt{\frac{6nx(1-x)+7}{2n}}\|f'''\|,\ f \in C^3[0,1];$$

$$ii)|(B_{n+1}-\overline{\mathbb{B}}_n)(f;x)| \le c\omega_3\left(f;\ \sqrt[3]{\frac{1}{2}(B_{n+1}+\overline{\mathbb{B}}_n)(|e_1-x|^3;x)}\right)$$

$$\le c\omega_3\left(f;\ \sqrt[6]{\frac{x^2(1-x)^2}{n^3}\cdot\frac{6nx(1-x)+7}{n}}\right).$$

Proposition 7.3 ([102]) *The Bernstein operators and the genuine Bernstein–Durrmeyer operators satisfy the following property:*

$$|(B_n - U_n)(f;x)| \le c\omega_2\left(f;\sqrt{\frac{3x(1-x)}{2n}}\right).$$

Proposition 7.4 ([102]) *The composition of two Bernstein operators*

$$D_n = B_n \circ B_{n+1}$$

and the genuine Bernstein–Durrmeyer operators U_n satisfy the following properties:

$$i)|(D_n - U_n)(f;x)| \le \frac{2x(1-x)}{n+1}\tilde{\omega}\left(f'';\sqrt{\frac{(n+1)(8nx(1-x)+13)}{12n^3}}\right), f \in C^2[0,1],$$

$$\le \frac{x(1-x)}{n\sqrt{n+1}}\sqrt{\frac{8nx(1-x)+13}{3n}}\|f'''\|, \; f \in C^3[0,1];$$

$$ii)|(D_n - U_n)(f;x)| \le c\omega_3\left(f;\sqrt[3]{\frac{1}{2}(D_n+U_n)(|e_1-x|^3;x)}\right)$$

$$\le c\omega_3\left(f;\sqrt[6]{\frac{x^2(1-x)^2}{(n+1)n^3}(24nx(1-x)+39)}\right).$$

7.2 Inequalities for Positive Linear Functionals and Applications

In [15] based on some inequalities involving positive linear functionals, new inequalities for such differences in terms of moduli of continuity were obtained. Firstly such inequalities were derived for smooth functions and the norms of their derivatives. Then, using some deep results from [99, 100], inequalities for continuous functions in terms of moduli of smoothness were obtained.

Let $I \subset \mathbb{R}$ be an interval and $E(I)$ a space of real-valued continuous functions on I containing the polynomials. $E_b(I)$ will be the space of all $f \in E(I)$ with

$$\|f\| := \sup\{|f(x)| : x \in I\} < \infty.$$

For $i \in \mathbb{N}$ let $e_i(x) := x^i, x \in I$. Let also $F : E(I) \to \mathbb{R}$ be a positive linear functional such that $F(e_0) = 1$. Set $b^F := F(e_1)$ and

$$\mu_i^F := \frac{1}{i!}F(e_1 - b^F e_0)^i, \; i \in \mathbb{N}.$$

Then

$$\mu_0^F = 1,\ \mu_1^F = 0,\ \mu_2^F = \frac{1}{2}\left(F(e_2) - (b^F)^2\right) \geq 0.$$

Lemma 7.3 ([15]) *Let $f \in E(I)$ with $f'' \in E_b(I)$. Then*

$$|F(f) - f(b^F)| \leq \mu_2^F \|f''\|. \tag{7.2.1}$$

Lemma 7.4 ([15]) *Let $f \in E(I)$ with $f^{IV} \in E_b(I)$. Then*

$$\left|F(f) - f(b^F) - \mu_2^F f''(b^F) - \mu_3^F f'''(b^F)\right| \leq \mu_4^F \|f^{IV}\|. \tag{7.2.2}$$

Proposition 7.5 ([15]) *Let $I = [0, 1]$, $f \in C[0, 1]$, $\lambda \geq 2\sqrt{\mu_2^F} > 0$. Then*

$$\left|F(f) - f(b^F)\right| \leq \frac{3}{2}\omega_2\left(f, \frac{\sqrt{\mu_2^F}}{\lambda}\right)(1 + \lambda^2). \tag{7.2.3}$$

Let K be a set of nonnegative integers and $p_k \in C(I)$, $p_k \geq 0$, $k \in K$, such that $\sum\limits_{k\in K} p_k = e_0$. For each $k \in K$ let $F_k : E(I) \to \mathbb{R}$ and $G_k : E(I) \to \mathbb{R}$ be positive linear functionals such that $F_k(e_0) = G_k(e_0) = 1$. Let $D(I)$ be the set of all $f \in E(I)$ for which

$$\sum_{k\in K} F_k(f)p_k \in C(I) \text{ and } \sum_{k\in K} G_k(f)p_k \in C(I).$$

Consider the positive linear operators

$$V : D(I) \to C(I) \text{ and } W : D(I) \to C(I)$$

defined, for $f \in D(I)$, by

$$V(f; x) := \sum_{k\in K} F_k(f)p_k(x) \text{ and } W(f; x) := \sum_{k\in K} G_k(f)p_k(x).$$

Denote

$$\sigma(x) := \sum_{k\in K}\left(\mu_2^{F_k} + \mu_2^{G_k}\right)p_k(x) \text{ and } \delta := \sup_{k\in K}\left|b^{F_k} - b^{G_k}\right|.$$

Theorem 7.5 ([15]) *Let $f \in D(I)$ with $f'' \in E_b(I)$. Then*

$$|(V - W)(f; x)| \leq \|f''\|\sigma(x) + \omega_1(f, \delta). \tag{7.2.4}$$

Theorem 7.6 ([15]) *Suppose that* $b^{F_k} = b^{G_k} = b_k$, $k \in K$. *Let* $f \in D(I)$ *with* f'', f''', $f^{IV} \in E_b(I)$. *Then for each* $x \in I$,

$$|(V - W)(f; x)| \leq \|f''\|\alpha(x) + \|f'''\|\beta(x) + \|f^{IV}\|\gamma(x), \tag{7.2.5}$$

where

$$\alpha(x) := \sum_{k \in K} |\mu_2^{F_k} - \mu_2^{G_k}| p_k(x),\ \beta(x) := \sum_{k \in K} |\mu_3^{F_k} - \mu_3^{G_k}| p_k(x),$$

$$\gamma(x) := \sum_{k \in K} (\mu_4^{F_k} + \mu_4^{G_k}) p_k(x).$$

Theorem 7.7 ([15]) *Let* $I = [0, 1]$, $f \in C[0, 1]$, $0 < h \leq \frac{1}{2}$, $x \in [0, 1]$. *Then*

$$|(V - W)(f; x)| \leq \frac{3}{2}\left(1 + \frac{\sigma(x)}{h^2}\right)\omega_2(f, h) + \frac{5\delta}{h}\omega_1(f, h). \tag{7.2.6}$$

Theorem 7.8 ([15]) *Let* $I = [0, 1]$, $f \in C[0, 1]$, $0 < h < 1$, $x \in [0, 1]$ *and* $b^{F_k} = b^{G_k}$, $k \in K$. *Then*

$$|(V - W)(f; x)| \leq c\left[\left(2 + \frac{\gamma(x)}{h^4}\right)\omega_4(f, h) + \frac{\beta(x)}{h^3}\omega_3(f, h) + \frac{\alpha(x)}{h^2}\omega_2(f, h)\right], \tag{7.2.7}$$

where c is an absolute constant.

In the following we apply the general results to certain positive linear operators. If we denote

$$F_k(f) := f\left(\frac{k}{n}\right), \quad G_k(f) := (n + 1)\int_0^1 p_{n,k}(t) f(t) dt,$$

the Bernstein operators and the Durrmeyer operators can be written as

$$B_n(f; x) = \sum_{k=0}^{n} F_k(f) p_{n,k}(x), \quad M_n(f; x) = \sum_{k=0}^{n} G_k(f) p_{n,k}(x).$$

Proposition 7.6 ([15]) *For Bernstein operators and Durrmeyer operators the following properties hold:*

i) $|(B_n - M_n)(f; x)| \leq \sigma(x)\|f''\| + \omega_1\left(f, \frac{1}{n+2}\right)$, *for* $f'' \in C[0, 1]$;

ii) $|(B_n - M_n)(f; x)| \leq 3\omega_2(f, \sqrt{\sigma(x)}) + \frac{5}{(n+2)\sqrt{\sigma(x)}}\omega_1\left(f, \sqrt{\sigma(x)}\right)$, *for* $f \in C[0, 1]$,

where

$$\sigma(x) = \frac{1}{2(n+3)(n+2)^2}\{x(1-x)n(n-1)+n+1\} \le \frac{1}{8(n+3)}.$$

Let U_n be the genuine Bernstein–Durrmeyer operators defined in (5.6.3).

Proposition 7.7 ([15]) *The Bernstein operators and the genuine Bernstein–Durrmeyer operators satisfy the following properties*

i) $|(B_n - U_n)(f;x)| \le \sigma(x)\|f''\|, \; f'' \in C[0,1]$,
ii) $|(B_n - U_n)(f;x)| \le 3\omega_2(f, \sqrt{\sigma(x)}), \; f \in C[0,1]$,
where

$$\sigma(x) = \frac{x(1-x)(n-1)}{2n(n+1)} \le \frac{1}{8(n+1)}.$$

Let $D_n := B_n \circ B_{n+1}$ be the composition of two Bernstein operators. Since this mapping has some similarities with U_n, in [15] was calculated the differences of these two operators.

Proposition 7.8 ([15]) *The following properties are satisfied*

i) $|(D_n - U_n)(f;x)| \le \dfrac{n-1}{n(n+1)}x(1-x)\|f''\|, \; f'' \in C[0,1]$;
ii) $|(D_n - U_n)(f;x)| \le \dfrac{x(1-x)}{2(n+1)^2}\left(\dfrac{1}{3}\|f^{(3)}\| + \dfrac{1}{8}\|f^{(4)}\|\right), \; f^{(4)} \in C[0,1]$;
iii) $|(D_n - U_n)(f;x)| \le 3\omega_2\left(f, \sqrt{\dfrac{(n-1)x(1-x)}{n(n+1)}}\right), \; f \in C[0,1]$;
iv) $|(D_n - U_n)(f;x)| \le c\left[\dfrac{33}{16}\omega_4\left(f, \sqrt[4]{\dfrac{x(1-x)}{(n+1)^2}}\right)\right.$
$\left. + \dfrac{\sqrt[4]{x(1-x)}}{6\sqrt{n+1}}\omega_3\left(f, \sqrt[4]{\dfrac{x(1-x)}{(n+1)^2}}\right)\right]$,
$f \in C[0,1]$, *where c is an absolute constant and $n \ge 6$.*

Let K_n be the Kantorovich operators defined in [148] as follows

$$K_n(f;x) = (n+1)\sum_{k=0}^{n} p_{n,k}(x)\int_{\frac{k}{n+1}}^{\frac{k+1}{n+1}} f(t)dt. \tag{7.2.8}$$

Proposition 7.9 ([15]) *The Bernstein operators and the Kantorovich operators satisfy the following properties*

i) $|(K_n - B_n)(f;x)| \le \dfrac{1}{24(n+1)^2}\|f''\| + \omega_1\left(f, \dfrac{1}{2(n+1)}\right), \; f'' \in C[0,1]$;

ii) $|(K_n - B_n)(f;x)| \leq 3\omega_2\left(f, \frac{1}{2\sqrt{6}(n+1)}\right) + 5\sqrt{6}\,\omega_1\left(f, \frac{1}{2\sqrt{6}(n+1)}\right)$, $f \in C[0,1]$.

In [15] this result was extended for a generalized class of Kantorovich-type operators. Let $C_b[0,\infty)$ be the space of all real-valued continuous function on $[0,\infty)$ with $\|f\| < \infty$ and

$$V_n : C_b[0,\infty) \to C_b[0,\infty),\ V_n(f;x) = \sum_{k=0}^{\infty} f\left(\frac{k}{n}\right)\varphi_{n,k}(x)$$

be a positive linear operator, where $(\varphi_{n,k})_{k\geq 0}$ is a sequence of real-valued functions which verify:

i) $\varphi_{n,k}(x) \geq 0,\ k \geq 0,\quad x \in [0,\infty)$,
ii) $\varphi_{n,k} \in C[0,\infty)$;
iii) $\sum_{k=0}^{\infty} \varphi_{n,k}(x) = 1$.

Let $W_n : C_b[0,\infty) \to C_b[0,\infty)$ be the Kantorovich generalized variant of the operator V_n. Therefore,

$$W_n(f;x) = n\sum_{k=0}^{\infty} \varphi_{n,k}(x)\int_{\frac{k}{n}}^{\frac{k+1}{n}} f(t)dt. \tag{7.2.9}$$

Some examples of the operators of the form (7.2.9) are the Kantorovich variants of the Szász–Mirakjan operators and the Baskakov operators. These operators are obtained choosing

$$\varphi_{n,k}(x) = e^{-nx}\frac{(nx)^k}{k!},$$

respectively

$$\varphi_{n,k}(x) = (1+x)^{-n}\binom{n+k-1}{k}\left(\frac{x}{1+x}\right)^k.$$

Proposition 7.10 ([15]) *Using the above notation, we have*

$$|W_n(f;x) - V_n(f;x)| \leq \frac{1}{24n^2}\|f''\| + \omega_1\left(f, \frac{1}{2n}\right),\ f^{(i)} \in C_b[0,\infty),\ i \in \{0,1,2\}.$$

7.3 The Class of Operators U_n^ρ

The class of operators U_n^ρ was introduced in [191] by Păltănea and further investigated by Păltănea and Gonska in [103, 104].

Let $\rho > 0$ and $n \in \mathbb{N}$. The operators $U_n^\rho : C[0, 1] \to \prod_n$ are defined by

$$U_n^\rho(f; x) := \sum_{k=0}^{n} F_k^\rho(f) p_{n,k}(x)$$
$$:= \sum_{k=1}^{n-1} \left(\int_0^1 \frac{t^{k\rho-1}(1-t)^{(n-k)\rho-1}}{B(k\rho, (n-k)\rho)} f(t)dt \right) p_{n,k}(x) + f(0)(1-x)^n + f(1)x^n,$$

for $f \in C[0, 1]$, $x \in [0, 1]$.

Remark 7.1 For $\rho = 1$ and $f \in C[0, 1]$, we obtain the genuine Bernstein–Durrmeyer operators, while for $\rho \to \infty$, for each $f \in C[0, 1]$ the sequence $U_n^\rho(f; x)$ converges uniformly to the Bernstein polynomials $B_n(f; x)$.

In [223] the following result for the images of the monomials under U_n^ρ is proved.

Theorem 7.9 *The images of the monomials under U_n^ρ can be written as*

$$U_n^\rho(e_m) = \frac{1}{(n\rho)_m} \sum_{l=0}^{m} c_{m-l}^{(m)} (n\rho)^l B_n(e_l) \tag{7.3.1}$$

where the coefficients $c_j^{(m)}$, $j = 0, 1, \ldots, m$ are given by the elementary symmetric sums:

$$\begin{aligned} c_0^{(m)} &:= 1, \quad c_m^{(m)} := 0, \\ c_1^{(m)} &= 1 + 2 + \cdots + (m-1) = \frac{m(m-1)}{2}, \\ c_2^{(m)} &= 1 \cdot 2 + 1 \cdot 3 + \cdots + 1 \cdot (m-1) + 2 \cdot 3 + \cdots + (m-2) \cdot (m-1), \\ &\ldots \\ c_{m-1}^{(m)} &= 1 \cdot 2 \cdot 3 \cdot \cdots \cdot (m-1) = (m-1)!. \end{aligned} \tag{7.3.2}$$

The moments of U_n^ϱ were calculated in [104] as follows:

Theorem 7.10 ([104]) *For $x, y \in [0, 1]$, we have*

$$U_n^\varrho(e_0; x) = 1, \quad U_n^\varrho(e_1 - ye_0; x) = x - y$$

and for $r \geq 1$ and $\Psi(x) = x(1-x)$

$$U_n^{\varrho}((e_1 - ye_0)^{r+1}; x) = \frac{\varrho\Psi(x)}{n\varrho + r}(U_n^{\varrho}((e_1 - ye_0)^r; x))_x' +$$
$$+ \frac{(1-2y)r + n\varrho(x-y)}{n\varrho + r}(U_n^{\varrho}((e_1 - ye_0)^r; x)) + \frac{r\Psi(y)}{n\varrho + r}(U_n^{\varrho}((e_1 - ye_0)^{r-1}; x)).$$

Let

$$M_{n,r}^{\varrho}(x) := U_n^{\varrho}((e_1 - xe_0)^r; x), n \geq 1, r \geq 0, x \in [0, 1] .$$

Then

$$(M_{n,r}^{\varrho}(x))' = (U_n^{\varrho}((e_1 - ye_0)^r; x))_x'|_{y=x} - rM_{n,r-1}^{\varrho}(x). \tag{7.3.3}$$

Using (7.3.3) and setting $y = x$ in Theorem 7.10, the following recursion for the central moments can be obtained (see [223]):

Corollary 7.1 *The following relations are true*

$$M_{n,0}^{\varrho}(x) = 1, M_{n,1}^{\varrho}(x) = 0,$$

and for $r \geq 1$

$$M_{n,r+1}^{\varrho}(x) = \frac{r(\varrho+1)\Psi(x)}{n\varrho + r}M_{n,r-1}^{\varrho}(x) + \frac{(1-2x)r}{n\varrho + r}M_{n,r}^{\varrho}(x) + \frac{\varrho\Psi(x)}{n\varrho + r}(M_{n,r}^{\varrho}(x))'. \tag{7.3.4}$$

In particular

$$M_{n,2}^{\varrho}(x) = \frac{(\varrho+1)\Psi(x)}{n\varrho + 1}, \tag{7.3.5}$$

$$M_{n,3}^{\varrho}(x) = \frac{(\varrho+1)(\varrho+2)\Psi(x)\Psi'(x)}{(n\varrho+1)(n\varrho+2)},$$

$$M_{n,4}^{\varrho}(x) = \frac{3\varrho(\varrho+1)^2\Psi^2(x)n}{(n\varrho+1)(n\varrho+2)(n\varrho+3)}$$
$$+ \frac{-6(\varrho+1)(\varrho^2+3\varrho+3)\Psi^2(x) + (\varrho+1)(\varrho+2)(\varrho+3)\Psi(x)}{(n\varrho+1)(n\varrho+2)(n\varrho+3)}.$$

Gonska et al. [103] proved that for for $n \geq 1$ and $f \in C[0, 1]$,

$$\lim_{\varrho \to 0^+} U_n^{\varrho} f = B_1 f, \text{ uniformly on } [0, 1]. \tag{7.3.6}$$

Moreover, the following result was obtained:

Theorem 7.11 ([103]) *For* $U_n^\varrho, 0<\varrho<\infty, n\ge 1$, *we have*

$$|U_n^\varrho f(x)-B_1f(x)|\le \frac{9}{4}\omega_2\left(f;\sqrt{\frac{n\varrho-\varrho}{n\varrho+1}x(1-x)}\right).$$

The following result was obtained with the method presented in [106].

Proposition 7.11 ([200, 223]) *Let* $f\in C[0,1]$, $n\ge 1$, $\rho, r>0$, $x\in[0,1]$. *The following inequality is satisfied*

$$|(U_n^\rho-U_n^r)(f;x)|\le c_1\omega_2\left(f;\sqrt{\frac{1}{2}(U_n^\rho+U_n^r)(|e_1-x|^2;x)}\right)$$

$$\le c_1\omega_2\left(f;\sqrt{\frac{1}{2}\frac{2n\rho r+(n+1)(\rho+r)+2}{(n\rho+1)(nr+1)}x(1-x)}\right).$$

Here c_1 *is an absolute constant independent of* f, x, ρ *and* r.

Another result in this direction was obtained in [200] and [223]:

Theorem 7.12 ([200, 223]) *Let* $f\in C[0,1]$, $n\ge 1$, $\rho, r>0$, $x\in[0,1]$. *Then*

$$|(U_n^\rho-U_n^r)(f;x)|\le \frac{9}{4}\omega_2\left(f;\sqrt{\frac{(n-1)|\rho-r|}{(n\rho+1)(nr+1)}x(1-x)}\right).$$

In the next statement we give some estimates of the difference $U_n^\rho-U_n^r$ using the results proved in the Section 7.2:

Proposition 7.12 *The following properties hold*

i) $|(U_n^\rho-U_n^r)(f;x)|\le \dfrac{(n-1)[2+(\rho+r)n]}{2n(1+\rho n)(1+rn)}x(1-x)\|f''\|,\ f''\in C[0,1];$

ii) $|(U_n^\rho-U_n^r)(f;x)|\le \dfrac{(n-1)|r-\rho|}{2(1+\rho n)(1+rn)}x(1-x)\,\|f''\|$

$$+\frac{1}{3}x(1-x)(n-1)\frac{|r-\rho|[3+n(r+\rho)]}{(1+\rho n)(2+\rho n)(1+rn)(2+rn)}\,\|f'''\|$$

$$+\frac{1}{32}x(1-x)\frac{n^2(\rho^2+r^2)+4n(\rho+r)+6}{(1+\rho n)(3+\rho n)(1+rn)(3+rn)}\|f^{IV}\|,\ f^{(4)}\in C[0,1],$$

$n\rho\ge 6,\quad nr\ge 6;$

iii) $|(U_n^\rho-U_n^r)(f;x)|\le 3\omega_2\left(f,\sqrt{\dfrac{2+(\rho+r)n}{2(1+\rho n)(1+rn)}x(1-x)}\right),\ f\in C[0,1].$

Denote $[A; B] := AB - BA$ the commutator of two positive linear operators A and B. In [200] the following theorem concerning the commutator $[U_n^{\varrho}; U_n^{\sigma}]$ was obtained:

Theorem 7.13 ([200]) *For each $f \in C^6[0, 1]$ one has*

$$\lim_{n\to\infty} n^3(U_n^{\varrho}U_n^{\sigma} - U_n^{\sigma}U_n^{\varrho})f(x) = \frac{(\sigma-\varrho)(\varrho+1)(\sigma+1)}{\varrho^2\sigma^2}x(1-x)f^{(4)}(x),$$

uniformly with respect to $x \in [0, 1]$.

7.4 Discrete Operators Associated with Certain Integral Operators

In 2015, Birou [58] introduced positive linear operators of discrete type associated with the classical Durrmeyer operator using some quadrature formulas with positive coefficients.

The classical Durrmeyer operators are defined as

$$M_n(f; x) = (n+1)\sum_{k=0}^{n} p_{n,k}(x)\int_0^1 p_{n,k}(t)f(t)dt, \; x \in [0, 1],$$

where

$$p_{n,k}(x) = \binom{n}{k}x^k(1-x)^{n-k}$$

and f is an integrable function on $[0, 1]$.

Approximating the Riemann integrals by the following quadrature formula

$$(n+1)\int_0^1 p_{n,k}(t)f(t)dt \simeq \sum_{j=0}^{m} A_{j,m}^{n,k} f(t_{j,m}^{n,k}), \; k = 0, 1, \ldots, n, \tag{7.4.1}$$

with $A_{j,m}^{n,k} \geq 0, \; j = 0, \ldots, m, \; k = 0, \ldots, n$, the associated discrete operator is obtained (see [58])

$$D_{n,m}(f; x) = \sum_{k=0}^{n} p_{n,k}(x)\sum_{j=0}^{m} A_{j,m}^{n,k} f(t_{j,m}^{n,k}), \; x \in [0, 1].$$

If

$$R^k_{n,m}(f) = (n+1)\int_0^1 p_{n,k}(t)f(t)dt - \sum_{j=0}^{m} A^{n,k}_{j,m} f(t^{n,k}_{j,m}),\ k = 0, 1, \ldots, n,$$

is the remainder term of the quadrature formula (7.4.1), then

$$|(M_n - D_{n,m})(f;x)| \le \max_{k\in\{0,\ldots,n\}} |R^k_{n,m}(f)|,\ x \in [0,1].$$

Using Korovkin theorem the following result follows immediately:

Theorem 7.14 ([58]) *If the quadrature formula (7.4.1) has the degree of exactness at least two, i.e.,* $R^k_{n,m}(e_i) = 0, i = 0, 1, 2,$ *then the sequence* $(D_{n,m}f)_{n\ge 1}$ *converges uniformly to the function* f*, for every* $f \in C[0,1]$.

Using quadrature formulas of Gauss–Jacobi type the following associated discrete operators can be obtained (see [58])

$$D^{GJ}_{n,m}(f;x) = \frac{n+1}{2^{n+1}} \sum_{k=0}^{n} p_{n,k}(x)\binom{n}{k} \sum_{j=0}^{m} B^{n,k}_{j,m} f\left(\frac{1+u^{n,k}_{j,m}}{2}\right),\ x \in [0,1],$$

where $u^{n,k}_{j,m}$, $j = 0, \ldots, m$, $k = 0, \ldots, n$ are the roots of the Jacobi orthogonal polynomial of degree $m+1$, and the coefficients are given as

$$B^{n,k}_{j,m} = \frac{2^n(2m+n+2)(m+k)!(m+n-k)!}{(m+1)!(m+n+1)!J^{(k,n-k)}_m(u^{n,k}_{j,m})\frac{d}{du}\left[J^{(k,n-k)}_{m+1}(u)\right]_{u=u^{n,k}_{j,m}}},$$

for $j = 0, \ldots m$ and $k = 0, \ldots, n$.

Using quadrature formulas of Gauss Legendre type the following associated discrete operators can be obtained (see [58])

$$D^{GL}_{n,m}(f;x) = \frac{n+1}{2^{n+1}} \sum_{k=0}^{n} p_{n,k}(x)\binom{n}{k} \sum_{j=0}^{m} B_{j,m}(1+u_{j,m})^k(1-u_{j,m})^{n-k} f\left(\frac{1+u_{j,m}}{2}\right),$$

$x \in [0,1]$, where the nodes $u_{j,m}$, $j = 0, \ldots, m$ are roots of the Legendre orthogonal polynomial $J^{(0,0)}_{m+1}(u)$ and the coefficients are given by

$$B_{j,m} = \frac{2}{(m+1)J^{(0,0)}_m(u_{j,m})\frac{d}{du}\left[J^{(0,0)(u)}_{m+1}\right]_{u=u_{j,m}}},$$

for $j = 0, \ldots, m$ and $k = 0, \ldots, n$.

Let $L_n : C[0,1] \to C[0,1]$, $n \geq 1$ be a linear positive operator, defined as follows

$$L_n(f;t) = \sum_{j=0}^{n} w_{n,j}(t) f(t_{j,n}),\ f \in C[0,1],\ t \in [0,1],$$

where $w_{n,j} \in C[0,1]$, $w_{n,j} \geq 0$.

Using the quadrature generated by some positive linear operators, Biro [58] introduced the following associated discrete operator

$$D_n^{PL}(f;x) = \sum_{k=0}^{n} p_{n,k}(x) \sum_{j=0}^{n} A_j^{n,k} f(t_{j,n}),\ x \in [0,1]$$

where

$$A_j^{n,k} = (n+1) \int_0^1 p_{n,k}(t) w_{n,j}(t) dt,\ j,k = 0,\ldots,n.$$

Theorem 7.15 ([58]) *If the sequence $(L_n f)_{n \geq 1}$ converges uniformly to the function $f \in C[0,1]$, then the sequence $(D_n^{PL} f)_{n \geq 1}$ converges uniformly to the function f and*

$$|(D_n^{PL} - M_n)(f;x)| \leq \sup_{x \in [0,1]} |R_n(f;x)|, x \in [0,1],$$

where

$$R_n(f;x) = f(x) - L_n(f;x).$$

In 2011, Raşa [199] constructed discrete operators associated with certain integral operators using a probabilistic approach.

Let $I_n : C[a,b] \to C[a,b]$, $n \geq 1$, be a sequence of positive linear operators of the form

$$I_n(f;x) = \sum_{k=0}^{n} h_{n,k}(x) A_{n,k}(f),\ f \in C[a,b],\ x \in [a,b],$$

where $h_{n,k} \in C[a,b]$, $h_{n,k} \geq 0$, and

$$A_{n,k}(f) = \int_a^b f(t) d\mu_{n,k}(t),$$

with $\mu_{n,k}$ probability Borel measures on $[a,b]$, $n \geq 1$, $k = 0, 1, \ldots, n$.

In [199] the following discrete operator was introduced, associated with the sequence (I_n)

$$D_n(f;x) = \sum_{k=0}^{n} h_{n,k}(x) f(x_{n,k}),$$

where

$$x_{n,k} = \int_a^b t d\mu_{n,k}(t) .$$

For example, the associated operators of the genuine Bernstein–Durrmeyer operators are Bernstein operators.

Mache introduced the sequence of positive linear operators (see [160, 161])

$$P_n(f;x) := \sum_{k=0}^{n} p_{n,k}(x) T_{n,k}(f),\ n \geq 1,$$

where

$$T_{n,k}(f) := \frac{\int_0^1 f(t) t^{ck+a}(1-t)^{c(n-k)+b} dt}{B(ck+a+1, c(n-k)+b+1)},$$

for $a, b > -1$, $\alpha \geq 0$, $c := c_n := [n^\alpha]$ and B is the Beta function.

Remark 7.2 ([199]) The sequence of positive linear operators (P_n) represents a link between the Durrmeyer operators with Jacobi weights ($\alpha = 0$) and the Bernstein operators ($\alpha \to \infty$).

Raşa [199] considered the sequence (P_n) of positive linear operators (V_n), defined by

$$V_n(f;x) := \sum_{k=0}^{n} p_{n,k}(x) f\left(\frac{ck+a+1}{cn+a+b+2}\right),\ f \in C[0,1],\ x \in [0,1].$$

Remark 7.3 ([199]) For $a = b = -1$, or $\alpha \to \infty$, we get the classical Bernstein operators. For $\alpha = 0$, the operators V_n reduce to the operators considered by Stancu in [222].

Theorem 7.16 ([199]) *For $n \geq 1$, $x \in [0,1]$, and $f \in C^2[0,1]$ we have*

$$|P_n(f;x) - V_n(f;x)| \leq$$

$$\frac{c^2 n(n-1)x(1-x) + cn(b-a)x + cn(a+1) + (a+1)(b+1)}{2(cn+a+b+2)^2(cn+a+b+3)} \|f''\|.$$

Chapter 8
Bivariate Operators of Discrete and Integral Type

8.1 Bivariate Operators of Bernstein Type

In [204] problems concerning polynomials of one and several variables and their applications have been discussed. If $f(x)$ is a bounded function on the interval $[0, 1]$, then the mth degree Bernstein polynomial is defined as

$$B_m(f; x) = \sum_{i=0}^{m} \lambda_{m,i}(x) f\left(\frac{i}{m}\right),$$

where

$$\lambda_{m,i}(x) = \binom{m}{i} x^i (1-x)^{m-i}, x \in [0, 1].$$

In 1951, Kingsley [151] initiated the study of approximation of functions of two variables by defining the Bernstein polynomials for functions of two variables of class $C^{(k)}$. Let us assume that $\phi(x, y)$ is a continuous function in a closed region $R : 0 \le x \le 1, 0 \le y \le 1$. Then, the Bernstein polynomials $B_{m,n}(x, y)$ associated with the function $\phi(x, y)$ is given by:

$$B_{m,n}(\phi; x, y) = \sum_{p=0}^{n} \sum_{q=0}^{m} \phi\left(\frac{p}{n}, \frac{q}{m}\right) \lambda_{n,p}(x) \lambda_{m,q}(y). \tag{8.1.1}$$

For fixed $m \in \mathbb{N}$, let $C^m(R)$ denote the space of all continuous functions f whose partial derivatives $\dfrac{\partial^k f}{\partial x^s \partial y^{k-s}}$ be continuous in $\mathbb{R}$, for all $s = 1, 2, \ldots, k; k = 1, 2, \ldots, m$.

V. Gupta et al., *Recent Advances in Constructive Approximation Theory*, Springer Optimization and Its Applications 138,
https://doi.org/10.1007/978-3-319-92165-5_8

It is easily seen that

$$\sum_{p=0}^{n} \lambda_{n,p}(x) = 1 \quad \text{and} \quad \sum_{p=0}^{n} (nx - p)^2 \lambda_{n,p}(x) = nx(1-x).$$

If, for $k \geq 0, i = 0, 1, 2 \ldots k$, we define

$$A_{p,q}^{i,k-i} = \sum_{\alpha=0}^{i} \sum_{\beta=0}^{k-i} (-1)^{\alpha+\beta} \binom{i}{\alpha} \binom{k-i}{\beta} \phi\left(\frac{p+(i-\alpha)}{n}, \frac{q+(k-i-\alpha)}{m}\right),$$

then by mathematical induction, the following lemmas hold.

Lemma 8.1 ([151]) *If $0 \leq i \leq k$, $i \leq n$, $k \leq m$ and $x, y \in R$, then the kth partial derivatives of Bernstein polynomials (8.1.1) are given by*

$$B_{m,n}^{(i,k-i)}(\phi; x, y) = \frac{n!m!}{(n-i)!(m-k+i)!} \sum_{p=0}^{n-i} \sum_{q=0}^{m-k+i} A_{p,q}^{i,k-i} \lambda_{n-i,p} \lambda_{m-k+i,q}.$$

Lemma 8.2 ([151]) *If $0 \leq i \leq k, 0 \leq p \leq n-i, 0 \leq q \leq m-k+i$, and if $\phi(x, y)$ is of class $C^{(k)}$ for x and y in R, then there exist two real numbers $\xi = \xi(p), \gamma = \gamma(q)$ such that $0 < \xi < 1, 0 < \gamma < 1$ and such that*

$$A_{p,q}^{i,k-i} = \frac{1}{n^i m^{k-i}} \phi^{(i,k-i)}\left(\frac{p+\xi i}{n}, \frac{q+\gamma(k-i)}{m}\right).$$

Lemma 8.3 ([151]) *For fixed x and y in R and for fixed positive integers M and N, let d be arbitrary positive number and let $a(p, q)$ be a quantity dependent upon p and q and such that*

$$|a(p,q)| \leq \pi_1 \quad for \quad \left|x - \frac{p}{N}\right| \leq d \quad and \quad \left|y - \frac{q}{M}\right| \leq d,$$

$$|a(p,q)| \leq \pi_2 \quad for \quad \left|x - \frac{p}{N}\right| > d \quad and \quad \left|y - \frac{q}{M}\right| > d.$$

Furthermore, assume that it is possible to split off from $a(p, q)$ terms $a'(p)$ independent of q or terms $b'(q)$ independent of p, that is

$$a(p,q) = a''(p,q) + a'(p) = b''(p,q) + b'(q)$$

such that

$$|a''(p,q)| \leq \pi_3 \quad for \quad \left|x - \frac{p}{N}\right| \leq d \quad and \quad \left|y - \frac{q}{M}\right| > d,$$

$$|a'(p,q)| \leq \pi_4 \quad for \quad \left|x - \frac{p}{N}\right| \leq d.$$

$$|b''(p,q)| \leq \pi_5 \quad for \quad \left|x-\frac{p}{N}\right|>d \quad and \quad \left|y-\frac{q}{M}\right|\leq d,$$

$$|b'(q)| \leq \pi_6 \quad for \quad \left|y-\frac{q}{M}\right|\leq d.$$

Then,

$$\left|\sum_{p=0}^{N}\sum_{q=0}^{M} a(p,q)\lambda_{N,p}\lambda_{M,q}\right| \leq \pi_1+\pi_4+\pi_6+\frac{\pi_2(M+N)}{8MNd^2}+\frac{\pi_3}{4Md^2}+\frac{\pi_5}{4Nd^2}.$$

Theorem 8.1 ([151]) *If $\phi(x,y)$ is of class $C^{(k)}(R)$, then*

$$\lim_{m,n\to\infty} B_{m,n}^{(s,k-s)}(\phi;x,y)=\phi^{(s,k-s)}(x,y)$$

and the convergence is uniform in R.

In the year 2008, Pop [198] obtained the rate of convergence by means of the modulus of continuity and proved the Voronovskaja type asymptotic theorem for the bivariate case of Bernstein polynomials :

Theorem 8.2 *Let $f:[0,1]\times[0,1]\to\mathbb{R}$ be a bivariate function.*

(i) If $(x,y)\in[0,1]\times[0,1]$ and f admits partial derivatives of second order continuous in a neighborhood of the point (x,y), then

$$\lim_{m\to\infty} m(B_{m,m}(f;x,y)-f(x,y))=\frac{x(1-x)}{2}f''_{xx}(x,y)+\frac{y(1-y)}{2}f''_{yy}(x,y).$$

If f admits partial derivatives of second order continuous on $[0,1]\times[0,1]$, then the convergence is uniform on $[0,1]\times[0,1]$.

(ii) If f is continuous on $[0,1]\times[0,1]$, then

$$|B_{m,m}(f;x,y)-f(x,y)|\leq\frac{25}{16}\omega\left(f;\frac{1}{\sqrt{m}},\frac{1}{\sqrt{m}}\right)$$

for any $(x,y)\in[0,1]\times[0,1]$, any $m\in\mathbb{N}$.

Stancu [220] introduced bivariate Bernstein polynomials on the triangle

$$\Delta:=S=\{(x,y):x+y\leq 1,\ 0\leq x,y\leq 1\}.$$

For the functions $f:S\to\mathbb{R}$ and $(x,y)\in S$, Stancu considered the bivariate Bernstein operators $G_n(f;x,y)$ as follows:

$$G_n(f; x, y) = \sum_{i=0}^{m} \sum_{j=0}^{m-i} \binom{m}{i} \binom{m-i}{j} x^i y^j (1 - x - y)^{m-i-j} f\left(\frac{i}{m}, \frac{j}{m}\right).$$

In 1989, Martinez [169] obtained the order of approximation of two-dimensional Bernstein polynomials. He established the convergence of the polynomials with integral coefficients $B_{m,n}^{(i,k-i),e} f$ to $f_{i,k-i}^{(k)}$ both in the uniform and L_p norms where the superscript "e" denotes a polynomial with integral coefficients in the above sense and $\| \cdot \|$ will denote the $L_p[\mathbb{R}]$ norm $(1 \leq p \leq \infty)$.

For $f \in C(\mathbb{R})$, the complete modulus of continuity for the bivariate case is defined as follows:

$$\bar{\omega}(f; \delta, \epsilon) = \sup_{(t,s),(x,y) \in I} \{|f(t, s) - f(x, y)| : |t - x| \leq \delta, \ |s - y| \leq \epsilon\},$$

where $\bar{\omega}(f, \delta, \epsilon)$ satisfies the following properties:

1. $\bar{\omega}(f, \delta, \epsilon) \to 0$, if $\delta \to 0$ and $\epsilon \to 0$;
2. $|f(t, s) - f(x, y)| \leq \bar{\omega}(f, \delta, \epsilon) \left(1 + \frac{|t - x|}{\delta_1}\right) \left(1 + \frac{|s - y|}{\delta_2}\right)$.

The details of the complete modulus of continuity for the bivariate case can be found in [30].

Furthermore, the partial moduli of continuity of f are defined as:

$$\omega^{(1)}(f; \delta) = \omega(f; \delta, 0) = \sup_{y} \sup_{|x_1 - x_2| \leq \delta} |f(x_1, y) - f(x_2, y)|$$

$$\omega^{(2)}(f; \delta) = \omega(f; 0, \delta) = \sup_{x} \sup_{|y_1 - y_2| \leq \delta} |f(x, y_1) - f(x, y_2)|.$$

It is known that the complete modulus of continuity $\omega(f; \delta, \epsilon)$ of the function f is connected with its partial moduli of continuity $\omega(f; \delta, 0)$ and $\omega(f; 0, \epsilon)$ by the inequalities

$$\omega(f; \delta, \epsilon) \leq \omega(f; \delta, 0) + \omega(f; 0, \epsilon) \leq 2\omega(f; \delta, \epsilon).$$

First, we state some basic lemmas which are required to prove the main theorem:

Lemma 8.4 *If f is continuous in R and $B_{m,n}(f; x)$ is the Bernstein polynomial of f, then*

$$\begin{aligned} |B_{m,n}(f; x, y) - f(x, y)| &\leq \frac{3}{2} \left(\omega^{(1)}(f; n^{-1/2}) + \omega^{(2)}(f; m^{-1/2})\right) \\ &\leq 3\omega(f; n^{-1/2}, m^{-1/2}), \end{aligned}$$

where $\omega^{(i)}(f; \delta), i = 1, 2$, are the partial moduli of continuity of f.

We now define the polynomial

$$\overline{B}_{m,n}^{(i,k-i)} f := \frac{(m-i)!}{m!} \frac{(n-k+i)!}{n!} m^i n^{k-i} B_{m,n}^{(i,k-i)} f$$

and let $\overline{B}_{m,n}^{(i,k-i),e} f$ be its corresponding polynomial with integral coefficients.

Lemma 8.5 *Let f be a function such that its partial derivatives of the first k orders exist and are continuous in R.*

1. If $1 \leq p < \infty$, then

$$\left\| f_{i,k-i}^{(k)} - \overline{B}_{m,n}^{(i,k-i),e} f \right\|_p \leq A + B + O((mn)^{-1/p})$$

where

$$A = 3\,\omega(f_{i,k-i}^{(k)}; (m-i)^{-1/2}, (n-k+i)^{-1/2}),$$

$$B = \omega(f_{i,k-i}^{(k)}; i/m, (k-i)/n);$$

2. If $p = \infty$ and the numbers $m^i n^{(k-i)} \Delta_{x,m^{-1}}^i \Delta_{y,n^{-1}}^{k-i} f(u_1, u_2)$ are integers, where u_1 is either zero or $(m-i)/m$ and u_2 is either zero or $(n-k+i)/n$, then

$$\left\| f_{i,k-i}^{(k)} - \overline{B}_{m,n}^{(i,k-i),e} f \right\|_\infty \leq A + B + O((mn)^{-1})$$

where A and B are as in (a).

Theorem 8.3 *Let* $\max(q, r-q) > 1$, $n_1 = \min(m, n)$, *and let f be a function such that its partial derivatives of the first k orders exist and are continuous in R.*

1. If $1 \leq p < \infty$, then

$$\left\| f_{i,k-i}^{(k)} - \overline{B}_{m,n}^{(i,k-i),e} f \right\|_p \leq A + B + O((n_1)^{-\lambda(p)}),$$

where A and B are as in Lemma 8.5 and

$$\lambda(p) = \begin{cases} 1\ , & if\ p = 1 \\ 2/p\ , if & 2 \leq p < \infty. \end{cases}$$

2. If $p = \infty$ and the numbers

$$\frac{m!n!}{(m-i)!(n-k+i)!} \Delta_{x,m^{-1}}^i \Delta_{y,n^{-1}}^{k-i} f(u_1, u_2)$$

are integers, where u_1 is either zero or $(m-i)/m$, and u_2 is either zero or $(n-k+i)/n$, then

$$\left\| f^{(k)}_{i,k-i} - \overline{B}^{(i,k-i),e}_{m,n} f \right\|_\infty \leq A + B + O((n_1)^{-1}),$$

where A and B are as in Lemma 8.5.

8.1.1 Bivariate Case of q-Bernstein–Schurer–Stancu

Let p be a nonnegative integer and α, β be real parameters satisfying $0 \leq \alpha \leq \beta$. In 2003, Bărbosu [49] defined the Bernstein–Schurer–Stancu operators as demonstrated below:

$\tilde{S}^{\alpha,\beta}_{m,p} : C[0, 1+p] \to C[0, 1]$ given by

$$\tilde{S}^{\alpha,\beta}_{m,p}(f; x) = \sum_{k=0}^{m+p} \tilde{p}_{m,k}(x) f\left(\frac{k+\alpha}{m+\beta}\right), \tag{8.1.2}$$

where

$$\tilde{p}_{m,k}(x) = \binom{m+p}{k} x^k (1-x)^{m+p-k}, x \in [0, 1]$$

are the fundamental Bernstein–Schurer polynomials . In particular, the operators (8.1.2) include the Bernstein–Stancu operator for $p = 0$, the Bernstein–Schurer operator $\alpha = \beta = 0$, and the Bernstein polynomial for $p = 0$ and $\alpha = \beta = 0$.

Subsequently, in [52] Bărbosu and Pop introduced bivariate Bernstein–Schurer–Stancu operators as

$$\tilde{S}^{\alpha_1,\beta_1,\alpha_2,\beta_2}_{m,n,p_1,p_2}(f; x, y) = \sum_{k=0}^{m+p_1} \sum_{j=0}^{n+p_2} \tilde{p}_{m,k}(x) \tilde{p}_{n,j}(y) f\left(\frac{k+\alpha_1}{m+\beta_1}, \frac{j+\alpha_2}{n+\beta_2}\right)$$

and studied the degree of approximation of these operators. Recently, Bărbosu and Muraru [55] defined the q-Bernstein–Schurer–Stancu operators for the bivariate case as follows:

Let p_1, p_2 be nonnegative integers such that,

$$I = [0, 1+p_1] \times [0, 1+p_2] \text{ and } J = [0, 1] \times [0, 1].$$

Let $\{q_m\}$ and $\{q_n\}$ be sequences in $(0, 1)$ such that

$$q_m \to 1, \; q_m^m \to a \,(0 \leq a < 1), \; \text{as } m \to \infty$$

and

$$q_n \to 1,\ q_n^n \to b\,(0 \le b < 1),\ \text{as } n \to \infty.$$

Further, let $0 \le \alpha_1 \le \beta_1$, $0 \le \alpha_2 \le \beta_2$ and $S_{m,n,p_1,p_2}^{(\alpha_1,\beta_1,\alpha_2,\beta_2)} : C(I) \to C(J)$, then for any $f \in C(I)$ we have

$$\begin{aligned} &S_{m,n,p_1,p_2}^{(\alpha_1,\beta_1,\alpha_2,\beta_2)}(f; q_m, q_n, x, y) \\ &\quad = \sum_{k_1=0}^{m+p_1}\sum_{k_2=0}^{n+p_2} \begin{bmatrix} m+p_1 \\ k_1 \end{bmatrix}_{q_m} \begin{bmatrix} n+p_2 \\ k_2 \end{bmatrix}_{q_n} \prod_{s=0}^{m+p_1-k_1-1}(1-q_m^s x) \\ &\quad\quad \times \prod_{r=0}^{n+p_2-k_2-1}(1-q_n^r y) x^{k_1} y^{k_2} f_{k_1,k_2}, \end{aligned} \tag{8.1.3}$$

where

$$f_{k_1,k_2} = f\left(\frac{[k_1]_{q_m}+\alpha_1}{[m]_{q_m}+\beta_1}, \frac{[k_2]_{q_n}+\alpha_2}{[n]_{q_n}+\beta_2}\right).$$

In order to study the approximation properties of the operators defined by (8.1.3), the following lemmas are necessary.

Lemma 8.6 ([55]) *Let $e_{i.j} : I \to \mathbb{R}, e_{i,j}(x,y) = x^i y^j (0 \le i+j \le 2, i, j(integers))$ be the test functions. Then the following equalities hold for the operators given by (8.1.3):*

1. $S_{m,n,p_1,p_2}^{(\alpha_1,\beta_1,\alpha_2,\beta_2)}(e_{0,0}; q_m, q_n, x, y) = e_{0,0}(x,y)$,
2. $S_{m,n,p_1,p_2}^{(\alpha_1,\beta_1,\alpha_2,\beta_2)}(e_{1,0}; q_m, q_n, x, y) = \frac{[m+p_1]_{q_m}x+\alpha_1}{[m]_{q_m}+\beta_1}$,
3. $S_{m,n,p_1,p_2}^{(\alpha_1,\beta_1,\alpha_2,\beta_2)}(e_{0,1}; q_m, q_n, x, y) = \frac{[n+p_2]_{q_n}y+\alpha_2}{[n]_{q_n}+\beta_2}$,
4. $S_{m,n,p_1,p_2}^{(\alpha_1,\beta_1,\alpha_2,\beta_2)}(e_{2,0}; q_m, q_n, x, y) = \frac{1}{([m]_{q_m}+\beta_1)^2}\Big([m+p_1]_{q_m}^2 x^2 + [m+p_1]_{q_m} x(1-x) + 2\alpha_1[m+p_1]_{q_m}x + \alpha_1^2\Big)$,
5. $S_{m,n,p_1,p_2}^{(\alpha_1,\beta_1,\alpha_2,\beta_2)}(e_{0,2}; q_m, q_n, x, y) = \frac{1}{([n]_{q_n}+\beta_2)^2}\Big([n+p_2]_{q_n}^2 y^2 + [n+p_2]_{q_n} y(1-y) + 2\alpha_2[n+p_2]_{q_n}y + \alpha_2^2\Big)$.

Lemma 8.7 ([216]) *For $(x, y) \in J$, we have*

1. $S_{m,n,p_1,p_2}^{(\alpha_1,\beta_1,\alpha_2,\beta_2)}((t-x)^2; q_m, q_n, x, y) = \frac{1}{([m+\beta_1]_{q_m})^2}\{((q_m^m[p_1]_{q_m}-\beta_1)x+\alpha_1)^2 + [m+p_1]_{q_m}x(1-x)\}$,

2. $S_{m,n,p_1,p_2}^{(\alpha_1,\beta_1,\alpha_2,\beta_2)}((s-y)^2; q_m, q_n, x, y) = \frac{1}{([n+\beta_2]_{q_n})^2}\{((q_n^n[p_2]_{q_n} - \beta_2)y + \alpha_2)^2 + [n+p_2]_{q_n} y(1-y)\}$.

Lemma 8.8 ([216]) *For* $(x, y) \in J$, *we have*

1. $\lim_{m\to\infty} [m]_{q_m} S_{m,n,p_1,p_2}^{(\alpha_1,\beta_1,\alpha_2,\beta_2)}((t-x); q_m, q_n, x, y) = \alpha_1 - \beta_1 x,$
2. $\lim_{n\to\infty} [n]_{q_n} S_{m,n,p_1,p_2}^{(\alpha_1,\beta_1,\alpha_2,\beta_2)}((s-y); q_m, q_n, x, y) = \alpha_2 - \beta_2 y,$
3. $\lim_{m\to\infty} [m]_{q_m} S_{m,n,p_1,p_2}^{(\alpha_1,\beta_1,\alpha_2,\beta_2)}((t-x)^2; q_m, q_n, x, y) = x(1-x),$
4. $\lim_{n\to\infty} [n]_{q_n} S_{m,n,p_1,p_2}^{(\alpha_1,\beta_1,\alpha_2,\beta_2)}((s-y)^2; q_m, q_n, x, y) = y(1-y).$

Similarly,

$$S_{m,n,p_1,p_2}^{(\alpha_1,\beta_1,\alpha_2,\beta_2)}((t-x)^4; q_m, q_n, x, y) = O\left(\frac{1}{[m]_{q_m}^2}\right), \quad \text{as} \quad m \to \infty \quad (8.1.4)$$

uniformly in $x \in [0, 1]$, and

$$S_{m,n,p_1,p_2}^{(\alpha_1,\beta_1,\alpha_2,\beta_2)}((s-y)^4; q_m, q_n, x, y) = O\left(\frac{1}{[n]_{q_n}^2}\right), \quad \text{as} \quad n \to \infty \quad (8.1.5)$$

uniformly in $y \in [0, 1]$.

Let $C^2(I) := \{f \in C(I) : f_{xx}, f_{xy}, f_{yx}, f_{yy} \in C(I)\}$. The norm on the space $C^2(I)$ is defined as

$$\|f\|_{C^{(2)}(I)} = \|f\| + \sum_{i=1}^{2}\left(\left\|\frac{\partial^i f}{\partial x^i}\right\| + \left\|\frac{\partial^i f}{\partial y^i}\right\|\right).$$

For $f \in C(I)$, let us consider the following Peetre's K-functional :

$$K_2(f, \delta) = inf\{||f - g|| + \delta||g||_{C^2(I)} : g \in C^2(I)\},$$

where $\delta > 0$.

By [78], there exists an absolute constant $C > 0$ such that

$$K_2(f, \delta) \le C\bar{\omega_2}(f, \sqrt{\delta}),$$

where $\bar{\omega_2}(f, \sqrt{\delta})$ denotes the second order modulus of continuity for the bivariate case.

Let δ_m and δ_n be defined as

$$\delta_m = \max_{x\in[0,1]}\left\{S_{m,p_1}^{(\alpha_1,\beta_1)}\left((t-x)^2; q_m, x\right)\right\}^{1/2}$$

$$= \frac{1}{[m]_{q_m} + \beta_1} \sqrt{4 \max_{x\in[0,1]} \left(((q_m^m [p_1]_{q_m} - \beta)x + \alpha_1)^2 + [m + p_1]_{q_m} \right)},$$

and

$$\delta_n = \max_{y\in[0,1]} \{S_{n,p_2}^{(\alpha_2,\beta_2)}((s-y)^2; q_n, y\}^{1/2}$$

$$= \frac{1}{[n]_{q_n} + \beta_2} \sqrt{4 \max_{y\in[0,1]} \left(((q_n^n [p_2]_{q_n} - \beta_2)y + \alpha_2)^2 + [n + p_2]_{q_n} \right)}.$$

Theorem 8.4 *Let $f \in C(I)$. Then we have the inequality*

$$\| S_{m,n,p_1,p_2}^{(\alpha_1,\beta_1,\alpha_2,\beta_2)}(f; q_m, q_n, ., .) - f \| \leq 2\left(\omega_1(f;\delta_m) + \omega_2(f;\delta_n)\right).$$

Proof Applying Lemma 8.6 and Cauchy–Schwarz inequality, we have

$$\left| S_{m,n,p_1,p_2}^{(\alpha_1,\beta_1,\alpha_2,\beta_2)}(f; q_m, q_n, x, y) - f(x,y) \right|$$

$$\leq S_{m,n,p_1,p_2}^{(\alpha_1,\beta_1,\alpha_2,\beta_2)}(|f(t,s) - f(x,y)|; q_m, q_n, x, y)$$

$$\leq S_{m,n,p_1,p_2}^{(\alpha_1,\beta_1,\alpha_2,\beta_2)}(|f(t,s) - f(t,y)|; q_m, q_n, x, y)$$
$$+ S_{m,n,p_1,p_2}^{(\alpha_1,\beta_1,\alpha_2,\beta_2)}(|f(t,y) - f(x,y)|; q_m, q_n, x, y)$$

$$\leq S_{m,n,p_1,p_2}^{(\alpha_1,\beta_1,\alpha_2,\beta_2)}(\omega_2(f; |s-y|); q_m, q_n, x, y)$$
$$+ S_{m,n,p_1,p_2}^{(\alpha_1,\beta_1,\alpha_2,\beta_2)}(\omega_1(f; |t-x|); q_m, q_n, x, y)$$

$$\leq \omega_2(f;\delta_n)\left[1 + \frac{1}{\delta_n} S_{m,n,p_1,p_2}^{(\alpha_1,\beta_1,\alpha_2,\beta_2)}(|s-y|; q_m, q_n, x, y)\right]$$
$$+ \omega_1(f;\delta_m)\left[1 + \frac{1}{\delta_m} S_{m,n,p_1,p_2}^{(\alpha_1,\beta_1,\alpha_2,\beta_2)}(|t-x|; q_m, q_n, x, y)\right]$$

$$\leq \omega_2(f;\delta_n)\left[1 + \frac{1}{\delta_n}\left(S_{m,n,p_1,p_2}^{(\alpha_1,\beta_1,\alpha_2,\beta_2)}((s-y)^2; q_m, q_n, x, y)\right)^{1/2}\right]$$
$$+ \omega_1(f;\delta_m)\left[1 + \frac{1}{\delta_m}\left(S_{m,n,p_1,p_2}^{(\alpha_1,\beta_1,\alpha_2,\beta_2)}((t-x)^2; q_m, q_n, x, y)\right)^{1/2}\right]$$

$$\leq \omega_2(f;\delta_n)\left(1 + \frac{1}{\delta_n}\frac{1}{[n]_{q_n} + \beta_2}\left\{4\max_{y\in[0,1]}\left(((q_n^n [p_2]_{q_n} - \beta_2)y + \alpha_2)^2\right.\right.\right.$$
$$\left.\left.\left. + [n + p_2]_{q_n}\right)\right\}^{1/2}\right)$$

$$+\omega_1(f;\delta_m)\left(1+\frac{1}{\delta_m}\frac{1}{[m]_{q_m}+\beta_1}\left\{4\max_{x\in[0,1]}\left(((q_m^m[p_1]_{q_m}-\beta_1)x+\alpha_1)^2\right.\right.\right.$$
$$\left.\left.\left.+[m+p_1]_{q_m}\right)\right\}^{1/2}\right).$$

Hence, we have proved the desired result. ■

Theorem 8.5 *Let $f\in C(I)$ and $0<q_m,q_n<1$. Then for all $(x,y)\in J$, we have*

$$\left\|S_{m,n,p_1,p_2}^{(\alpha_1,\beta_1,\alpha_2,\beta_2)}(f;q_m,q_n,.,.)-f\right\|\leq 4\bar{\omega}(f,\delta_m,\delta_n).$$

Proof Using the linearity and positivity of the operator

$$S_{m,n,p_1,p_2}^{(\alpha_1,\beta_1,\alpha_2,\beta_2)}(f;q_m,q_n,x,y),$$

we have

$$\left|S_{m,n,p_1,p_2}^{(\alpha_1,\beta_1,\alpha_2,\beta_2)}(f;q_m,q_n,x,y)-f(x,y)\right|$$

$$\leq S_{m,n,p_1,p_2}^{(\alpha_1,\beta_1,\alpha_2,\beta_2)}(|f(t,s)-f(x,y)|;q_m,q_n,x,y|)$$
$$\leq \omega(f;\delta_m,\delta_n)\left(S_{m,p_1}^{(\alpha_1,\beta_1)}(f_0;q_{m_1},x)+\frac{1}{\delta_m}S_{m,p_1}^{(\alpha_1,\beta_1)}(|t-x|;q_m,x)\right)$$
$$\times\left(S_{n,p_2}^{(\alpha_2,\beta_2)}(f_0;q_n,y)+\frac{1}{\delta_n}S_{n,p_2}^{(\alpha_2,\beta_2)}(|s-y|;q_n,y)\right).$$

Applying Cauchy–Schwarz inequality, we have

$$|S_{m,n,p_1,p_2}^{(\alpha_1,\beta_1,\alpha_2,\beta_2)}(f;q_m,q_n,x,y)-f(x,y)|$$
$$\leq\bar{\omega}(f;\delta_m,\delta_n)\left\{\left(1+\frac{1}{\delta_m}\sqrt{S_{m,p_1}^{(\alpha_1,\beta_1)}((t-x)^2;q_m,x)}\right)\right.$$
$$\left.\times\left(1+\frac{1}{\delta_n}\sqrt{S_{n,p_2}^{(\alpha_2,\beta_2)}((s-y)^2;q_n,y)}\right)\right\}$$
$$\leq 4\bar{\omega}(f;\delta_m,\delta_n).$$

This completes the proof. ■

Example 1 For $n,m=10$, $p_1,p_2=2$, $\alpha_1=3$, $\beta_1=4$, $\alpha_2=5$, $\beta_2=7$, $q_1,q_2=0.5$, $q_1,q_2=0.7$ and $q_1,q_2=0.9$ the convergence of the operators

$$S_{10,10,2,2}^{(3,4,5,7)}(f;.5,.5,x,y)\ \text{(yellow)},$$

$$S_{10,10,2,2}^{(3,4,5,7)}(f;.7,.7,x,y) \text{ (pink)},$$

$$S_{10,10,2,2}^{(3,4,5,7)}(f;.9,.9,x,y) \text{ (blue)}$$

to

$$f(x,y) = x\left(x-\frac{1}{4}\right)\left(y-\frac{3}{7}\right) \quad (red)$$

is illustrated by Figure 8.1.

Fig. 8.1

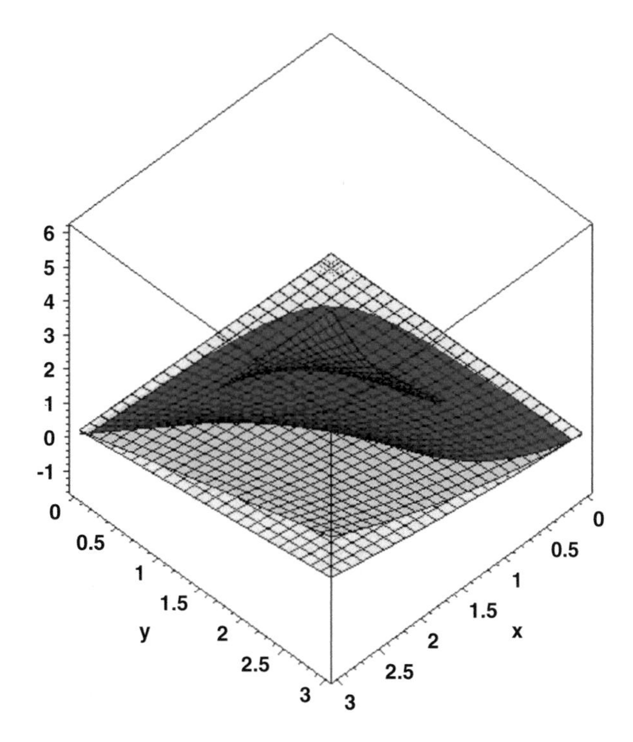

Example 2 For $m,n=10$, $\alpha_1,\alpha_2=1$, $\beta_1,\beta_2=2$, $p_1,p_2=1$, the comparison of the convergence of q -Bernstein–Schurer–Stancu (blue) given by

$$S_{m,n,p_1,p_2}^{(\alpha_1,\beta_1,\alpha_2,\beta_2)}(f;q_m,q_n,x,y)$$

and the operators bivariate q-Bernstein–Schurer (green), q-Bernstein–Stancu (red), to

$$f(x,y) = 2x\cos(\pi x)\,y^3 \text{ (yellow)}$$

with $q_m = m/(m+1)$, $q_n = 1-1/\sqrt{n}$ are illustrated in Figure 8.2.

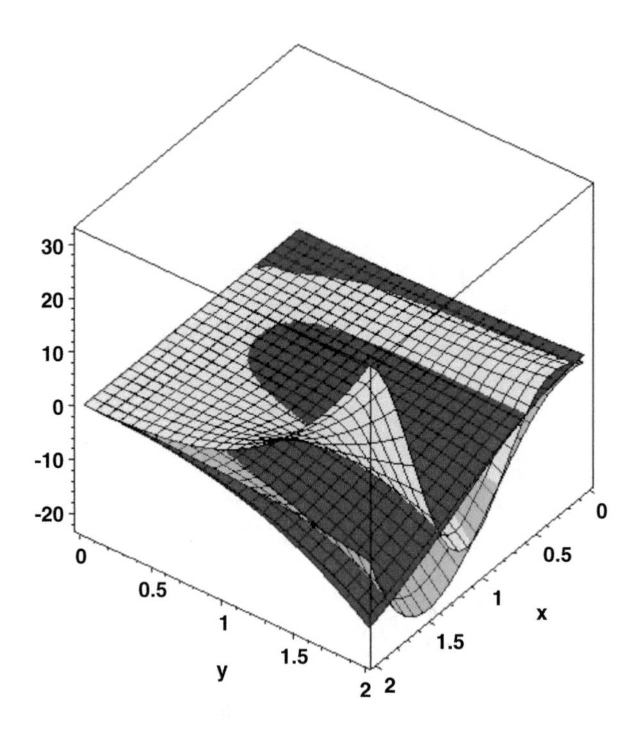

Fig. 8.2

For $0 < \xi \leq 1$ and $0 < \gamma \leq 1$, for $f \in C(I)$ we define the Lipschitz class $Lip_M(\xi, \gamma)$ for the bivariate case as follows:

$$|f(t,s) - f(x,y)| \leq M|t-x|^{\xi}|s-y|^{\gamma}.$$

Theorem 8.6 *Let $f \in Lip_M(\xi, \gamma)$. Then, we have*

$$\|S_{m,n,p_1,p_2}^{(\alpha_1,\beta_1,\alpha_2,\beta_2)}(f; q_m, q_n, ., .) - f\| \leq M\delta_m^{\xi}\delta_n^{\gamma}.$$

Proof By our hypothesis, we may write

$$\begin{aligned}
&\left|S_{m,n,p_1,p_2}^{(\alpha_1,\beta_1,\alpha_2,\beta_2)}(f; q_m, q_n, x, y) - f(x,y)\right| \\
&\qquad \leq S_{m,n,p_1,p_2}^{(\alpha_1,\beta_1,\alpha_2,\beta_2)}(|f(t,s) - f(x,y)|; q_m, q_n, x, y) \\
&\qquad \leq M S_{m,n,p_1,p_2}^{(\alpha_1,\beta_1,\alpha_2,\beta_2)}(|t-x|^{\xi}|s-y|^{\gamma}; q_m, q_n, x, y) \\
&\qquad = M S_{m,p_1}^{(\alpha_1,\beta_1)}(|t-x|^{\xi}; q_m, x) S_{n,p_2}^{(\alpha_2,\beta_2)}(|s-y|^{\gamma}; q_n, y).
\end{aligned}$$

Now, using the Hölder's inequality with

$$u_1 = \frac{2}{\xi},\ v_1 = \frac{2}{2-\xi} \text{ and } u_2 = \frac{2}{\gamma} \text{ and } v_2 = \frac{2}{2-\gamma},$$

we have

$$\begin{aligned}
&\left|S_{m,n,p_1,p_2}^{(\alpha_1,\beta_1,\alpha_2,\beta_2)}(f; q_m, q_n, x, y) - f(x, y)\right| \\
&\quad \leq M S_{m,p_1}^{(\alpha_1,\beta_1)}((t-x)^2; q_m, x)^{\frac{\xi}{2}} S_{m,p_1}^{(\alpha_1,\beta_1)}(f_0; q_m, x)^{\frac{2-\xi}{2}} \\
&\qquad \times S_{n,p_2}^{(\alpha_2,\beta_2)}((s-y)^2; q_n, y)^{\frac{\gamma}{2}} S_{n,p_2}^{(\alpha_2,\beta_2)}(f_0; q_n, y)^{\frac{2-\gamma}{2}} \\
&\quad \leq M\delta_m^{\xi}\delta_n^{\gamma}.
\end{aligned}$$

Hence, the proof is completed. ■

Theorem 8.7 *Let $f \in C^1(I)$ and $(x, y) \in J$. Then, we have*

$$\left|S_{m,n,p_1,p_2}^{(\alpha_1,\beta_1,\alpha_2,\beta_2)}(f; q_m, q_n, x, y) - f(x, y)\right| \leq \|f_x\|\delta_m + \|f_y\|\delta_n.$$

Proof Let $(x, y) \in J$ be a fixed point. Then, we can write

$$f(t, s) - f(x, y) = \int_x^t f_u(u, s)d_q u + \int_y^s f_v(x, v)d_q v.$$

Now applying $S_{m,n,p_1,p_2}^{(\alpha_1,\beta_1,\alpha_2,\beta_2)}(.; q_m, q_n, x, y)$ on both sides, we have

$$\begin{aligned}
&\left|S_{m,n,p_1,p_2}^{(\alpha_1,\beta_1,\alpha_2,\beta_2)}(f; q_m, q_n, x, y) - f(x, y)\right| \\
&\quad \leq S_{m,n,p_1,p_2}^{(\alpha_1,\beta_1,\alpha_2,\beta_2)}\left(\int_x^t f_u(u, s)d_q u; q_m, q_n, x, y\right) \\
&\qquad + S_{m,n,p_1,p_2}^{(\alpha_1,\beta_1,\alpha_2,\beta_2)}\left(\int_y^s f_v(x, v)d_q v; q_m, q_n, x, y\right).
\end{aligned}$$

Since

$$\left|\int_x^t f_u(u, s)d_q u\right| \leq \|f_x\||t-x| \text{ and } \left|\int_y^s f_v(x, v)d_q v\right| \leq \|f_y\||s-y|,$$

we have

$$\begin{aligned}
\left|S_{m,n,p_1,p_2}^{(\alpha_1,\beta_1,\alpha_2,\beta_2)}(f; q_m, q_n, x, y) - f(x, y)\right| &\leq \|f_x\| S_{m,p_1}^{(\alpha_1,\beta_1)}(|t-x|; q_m, x) \\
&\quad + \|f_y\| S_{n,p_2}^{(\alpha_2,\beta_2)}(|s-y|; q_n, y).
\end{aligned}$$

Now, applying the Cauchy–Schwarz inequality, we get

$$\left|S_{m,n,p_1,p_2}^{(\alpha_1,\beta_1,\alpha_2,\beta_2)}(f;q_m,q_n,x,y)-f(x,y)\right|$$

$$\begin{aligned}&\leq \|f_x\|S_{m,p_1}^{(\alpha_1,\beta_1)}((t-x)^2;q_m,x)^{1/2}S_{m,p_1}^{(\alpha_1,\beta_1)}(f_0;q_m,x)^{1/2}\\&+\|f_y\|S_{n,p_2}^{(\alpha_2,\beta_2)}((s-y)^2;q_n,y)^{1/2}S_{n,p_2}^{(\alpha_2,\beta_2)}(f_0;q_n,y)^{1/2}\\&\leq \|f_x\|\delta_m+\|f_y\|\delta_n.\end{aligned}$$

This completes the proof of the theorem. ■

Theorem 8.8 *For the function $f\in C(I)$, we have the following inequality*

$$\left|S_{m,n,p_1,p_2}^{(\alpha_1,\beta_1,\alpha_2,\beta_2)}(f;q_m,q_n,x,y)-f(x,y)\right|$$

$$\leq M\left\{\bar{\omega_2}(f;\sqrt{C_{m,n}})+\min\{1,C_{m,n}\}\|f\|_{C(I)}\right\}+\omega(f;\psi_{m,n}),$$

where

$$\psi_{m,n}=\sqrt{\max_{(x,y)\in J}\left\{\left(\frac{[m+p_1]_{q_m}x+\alpha_1}{[m]_{q_m}+\beta_1}-x\right)^2+\left(\frac{[n+p_2]_{q_n}y+\alpha_2}{[n]_{q_n}+\beta_2}-y\right)^2\right\}},$$

$C_{m,n}=\delta_m^2+\delta_n^2+\psi_{m,n}^2$ and the constant $M>0$, is independent of f and $C_{m,n}$.

Proof We introduce the auxiliary operators as follows:

$$\begin{aligned}S_{m,n,p_1,p_2}^{*(\alpha_1,\beta_1,\alpha_2,\beta_2)}(f;q_m,q_n,x,y)&=S_{m,n,p_1,p_2}^{(\alpha_1,\beta_1,\alpha_2,\beta_2)}(f;q_m,q_n,x,y)\\&-f\left(\frac{[m+p_1]_{q_m}x+\alpha_1}{[m]_{q_m}+\beta_1},\frac{[n+p_2]_{q_n}y+\alpha_2}{[n]_{q_n}+\beta_2}\right)\\&+f(x,y).\end{aligned}$$

Then using Lemma 8.6, we have

$$S_{m,n,p_1,p_2}^{*(\alpha_1,\beta_1,\alpha_2,\beta_2)}((t-x);q_m,q_n,x,y)=0$$

and

$$S_{m,n,p_1,p_2}^{*(\alpha_1,\beta_1,\alpha_2,\beta_2)}((s-y);q_m,q_n,x,y)=0.$$

Let $g\in C^2(I)$ and $t,s\in I$. Using the Taylor's theorem, we may write

$$g(t,s)-g(x,y)=g(t,y)-g(x,y)+g(t,s)-g(t,y)$$

$$= \frac{\partial g(x,y)}{\partial x}(t-x) + \int_x^t (t-u)\frac{\partial^2 g(u,y)}{\partial u^2}du + \frac{\partial g(x,y)}{\partial y}(s-y) + \int_y^s (s-v)\frac{\partial^2 g(x,v)}{\partial v^2}dv.$$

Applying the operator $S^{*(\alpha_1,\beta_1,\alpha_2,\beta_2)}_{m,n,p_1,p_2}(.,q_m,q_n,x,y)$ on both sides, we get

$$S^{*(\alpha_1,\beta_1,\alpha_2,\beta_2)}_{m,n,p_1,p_2}(f;q_m,q_n,x,y) - f(x,y)$$

$$= S^{*(\alpha_1,\beta_1,\alpha_2,\beta_2)}_{m,n,p_1,p_2}\left(\int_x^t (t-u)\frac{\partial^2 g(u,y)}{\partial u^2}du;q_m,q_n,x,y\right) + S^{*(\alpha_1,\beta_1,\alpha_2,\beta_2)}_{m,n,p_1,p_2}\left(\int_y^s (s-v)\frac{\partial^2 g(x,v)}{\partial v^2}dv;q_m,q_n,x,y\right)$$

$$= S^{(\alpha_1,\beta_1,\alpha_2,\beta_2)}_{m,n,p_1,p_2}\left(\int_x^t (t-u)\frac{\partial^2 g(u,y)}{\partial u^2}du;q_m,q_n,x,y\right) - \int_x^{\frac{[m+p_1]_{q_m}x+\alpha_1}{[m]_{q_m}+\beta_1}}\left(\frac{[m+p_1]_{q_m}x+\alpha_1}{[m]_{q_m}+\beta_1}-u\right)\frac{\partial^2 g(u,y)}{\partial u^2}du$$
$$+ S^{(\alpha_1,\beta_1,\alpha_2,\beta_2)}_{m,n,p_1,p_2}\left(\int_y^s (s-v)\frac{\partial^2 g(x,v)}{\partial v^2}dv;q_m,q_n,x,y\right) - \int_x^{\frac{[n+p_2]_{q_n}y+\alpha_2}{[n]_{q_n}+\beta_2}}\left(\frac{[n+p_2]_{q_n}y+\alpha_2}{[n]_{q_n}+\beta_2}-v\right)\frac{\partial^2 g(x,v)}{\partial v^2}dv.$$

Hence

$$|S^{(\alpha_1,\beta_1,\alpha_2,\beta_2)}_{m,n,p_1,p_2}(f;q_m,q_n,x,y) - f(x,y)|$$

$$\leq S^{(\alpha_1,\beta_1,\alpha_2,\beta_2)}_{m,n,p_1,p_2}\left(\left|\int_x^t |t-u|\left|\frac{\partial^2 g(u,y)}{\partial u^2}\right|du\right|;x,y\right) + \left|\int_x^{\frac{[m+p_1]_{q_m}x+\alpha_1}{[m]_{q_m}+\beta_1}}\left|\frac{[m+p_1]_{q_m}x+\alpha_1}{[m]_{q_m}+\beta_1}-u\right|\left|\frac{\partial^2 g(u,y)}{\partial u^2}\right|du\right|$$
$$+ S^{(\alpha_1,\beta_1,\alpha_2,\beta_2)}_{m,n,p_1,p_2}\left(\left|\int_y^s |s-v|\left|\frac{\partial^2 g(x,v)}{\partial v^2}\right|dv\right|;x,y\right)$$

$$+\left|\int_x^{\frac{[n+p_2]_{q_n}y+\alpha_2}{[n]_{q_n}+\beta_2}}\left|\frac{[n+p_2]_{q_n}y+\alpha_2}{[n]_{q_n}+\beta_2}-v\right|\left|\frac{\partial^2 g(x,v)}{\partial v^2}\right|dv\right|$$

$$\leq\left\{S_{m,n,p_1,p_2}^{(\alpha_1,\beta_1,\alpha_2,\beta_2)}((t-x)^2;q_m,q_n,x,y)+\left(\frac{[m+p_1]_{q_m}x+\alpha_1}{[m]_{q_m}+\beta_1}-x\right)^2\right\}\|g\|_{C^2(I)}$$

$$+\left\{S_{m,n,p_1,p_2}^{(\alpha_1,\beta_1,\alpha_2,\beta_2)}((s-y)^2;q_m,q_n,x,y)+\left(\frac{[n+p_2]_{q_n}y+\alpha_2}{[n]_{q_n}+\beta_2}-y\right)^2\right\}\|g\|_{C^2(I)}$$

$$\leq(\delta_m^2+\delta_n^2+\psi_{m,n}^2)\|g\|_{C^2(I)}$$

$$=C_{m,n}\|g\|_{C^2(I)}. \tag{8.1.6}$$

Also, using Lemma 8.6

$$|S_{m,n,p_1,p_2}^{*(\alpha_1,\beta_1,\alpha_2,\beta_2)}(f;q_m,q_n,x,y)|\leq|S_{m,n,p_1,p_2}^{(\alpha_1,\beta_1,\alpha_2,\beta_2)}(f;q_m,q_n,x,y)|$$

$$+\left|f\left(\frac{[m+p_1]_{q_m}x+\alpha_1}{[m]_{q_m}+\beta_1},\frac{[n+p_2]_{q_n}y+\alpha_2}{[n]_{q_n}+\beta_2}\right)\right|$$

$$+|f(x,y)|$$

$$\leq 3\|f\|_{C(I)}. \tag{8.1.7}$$

Hence in view of (8.1.6) and (8.1.7), we get

$$|S_{m,n,p_1,p_2}^{(\alpha_1,\beta_1,\alpha_2,\beta_2)}(f;q_m,q_n,x,y)-f(x,y)|$$

$$=\left|S_{m,n,p_1,p_2}^{*(\alpha_1,\beta_1,\alpha_2,\beta_2)}(f;q_m,q_n,x,y)-f(x,y)\right.$$

$$\left.+f\left(\frac{[m+p_1]_{q_m}x+\alpha_1}{[m]_{q_m}+\beta_1},\frac{[n+p_2]_{q_n}y+\alpha_2}{[n]_{q_n}+\beta_2}\right)-f(x,y)\right|$$

$$\leq|S_{m,n,p_1,p_2}^{*(\alpha_1,\beta_1,\alpha_2,\beta_2)}(f-g;q_m,q_n,x,y)|+|S_{m,n,p_1,p_2}^{*(\alpha_1,\beta_1,\alpha_2,\beta_2)}(g;q_m,q_n,x,y)-g(x,y)|$$

$$+|g(x,y)-f(x,y)|+\left|f\left(\frac{[m+p_1]_{q_m}x+\alpha_1}{[m]_{q_m}+\beta_1},\frac{[n+p_2]_{q_n}y+\alpha_2}{[n]_{q_n}+\beta_2}\right)-f(x,y)\right|$$

$$\leq 4\|f-g\|_{C(I)}+\left|S_{m,n,p_1,p_2}^{(\alpha_1,\beta_1,\alpha_2,\beta_2)}(g;q_m,q_n,x,y)-g(x,y)\right|$$

$$+\left|f\left(\frac{[m+p_1]_{q_m}x+\alpha_1}{[m]_{q_m}+\beta_1},\frac{[n+p_2]_{q_n}y+\alpha_2}{[n]_{q_n}+\beta_2}\right)-f(x,y)\right|$$

$$\leq\left(4\|f-g\|_{C(I)}+C_{m,n}\|g\|_{C^2(I)}\right)+\omega(f;\psi_{m,n})$$

$$\leq 4K_2(f;C_{m,n})+\omega(f;\psi_{m,n})$$

$$\leq M\left\{\bar{\omega}_2(f;\sqrt{C_{m,n}})+\min\{1,C_{m,n}\}\|f\|_{C(I)}\right\}+\omega(f;\psi_{m,n}).$$

Therefore, we get the desired result. ■

Here, we obtain a Voronovskaja type asymptotic theorem for the bivariate operators $S_{m,n,p_1,p_2}^{(\alpha_1,\beta_1,\alpha_2,\beta_2)}$.

Theorem 8.9 *Let $f \in C^2(I)$. Then, we have*

$$\lim_{n\to\infty}[n]_{q_n}(S_{n,n,p_1,p_2}^{(\alpha_1,\beta_1,\alpha_2,\beta_2)}(f;q_n,x,y)-f(x,y))$$

$$=(\alpha_1-\beta_1 x)f_x(x,y)+(\alpha_2-\beta_2 y)f_y(x,y)$$
$$+\frac{f_{xx}(x,y)}{2}x(1-x)+\frac{f_{yy}(x,y)}{2}y(1-y),$$

uniformly in $(x,y)\in J$.

Proof Let $(x,y)\in J$. By the Taylor's theorem, we have

$$f(t,s)=f(x,y)+f_x(x,y)(t-x)+f_y(x,y)(s-y)+\frac{1}{2}\left\{f_{xx}(x,y)(t-x)^2\right.$$
$$\left.+2f_{xy}(x,y)(t-x)(s-y)+f_{yy}(x,y)(s-y)^2\right\}$$
$$+\varepsilon(t,s;x,y)\sqrt{(t-x)^4+(s-y)^4}, \tag{8.1.8}$$

for $t,s\in I$, where $\varepsilon(t,s;x,y)\in C(I)$ and $\varepsilon(t,s;x,y)\to 0$ as $(t,s)\to(x,y)$.

Applying $S_{n,n,p_1,p_2}^{(\alpha_1,\beta_1,\alpha_2,\beta_2)}(f;q_n,x,y)$ on both sides of (8.1.8), we get

$$S_{n,n,p_1,p_2}^{(\alpha_1,\beta_1,\alpha_2,\beta_2)}(f(t,s);q_n,x,y)$$
$$=f(x,y)+f_x(x,y)S_{n,p_1}^{(\alpha_1,\beta_1,)}((t-x);q_n,x)$$
$$+f_y(x,y)S_{n,p_2}^{(\alpha_2,\beta_2)}((s-y);q_n,y)$$
$$+\frac{1}{2}\{f_{xx}(x,y)S_{n,p_1}^{(\alpha_1,\beta_1)}((t-x)^2;q_n,x)$$
$$+2f_{xy}(x,y)S_{n,n,p_1,p_2}^{(\alpha_1,\beta_1,\alpha_2,\beta_2)}((t-x)(s-y);q_n,x,y)$$
$$+f_{yy}(x,y)S_{n,p_2}^{(\alpha_2,\beta_2)}((s-y)^2;q_n,y)\}$$
$$+S_{n,n,p_1,p_2}^{(\alpha_1,\beta_1,\alpha_2,\beta_2)}\left(\varepsilon(t,s;x,y)\sqrt{(t-x)^4+(s-y)^4};q_n,x,y\right).$$

By using Lemma 8.8, we may write,

$$\lim_{n\to\infty}[n]_{q_n}(S_{n,n,p_1,p_2}^{(\alpha_1,\beta_1,\alpha_2,\beta_2)}(f;q_n,x,y)-f(x,y))$$

$$\begin{aligned}&=(\alpha_1-\beta_1 x)f_x(x,y)+(\alpha_2-\beta_2 y)f_y(x,y)\\&\quad+\frac{f_{xx}(x,y)}{2}x(1-x)+\frac{f_{yy}(x,y)}{2}y(1-y)\\&\quad+\lim_{n\to\infty}[n]_{q_n}S_{n,n,p_1,p_2}^{(\alpha_1,\beta_1,\alpha_2,\beta_2)}\left(\varepsilon(t,s;x,y)\sqrt{(t-x)^4+(s-y)^4};q_n,x,y\right).\end{aligned}$$

Now, applying Hölder inequality, we have

$$\left|S_{n,n,p_1,p_2}^{(\alpha_1,\beta_1,\alpha_2,\beta_2)}(\varepsilon(t,s;x,y)\sqrt{(t-x)^4+(s-y)^4};q_n,x,y)\right|$$

$$\begin{aligned}&\leq\{S_{n,n,p_1,p_2}^{(\alpha_1,\beta_1,\alpha_2,\beta_2)}(\varepsilon(t,s;x,y);q_n,x,y)\}^{1/2}\\&\quad\{S_{n,n,p_1,p_2}^{(\alpha_1,\beta_1,\alpha_2,\beta_2)}((t-x)^4+(s-y)^4;q_n,x,y)\}^{1/2}\\&\leq\{S_{n,n,p_1,p_2}^{(\alpha_1,\beta_1,\alpha_2,\beta_2)}(\varepsilon^2(t,s;x,y)q_n,x,y)\}^{1/2}\\&\quad\{S_{n,p_1}^{(\alpha_1,\beta_1)}((t-x)^4;q_n,x)+S_{n,p_2}^{(\alpha_2,\beta_2)}((s-y)^4;q_n,y)\}^{1/2}.\end{aligned}$$

Since,

$$\varepsilon^2(t,s;x,y)\to 0 \text{ as } (t,s)\to(x,y),$$

applying Theorem 8.5, we get

$$\lim_{n\to\infty}S_{n,n,p_1,p_2}^{(\alpha_1,\beta_1,\alpha_2,\beta_2)}(\varepsilon^2(t,s;x,y),x,y)=0.$$

Further, in view of (8.1.4) and (8.1.5),

$$[n]_{q_n}\left\{S_{n,p_1}^{(\alpha_1,\beta_1)}((t-x)^4;q_n,x)+S_{n,p_1}^{(\alpha_2,\beta_2)}((s-y)^4;q_n,y)\right\}^{1/2}=O(1),$$

as $n\to\infty$ uniformly in $(x,y)\in J$. Hence

$$\lim_{n\to\infty}[n]_{q_n}S_{m,n,p_1,p_2}^{(\alpha_1,\beta_1,\alpha_2,\beta_2)}(\varepsilon(t,s;x,y)\sqrt{(t-x)^4+(s-y)^4};q_n,x,y)=0,$$

uniformly in $(x,y)\in J$. This completes the proof. ∎

8.2 Bivariate Operators of Kantorovich Type

8.2.1 Bivariate Case of q-Bernstein–Schurer–Kantorovich

Agrawal, Finta, and Kumar [20] introduced a new Kantorovich type generalization of the q-Bernstein–Schurer operators as follows:

$$K_{n,p}(f;q,x) = [n+1]_q \sum_{k=0}^{n+p} b^q_{n+p,k}(x) q^{-k} \int_{[k]_q/[n+1]_q}^{[k+1]_q/[n+1]_q} f(t) d^R_q t, \quad x \in [0,1]$$

and studied the rate of convergence by means of the modulus of continuity.

A Voronovskaja type asymptotic theorem and a Kantorovich type A-statistical theorem were also proved. Agrawal et al. [20] constructed a bivariate case of a new kind of Kantorovich type generalization of the operators $\bar{B}_{n,p}$ as follows:

Let $I = [0, 1+p] \times [0, 1+p]$ and $C(I)$ be the space of all real valued continuous functions on I endowed with the norm $\|f\| = \sup\limits_{(x,y)\in I} |f(x,y)|$. For $f \in C(I)$ and $0 < q_{n_1}, q_{n_2} < 1$, the bivariate generalization of Kantorovich type q-Bernstein–Schurer operators is defined by

$$\begin{aligned} &K_{n_1,n_2,p}(f; q_{n_1}, q_{n_2}, x, y) \\ &\quad = [n_1+1]_{q_{n_1}} [n_2+1]_{q_{n_2}} \sum_{k_1=0}^{n_1+p} \sum_{k_2=0}^{n_2+p} b^{q_{n_1},q_{n_2}}_{n_1+p,n_2+p,k_1,k_2}(x,y) q_{n_1}^{-k_1} q_{n_2}^{-k_2} \\ &\qquad \times \int_{[k_2]_{q_{n_2}}/[n_2+1]_{q_{n_2}}}^{[k_2+1]_{q_{n_2}}/[n_2+1]_{q_{n_2}}} \int_{[k_1]_{q_{n_1}}/[n_1+1]_{q_{n_1}}}^{[k_1+1]_{q_{n_1}}/[n_1+1]_{q_{n_1}}} f(t,s) d^R_{q_{n_1}}(t) d^R_{q_{n_2}}(s), \end{aligned} \tag{8.2.1}$$

where

$$b^{q_{n_1},q_{n_2}}_{n_1+p,n_2+p,k_1,k_2}(x,y) = \binom{n_1+p}{k_1}_{q_{n_1}} \binom{n_2+p}{k_2}_{q_{n_2}} x^{k_1} y^{k_2} (1-x)^{n_1+p-k_1}_{q_{n_1}} (1-y)^{n_2+p-k_2}_{q_{n_2}}$$

and $(x,y) \in J = [0,1] \times [0,1]$.

For $i = 1, 2$, let (q_{n_i}) be a sequence in $(0,1)$ satisfying $q_{n_i} \to 1$ and $q_{n_i}^{n_i} \to 0$, as $n \to \infty$. In the sequel, we need the following lemmas to discuss the rate of convergence of the operators (8.2.1).

Lemma 8.9 ([20]) *Let $e_{ij} = x^i y^j$, $(i,j) \in \mathbb{N}^0 \times \mathbb{N}^0$ with $i+j \leq 2$ be the two-dimensional test functions. Then the following equalities hold for the operators given by (8.2.1):*

1. $K_{n_1,n_2,p}(e_{00}; q_{n_1}, q_{n_2}, x, y) = 1;$
2. $K_{n_1,n_2,p}(e_{10}; q_{n_1}, q_{n_2}, x, y) = \frac{[n_1+p]_{q_{n_1}}}{[n_1+1]_{q_{n_1}}} \frac{2q_{n_1}}{[2]_{q_{n_1}}} x + \frac{1}{[2]_{q_{n_1}}[n_1+1]_{q_{n_1}}};$
3. $K_{n_1,n_2,p}(e_{01}; q_{n_1}, q_{n_2}, x, y) = \frac{[n_2+p]_{q_{n_2}}}{[n_2+1]_{q_{n_2}}} \frac{2q_{n_2}}{[2]_{q_{n_2}}} y + \frac{1}{[2]_{q_{n_2}}[n_2+1]_{q_{n_2}}};$
4. $K_{n_1,n_2,p}(e_{20}; q_{n_1}, q_{n_2}, x, y) = \frac{1}{[n_1+1]^2_{q_{n_1}}[3]_{q_{n_1}}} + \frac{q_{n_1}(3+5q_{n_1}+4q^2_{n_1})}{[2]_{q_{n_1}}[3]_{q_{n_1}}}$
$\frac{[n_1+p]_{q_{n_1}}}{[n_1+1]_{q^2_{n_1}}} x + \frac{q^2_{n_1}(1+q_{n_1}+4q^2_{n_1})}{[2]_{q_{n_1}}[3]_{q_{n_1}}} \frac{[n_1+p]_{q_{n_1}}[n_1+p-1]_{q_{n_1}}}{[n_1+1]_{q^2_{n_1}}} x^2;$
5. $K_{n_1,n_2,p}(e_{02}; q_{n_1}, q_{n_2}, x, y) = \frac{1}{[n_2+1]^2_{q_{n_2}}[3]_{q_{n_2}}} + \frac{q_{n_2}(3+5q_{n_2}+4q^2_{n_2})}{[2]_{q_{n_2}}[3]_{q_{n_2}}}$
$\frac{[n_2+p]_{q_{n_2}}}{[n_2+1]^2_{q_{n_2}}} y + \frac{q^2_{n_2}(1+q_{n_2}+4q^2_{n_2})}{[2]_{q_{n_2}}[3]_{q_{n_2}}} \frac{[n_2+p]_{q_{n_2}}[n_2+p-1]_{q_{n_2}}}{[n_2+1]_{q^2_{n_2}}} y^2.$

Remark 8.1 From Lemma 8.9, we get

1. $K_{n_1,n_2,p}((t-x); q_{n_1}, q_{n_2}, x, y) = \frac{1}{[2]_{q_{n_1}}[n_1+1]_{q_{n_1}}} + x\left(\frac{[n_1+p]_{q_{n_1}}}{[n_1+1]_{q_{n_1}}} \frac{2q_{n_1}}{[2]_{q_{n_1}}} - 1\right);$
2. $K_{n_1,n_2,p}((s-y); q_{n_1}, q_{n_2}, x, y) = \frac{1}{[2]_{q_{n_2}}[n_2+1]_{q_{n_2}}} + y\left(\frac{[n_2+p]_{q_{n_2}}}{[n_2+1]_{q_{n_2}}} \frac{2q_{n_2}}{[2]_{q_{n_2}}} - 1\right);$
3. $K_{n_1,n_2,p}((t-x)^2; q_{n_1}, q_{n_2}, x, y) =$

$$\left(\frac{q^2_{n_1}(1+q_{n_1}+4q^2_{n_1})}{[2]_{q_{n_1}}[3]_{q_{n_1}}} \frac{[n_1+p]_{q_{n_1}}[n_1+p-1]_{q_{n_1}}}{[n_1+1]^2_{q_{n_1}}} - \frac{[n_1+p]_{q_{n_1}}}{[n_1+1]_{q_{n_1}}} \frac{4q_{n_1}}{[2]_{q_{n_1}}} + 1\right)x^2 + \left(\frac{q_{n_1}(3+5q_{n_1}+4q^2_{n_1})}{[2]_{q_{n_1}}[3]_{q_{n_1}}} \frac{[n_1+p]_{q_{n_1}}}{[n_1+1]^2_{q_{n_1}}} - \frac{2}{[n_1+1]_{q_{n_1}}[2]_{q_{n_1}}}\right)x + \frac{1}{[n_1+1]^2_{q_{n_1}}[3]_{q_{n_1}}};$$

4. $K_{n_1,n_2,p}((s-y)^2; q_{n_1}, q_{n_2}, x, y) =$

$$\left(\frac{q^2_{n_2}(1+q_{n_2}+4q^2_{n_2})}{[2]_{q_{n_2}}[3]_{q_{n_2}}} \frac{[n_2+p]_{q_{n_2}}[n_2+p-1]_{q_{n_2}}}{[n_2+1]^2_{q_{n_2}}} - \frac{[n_2+p]_{q_{n_2}}}{[n_2+1]_{q_{n_2}}} \frac{4q_{n_2}}{[2]_{q_{n_2}}} + 1\right)y^2 + \left(\frac{q_{n_2}(3+5q_{n_2}+4q^2_{n_2})}{[2]_{q_{n_2}}[3]_{q_{n_2}}} \frac{[n_2+p]_{q_{n_2}}}{[n_2+1]^2_{q_{n_2}}} - \frac{2}{[n_2+1]_{q_{n_2}}[2]_{q_{n_2}}}\right)y + \frac{1}{[n_2+1]^2_{q_{n_2}}[3]_{q_{n_2}}}.$$

In order to prove the uniform convergence of the operators $K_{n_1,n_2,p}(f)$ to f for each $f \in C(I)$, we need some notions concerning the j_{th} projection and Korovkin subset according to [20].

From [20, p. 29], for locally compact Hausdorf spaces X_1 and X_2, let us denote by $pr_j : X_1 \times X_2 \to X_j$ $(j = 1, 2)$ the jth projection which is defined by

$$pr_j(x) = x_j, \ \forall x = (x_1, x_2) \in X_1 \times X_2.$$

Let X and Y be two locally compact Hausdorf spaces and $T : C(X) \to C(Y)$ be a positive linear operator. A subset H of $C(X)$ is called a Korovkin subset for T with respect to linear operators if it satisfies the following property: if $\{L_i\}_{i\in I}^{\leq}$ is an arbitrary set of positive linear operators from $C(X)$ into $C(Y)$ such that $\sup_{i\in I} ||L_i|| < +\infty$ and if $\lim_{i\in I\leq} L_i(h) = T(h)$ for all $h \in H$, then $\lim_{i\in I\leq} L_i(f) = T(f)$ for every $f \in C(X)$ (see [20], pp. 122–123).

The following lemma follows from [225]

Lemma 8.10 $\{1, pr_1, pr_2, pr_1^2 + pr_2^2\}$ *is a Korovkin subset in* $C(I)$ *for the identity operator with respect to positive linear operators.*

Theorem 8.10 ([20]) *For any* $f \in C(I)$ *and* $q_{n_1}, q_{n_2} \in (0, 1)$ *such that* $q_{n_1} \to 1$ *as* $n_1 \to \infty$ *and* $q_{n_2} \to 1$ *as* $n_2 \to \infty$, *we have*

$$\lim_{n_1,n_2\to\infty} ||K_{n_1,n_2,p}(f; q_{n_1}, q_{n_2}, ., .) - f||_j = 0.$$

Proof Now, by Lemma 8.9 (i), we have

$$\|K_{n_1,n_2,p}(f; q_{n_1}, q_{n_2}, \cdot, \cdot)\|_J \leq \|f\|_I \text{ for all } f \in C(I).$$

Hence $\|K_{n_1,n_2,p}\| \leq 1$. Further, because $q_{n_i} \to 1$ as $n_i \to \infty i = 1, 2)$, we find that

$$\frac{1}{[n_i+1]_{q_{n_i}}} \to 0, \ \frac{[n_i+p]_{q_{n_i}}}{[n_i+1]_{q_{n_i}}} \to 1 \text{ and } \frac{[n_i+p-1]_{q_{n_i}}}{[n_i+1]_{q_{n_i}}} \to 1 \text{ as } n_i \to \infty \ (i = 1, 2).$$

Hence, we have

$$K_{n_1,n_2,p}(1; q_{n_1}, q_{n_2}, x, y) = K_{n_1,n_2,p}(e_{00}; q_{n_1}, q_{n_2}, x, y) = 1,$$

$$K_{n_1,n_2,p}(pr_1; q_{n_1}, q_{n_2}, x, y) = K_{n_1,n_2,p}(e_{10}; q_{n_1}, q_{n_2}, x, y) \to x$$

uniformly on J, as $n_1 \to \infty$ and $n_2 \to \infty$,

$$K_{n_1,n_2,p}(pr_2; q_{n_1}, q_{n_2}, x, y) = K_{n_1,n_2,p}(e_{01}; q_{n_1}, q_{n_2}, x, y) \to y$$

uniformly on J, as $n_1 \to \infty$ and $n_2 \to \infty$, and

$$K_{n_1,n_2,p}(pr_1^2+pr_2^2; q_{n_1}, q_{n_2}, x, y) = K_{n_1,n_2,p}(e_{20}+e_{02}; q_{n_1}, q_{n_2}, x, y) \to x^2+y^2$$

uniformly on J, as $n_1 \to \infty$ and $n_2 \to \infty$, respectively. Thus, the hypotheses of Lemma 8.10 are satisfied and the proof is completed. ∎

The proofs of the Theorems 8.11–8.13 are similar to the proof of the Theorem 8.10 and hence are omitted.

Theorem 8.11 ([20]) *For any $f \in C(I)$ and $q_{n_1}, q_{n_2} \in (0,1)$ such that $q_{n_1} \to 1$ as $n_1 \to \infty$ and $q_{n_2} \to 1$ as $n_2 \to \infty$, and for all $(x,y) \in J$, we have*

$$|K_{n_1,n_2,p}(f; q_{n_1}, q_{n_2}, x, y) - f(x,y)| \leq 4\omega\left(f; \sqrt{\delta_{n_1}(x)}, \sqrt{\delta_{n_2}(y)}\right),$$

where $\delta_{n_1}(x) = K_{n_1,p}((t-x)^2; q_{n_1}, x)$ and $\delta_{n_2}(y) = K_{n_2,p}((s-y)^2; q_{n_2}, y)$.

Theorem 8.12 ([20]) *Let $f \in Lip_M(\alpha_1, \alpha_2)$ and $q_{n_1}, q_{n_2} \in (0,1)$ such that $q_{n_1} \to 1$ as $n_1 \to \infty$ and $q_{n_2} \to 1$ as $n_2 \to \infty$. Then, for all $(x,y) \in J$, we have*

$$|K_{n_1,n_2,p}(f; q_{n_1}, q_{n_2}, x, y) - f(x,y)| \leq 4M\delta_{n_1}^{\alpha_1/2}(x)\delta_{n_2}^{\alpha_2/2}(y),$$

where $\delta_{n_1}(x)$ and $\delta_{n_2}(y)$ are defined in Theorem 8.11.

In the next theorem, we shall use the following notations:

$$C^1(I) = \{f \in C(I) : f'_x, f'_y \in C(I)\}$$

and

$$C^2(I) = \{f \in C(I) : f''_{xx}, f''_{xy}, f''_{yx}, f''_{yy} \in C(I)\}$$

respectively.

Theorem 8.13 ([20]) *Let $f \in C^1(I)$, $(x,y) \in J$ and $0 < q_{n_1}, q_{n_2} < 1$ such that $q_{n_1} \to 1$ as $n_1 \to \infty$ and $q_{n_2} \to 1$ as $n_2 \to \infty$. Then, we have*

$$|K_{n_1,n_2,p}(f; q_{n_1}, q_{n_2}, x, y) - f(x,y)| \leq \|f'_x\|\sqrt{\delta_{n_1}(x)} + \|f'_y\|\sqrt{\delta_{n_2}(y)},$$

where $\delta_{n_1}(x)$ and $\delta_{n_2}(y)$ are defined as in Theorem 8.11.

8.3 Bivariate Operators of Durrmeyer Type

8.3.1 The Bivariate Generalization of q-Stancu–Durrmeyer Type Operators

In 2009, for any function

$$f \in C[0,1], q > 0, \alpha = \alpha(n) \geq 0 \ \ (\alpha(n) \to 0, \ as\ n \to \infty)$$

and each $n \in \mathbb{N}$ Nowak [184] defined the q-analogue for the operators defined by Stancu [221] as

$$B_n^{q,\alpha}(f;x) = \sum_{k=0}^{n} v_{n,k}^{q,\alpha}(x) f\left(\frac{[k]_q}{[n]_q}\right), \quad x \in [0,1], \tag{8.3.1}$$

where,

$$v_{n,k}^{q,\alpha}(x) = \begin{bmatrix} n \\ k \end{bmatrix}_q \frac{\prod_{\nu=0}^{k-1}(x+\alpha[\nu]_q)\prod_{\mu=0}^{n-k-1}(1-q^{\mu}x+\alpha[\mu]_q)}{\prod_{\lambda=0}^{n-1}(1+\alpha[\lambda]_q)}$$

and investigated the Korovkin type approximation properties for these operators. Recently, Neer and Agrawal [182] introduced the Durrmeyer type integral modification for the operators (8.3.1) as

$$D_n^{\alpha}(f;q;x) = [n+1]_q \sum_{k=0}^{n} v_{n,k}^{q,\alpha}(x) \int_0^1 p_{n,k}^q(t) f(t) d_q t, \tag{8.3.2}$$

where

$$p_{n,k}^q(t) = \begin{bmatrix} n \\ k \end{bmatrix}_q t^k (1-qt)_q^{n-k},$$

and obtained the basic convergence theorem, local approximation theorem, Korovkin type A-statistical approximation theorem, and the rate of A-statistical convergence for these operators (8.3.2). Subsequently, Neer et al. [183] proposed a bivariate case of the operators (8.3.2) as follows:

Let $I = [0,1] \times [0,1]$ and $C(I)$ denote the class of all real valued continuous functions on I endowed with the norm $\|f\| = \sup_{(x,y)\in I} |f(x,y)|$. Then, for $f \in C(I)$ and for all $(x,y) \in I$, the bivariate generalization of q-Stancu–Durrmeyer type operators (8.3.2) is defined as

$$D_{n_1,n_2}^{\alpha_{n_1},\alpha_{n_2}}(f;q_{n_1},q_{n_2},x,y) = [n_1+1]_{q_{n_1}}[n_2+1]_{q_{n_2}} \sum_{k_1=0}^{n_1}\sum_{k_2=0}^{n_2} v_{n_1,k_1}^{q_{n_1},\alpha_{n_1}}(x) v_{n_2,k_2}^{q_{n_2},\alpha_{n_2}}(y)$$
$$\times \int_0^1\int_0^1 p_{n_1,k_1}^{q_{n_1}}(t) p_{n_2,k_2}^{q_{n_2}}(s) f(t,s) d_{q_{n_1}}(t) d_{q_{n_2}}(s). \tag{8.3.3}$$

In order to discuss the main results, the following lemmas are required:

Lemma 8.11 ([182]) *For $D_n^\alpha(t^m; q; x)$, $m = 0, 1, 2$, one has*

1. $D_n^\alpha(1; q; x) = 1;$
2. $D_n^\alpha(t; q; x) = \frac{1}{[n+2]_q}(1 + q[n]_q x);$
3.

$$D_n^\alpha(t^2; q; x)$$
$$= \frac{1}{[n+2]_q[n+3]_q}\left\{[2]_q + q(1+2q)[n]_q x + \frac{q^3[n]_q^2}{1+\alpha}\left(x(x+\alpha) + \frac{x(1-x)}{[n]_q}\right)\right\}.$$

Consequently,

1. $D_n^\alpha(t - x; q; x) = \dfrac{1}{[n+2]_q} + \dfrac{1}{[n+2]_q}(q[n]_q - [n+2]_q)x;$
2.

$$D_n^\alpha((t-x)^2; q; x) = \frac{[2]_q}{[n+2]_q[n+3]_q}$$
$$+ \left\{\frac{1}{[n+2]_q[n+3]_q}\left(q(1+2q)[n]_q + \frac{q^3[n]_q([n]_q\alpha + 1)}{(1+\alpha)}\right) - \frac{2}{[n+2]_q}\right\}x$$
$$+ \left\{\frac{q^3[n]_q([n]_q - 1)}{[n+2]_q[n+3]_q(1+\alpha)} - \frac{2q[n]_q}{[n+2]_q} + 1\right\}x^2.$$

Lemma 8.12 ([183]) *For $D_{n_1,n_2}^{\alpha_{n_1},\alpha_{n_2}}(e_{ij}; q_1, q_2, x, y)$, $e_{ij} = x^i y^j$, $i, j \in \mathbb{N} \cup \{0\}$, $x, y \in [0, 1]$, we have*

1. $D_{n_1,n_2}^{\alpha_{n_1},\alpha_{n_2}}(e_{00}; q_{n_1}, q_{n_2}, x, y) = 1;$
2. $D_{n_1,n_2}^{\alpha_{n_1},\alpha_{n_2}}(e_{10}; q_{n_1}, q_{n_2}, x, y) = \dfrac{1}{[n_1+2]_{q_{n_1}}}(1 + q_{n_1}[n_1]_{q_{n_1}} x);$
3. $D_{n_1,n_2}^{\alpha_{n_1},\alpha_{n_2}}(e_{01}; q_{n_1}, q_{n_2}, x, y) = \dfrac{1}{[n_2+2]_{q_{n_2}}}(1 + q_{n_2}[n_2]_{q_{n_2}} y);$
4. $D_{n_1,n_2}^{\alpha_{n_1},\alpha_{n_2}}(e_{11}; q_{n_1}, q_{n_2}, x, y) = \dfrac{1}{[n_1+2]_{q_{n_1}}}(1 + q_{n_1}[n_1]_{q_{n_1}} x)\dfrac{1}{[n_2+2]_{q_{n_2}}}$ $(1 + q_{n_2}[n_2]_{q_{n_2}} y);$
5. $D_{n_1,n_2}^{\alpha_{n_1},\alpha_{n_2}}(e_{20}; q_{n_1}, q_{n_2}, x, y) = \dfrac{1}{[n_1+2]_{q_{n_1}}[n_1+3]_{q_{n_1}}}\Bigg\{[2]_{q_{n_1}}$

$$+ q_{n_1}(1+2q_{n_1})[n_1]_{q_{n_1}} x + \frac{q_{n_1}^3[n_1]_{q_{n_1}}^2}{1+\alpha_{n_1}}\left(x(x+\alpha_{n_1}) + \frac{x(1-x)}{[n_1]_{q_{n_1}}}\right)\Bigg\};$$

6. $D_{n_1,n_2}^{\alpha_{n_1},\alpha_{n_2}}(e_{02}; q_{n_1}, q_{n_2}, x, y) = \dfrac{1}{[n_2+2]_{q_{n_2}}[n_2+3]_{q_{n_2}}}\Bigg\{[2]_{q_{n_2}}$
$+ q_{n_2}(1+2q_{n_2})[n_2]_{q_{n_2}}y + \dfrac{q_{n_2}^3[n_2]_{q_{n_2}}^2}{1+\alpha_{n_2}}\left(y(y+\alpha_{n_2}) + \dfrac{y(1-y)}{[n_2]_{q_{n_2}}}\right)\Bigg\};$

7. $D_{n_1,n_2}^{\alpha_{n_1},\alpha_{n_2}}(e_{20} + e_{02}; q_1, q_2, x, y) = \dfrac{1}{[n_1+2]_{q_{n_1}}[n_1+3]_{q_{n_1}}}\Bigg\{[2]_{q_{n_1}} + q_{n_1}(1+$
$2q_{n_1})[n_1]_{q_{n_1}}x$
$+ \dfrac{q_{n_1}^3[n_1]_{q_{n_1}}^2}{1+\alpha_{n_1}}\left(x(x+\alpha_{n_1}) + \dfrac{x(1-x)}{[n_1]_{q_{n_1}}}\right)\Bigg\}\dfrac{1}{[n_2+2]_{q_{n_2}}[n_2+3]_{q_{n_2}}}\Bigg\{[2]_{q_{n_2}} +$
$q_{n_2}(1+2q_{n_2})[n_2]_{q_{n_2}}y$
$+ \dfrac{q_{n_2}^3[n_2]_{q_{n_2}}^2}{1+\alpha_{n_2}}\left(y(y+\alpha_{n_2}) + \dfrac{y(1-y)}{[n_2]_{q_{n_2}}}\right)\Bigg\}.$

Lemma 8.13 ([183])

1. $D_{n_1,n_2}^{\alpha_{n_1},\alpha_{n_2}}(t-x; q_{n_1}, q_{n_2}, x, y) = \dfrac{1}{[n_1+2]_{q_{n_1}}} + \dfrac{1}{[n_1+2]_{q_{n_1}}}(q_{n_1}[n_1]_{q_{n_1}} - [n+2]_{q_{n_1}})x;$
2. $D_{n_1,n_2}^{\alpha_{n_1},\alpha_{n_2}}(s-y; q_{n_1}, q_{n_2}, x, y) = \dfrac{1}{[n_2+2]_{q_{n_2}}} + \dfrac{1}{[n_2+2]_{q_{n_2}}}(q_{n_2}[n_2]_{q_{n_2}} - [n+2]_{q_{n_2}})y;$
3. $D_{n_1,n_2}^{\alpha_{n_1},\alpha_{n_2}}((t-x)^2; q_{n_1}, q_{n_2}, x, y)$
$= \dfrac{[2]_{q_{n_1}}}{[n_1+2]_{q_{n_1}}[n_1+3]_{q_{n_1}}} + \Bigg\{\dfrac{1}{[n_1+2]_{q_{n_1}}[n_1+3]_{q_{n_1}}}\Bigg(q_{n_1}(1+2q_{n_1})[n_1]_{q_{n_1}} +$
$\dfrac{q_{n_1}^3[n_1]_{q_{n_1}}([n_1]_{q_{n_1}}\alpha_{n_1}+1)}{(1+\alpha_{n_1})}\Bigg) - \dfrac{2}{[n_1+2]_{q_{n_1}}}\Bigg\}x + \Bigg\{\dfrac{q_{n_1}^3[n_1]_{q_{n_1}}([n_1]_{q_{n_1}}-1)}{[n_1+2]_{q_{n_1}}[n_1+3]_{q_{n_1}}(1+\alpha_{n_1})} -$
$\dfrac{2q_{n_1}[n_1]_{q_{n_1}}}{[n_1+2]_{q_{n_1}}} + 1\Bigg\}x^2;$
4. $D_{n_1,n_2}^{\alpha_{n_1},\alpha_{n_2}}((s-y)^2; q_{n_1}, q_{n_2}, x, y)$
$= \dfrac{[2]_{q_{n_2}}}{[n_2+2]_{q_{n_2}}[n_2+3]_{q_{n_2}}} + \Bigg\{\dfrac{1}{[n_2+2]_{q_{n_2}}[n_2+3]_{q_{n_2}}}\Bigg(q_{n_2}(1+2q_{n_2})[n_2]_{q_{n_2}} +$
$\dfrac{q_{n_2}^3[n_2]_{q_{n_2}}([n_2]_{q_{n_2}}\alpha_{n_2}+1)}{(1+\alpha_{n_2})}\Bigg) - \dfrac{2}{[n_2+2]_{q_{n_2}}}\Bigg\}y + \Bigg\{\dfrac{q_{n_2}^3[n_2]_{q_{n_2}}([n_2]_{q_{n_2}}-1)}{[n_2+2]_{q_{n_2}}[n_2+3]_{q_{n_2}}(1+\alpha_{n_2})} -$
$\dfrac{2q_{n_2}[n_2]_{q_{n_2}}}{[n_2+2]_{q_{n_2}}} + 1\Bigg\}y^2.$

The authors [183] obtained the rate of convergence of the operators (8.3.3) by using the complete and partial modulus of continuity and Peetre's K-functional. They proved the following results:

Let $0 < q_{n_i} < 1$ and $\alpha_{n_i} \geq 0$ be sequences such that

$$\lim_{n_i\to\infty} q_{n_i} = 1,\ \ \lim_{n_i\to\infty} q_{n_i}^{n_i} = a_i\ (0 \leq a_i < 1)$$

Table 8.1 Error of approximation for $D_{n_1,n_2}^{\alpha_{n_1},\alpha_{n_2}}$

$q_{n_1} = q_{n_2}$	Error of approximation
0.4	2.172377390
0.5	1.880657031
0.6	1.606452189
0.7	1.366444984
0.8	1.170958744
0.9	1.022298931

and $\lim_{n_i\to\infty} \alpha_{n_i} = 0,\ i = 1, 2$. Also, assume that

$$\delta_{n_1,q_{n_1}}^{\alpha_{n_1}}(x) = \sqrt{D_{n_1}^{\alpha_{n_1}}((t-x)^2; q_{n_1}, x)},$$

$$\delta_{n_2,q_{n_2}}^{\alpha_{n_2}}(y) = \sqrt{D_{n_2}^{\alpha_{n_2}}((s-y)^2; q_{n_2}, y)}. \tag{8.3.4}$$

Theorem 8.14 *The sequence of bivariate q-Stancu–Durrmeyer operators* $D_{n_1,n_2}^{\alpha_{n_1},\alpha_{n_2}}(f; q_{n_1}, q_{n_2}, x, y)$ *converges uniformly to* $f(x, y)$, *for any* $f \in C(I)$.

Theorem 8.15 *Let* $f \in C^1(I)$ *and* $(x, y) \in I$. *Then, we have*

$$|D_{n_1,n_2}^{\alpha_{n_1},\alpha_{n_2}}(f; q_{n_1}, q_{n_2}, x, y) - f(x, y)| \le \|f_x'\|\delta_{n_1,q_{n_1}}^{\alpha_{n_1}}(x) + \|f_y'\|\delta_{n_2,q_{n_2}}^{\alpha_{n_2}}(y),$$

where $\delta_{n_1,q_{n_1}}^{\alpha_{n_1}}(x)$ *and* $\delta_{n_2,q_{n_2}}^{\alpha_{n_2}}(x)$ *are defined in (8.3.4).*

Example 8.1 Let $f \in C^1(I^2)$. Considering $n_1 = n_2 = 10$ and $\alpha_{n_1} = \alpha_{n_2} = 0.2$, in Table 8.1 we compute the error of approximation of

$$f(x, y) = x^2y^2 + x^3y - 2x^4$$

by using the relation (5.1.5).

Theorem 8.16 *Let* $f \in C(I)$ *and* $(x, y) \in I$. *Then, we have*

$$|D_{n_1,n_2}^{\alpha_{n_1},\alpha_{n_2}}(f; q_{n_1}, q_{n_2}, x, y) - f(x, y)| \le 4\bar{\omega}\left(f, \delta_{n_1,q_{n_1}}^{\alpha_{n_1}}(x), \delta_{n_2,q_{n_2}}^{\alpha_{n_2}}(y)\right),$$

where $\delta_{n_1,q_{n_1}}^{\alpha_{n_1}}(x)$ *and* $\delta_{n_2,q_{n_2}}^{\alpha_{n_2}}(y)$ *are defined in (8.3.4).*

Theorem 8.17 *Let* $f \in C(I)$ *and* $(x, y) \in I$. *Then, we have*

$$|D_{n_1,n_2}^{\alpha_{n_1},\alpha_{n_2}}(f; q_{n_1}, q_{n_2}, x, y) - f(x, y)| \le 2\omega^1\left(f; \delta_{n_1,q_{n_1}}^{\alpha_{n_1}}(x)\right) + 2\omega^2\left(f; \delta_{n_2,q_{n_2}}^{\alpha_{n_2}}(y)\right),$$

where $\delta_{n_1,q_{n_1}}^{\alpha_{n_1}}(x)$ *and* $\delta_{n_2,q_{n_2}}^{\alpha_{n_2}}(y)$ *are defined in (8.3.4).*

8.3.2 *Bivariate of Lupaş–Durrmeyer Type Operators*

In [172], Miclaus studied some approximation properties of Bernstein–Stancu type operators based on Pòlya distribution defined by

$$P_n^{(1/n)}(f;x) = \sum_{k=0}^{n} f\left(\frac{k}{n}\right) p_{n,k}^{(1/n)}(x), \tag{8.3.5}$$

where $f \in C(I)$, with $I = [0, 1]$,

$$p_{n,k}^{(1/n)}(x) = \frac{2(n!)}{(2n)!}\binom{n}{k}(nx)_k(n-nx)_{n-k}$$

and $(n)_k = n(n+1)\cdots(n+k-1)$. These operators were introduced by Lupaş and Lupaş [159].

In [23], Kantorovich variant of the operators given by (8.3.5) was introduced and the properties of local and global approximation were investigated in univariate and bivariate cases. The authors also considered the bivariate form of these operators and discussed the degree of approximation, using the second order Ditzian–Totik modulus of smoothness and the corresponding Peeter's K-functional. In [127], the Durrmeyer type modification of the operator (8.3.5) was introduced and studied by Gupta and Rassias. The Durrmeyer–Lupaş-type operator based on Pòlya distribution was defined by

$$D_n^{(1/n)}(f;x) = (n+1)\sum_{k=0}^{n} p_{n,k}^{(1/n)}(x)\int_0^1 p_{n,k}(t)\, f(t)dt, \tag{8.3.6}$$

where $f \in C(I)$, $p_{n\,k}(t) = \binom{n}{k} t^k\,(1-t)^{n-k}$ and $p_{n,k}^{(1/n)}(x)$ is as given in (8.3.5).

For $I = [0, 1] \times [0, 1]$, let $C(I)$ be the space of all ordinary continuous functions on I, equipped with the norm given by $\|f\|_{C(I)} = \sup_{(x,y)\in I} |f(x, y)|$. For $f \in C(I)$, the bivariate case of the operator (8.3.6) is defined by

$$D_{n,m}^{(1/n,1/m)}(f;x,y) = (n+1)(m+1)\sum_{k=0}^{n}\sum_{j=0}^{m} p_{n,m,k,j}^{(1/n,1/m)}(x,y)$$
$$\times \int_0^1\int_0^1 p_{n,m,k,j}(t,s) f(t,s)\; dtds, \tag{8.3.7}$$

where

$$p_{n,m,k,j}\,(t,s) = \binom{n}{k} t^k(1-t)^{n-k}\binom{m}{j} s^j(1-s)^{m-j}$$

is the bivariate Bernstein basis element and

$$p_{n,m,k,j}^{(1/n,1/m)}(x,y) = \frac{2(n!)}{(2n)!}\frac{2(m!)}{(2m)!}\binom{n}{k}\binom{m}{j}(nx)_k(n-nx)_{n-k}(my)_j(m-my)_{m-j}.$$

Now we establish the following lemmas which will be useful in the sequel.

Lemma 8.14 ([127]) *Let* $e_i(t) = t^i, i = 0, 1, 2, 3, 4$ *we have*

1. $D_n^{(1/n)}(e_0; x) = 1;$
2. $D_n^{(1/n)}(e_1; x) = \dfrac{nx+1}{(n+2)};$
3. $D_n^{(1/n)}(e_2; x) = \dfrac{n^3x^2+5n^2x-n^2x^2+3nx+2n+2}{(n+1)(n+2)(n+3)};$
4. $D_n^{(1/n)}(e_3; x) = \frac{1}{(n+1)(n+2)^2(n+3)(n+4)}\Big\{n^3(n-1)(n-2)x^3+6n^2x^2(2n^2+n-2)+nx(17n^2+57n+22)+6(n+1)(n+2)\Big\};$
5. $D_n^{(1/n)}(e_4; x) = \frac{1}{n(n+1)(n+2)^2(n+3)^2(n+4)(n+5)}\Big\{n^5(n-1)(n-2)(n-3)x^4+2n^5x^3(11n^2-18n-23)+n^3x^2(166n^3+327n^2-143n-210)+n^2x(206n^3+828n^2+970n+300)+24\Big\}.$

Lemma 8.15 ([24]) *Let* $e_{i,j}(x,y) = x^iy^j, (i,j) \in \mathbb{N}^0 \times \mathbb{N}^0$, *with* $i+j \le 4$ *and* $\mathbb{N}^0 = \mathbb{N} \cup \{0\}$ *be the two-dimensional test functions. Then*

1. $D_{n,m}^{(1/n,1/m)}(e_{00}; x,y) = 1;$
2. $D_{n,m}^{(1/n,1/nm)}(e_{10}; x,y) = \dfrac{nx+1}{(n+2)};$
3. $D_{n,m}^{(1/n,1/m)}(e_{01}; x,y) = \dfrac{my+1}{(m+2)};$
4. $D_{n,m}^{(1/n,1/m)}(e_{20}; x,y) = \dfrac{n^3x^2+5n^2x-n^2x^2+3nx+2n+2}{(n+1)(n+2)(n+3)};$
5. $D_{n,m}^{(1/n,1/m)}(e_{02}; x,y) = \dfrac{m^3y^2+5m^2y-m^2y^2+3my+2m+2}{(m+1)(m+2)(m+3)},$
6. $D_{n,m}^{(1/n,1/m)}(e_{30}; x,y) = \frac{1}{(n+1)(n+2)^2(n+3)(n+4)}\Big\{n^3(n-1)(n-2)x^3+6n^2x^2(2n^2+n-2)+nx(17n^2+57n+22)+6(n+1)(n+2)\Big\};$
7. $D_{n,m}^{(1/n,1/m)}(e_{03}; x,y) = \frac{1}{(m+1)(m+2)^2(m+3)(m+4)}\Big\{m^3(m-1)(m-2)x^3+6m^2x^2(2m^2+m-2)+mx(17m^2+57m+22)+6(m+1)(m+2)\Big\};$

8. $D_{n,m}^{(1/n,1/m)}(e_{40};x,y)=\frac{1}{n(n+1)(n+2)^2(n+3)^2(n+4)(n+5)}\Big\{n^5(n-1)(n-2)(n-3)x^4+2n^5x^3(11n^2-18n-23)+n^3x^2(166n^3+327n^2-143n-210)+n^2x(206n^3+828n^2+970n+300)+24\Big\}$;
9. $D_{n,m}^{(1/n,1/m)}(e_{04};x,y)=\frac{1}{m(m+1)(m+2)^2(m+3)^2(m+4)(m+5)}\Big\{m^5(m-1)(m-2)(m-3)x^4+2m^5x^3(11m^2-18m-23)+m^3x^2(166m^3+327m^2-143m-210)+m^2x(206m^3+828m^2+970m+300)+24\Big\}$.

Remark 8.2 ([24]) From Lemma 8.15, we have

1. $D_{n,m}^{(1/n,1/m)}((t-x);x,y)=\frac{(1-2x)}{(n+1)}$;
2. $D_{n,m}^{(1/n,1/m)}((s-y);x,y)=\frac{(1-2y)}{(m+1)}$;
3. $D_{n,m}^{(1/n,1/m)}((t-x)^2;x,y)=\frac{x(1-x)(3n^2-5n-6)+2(n+1)}{(n+1)(n+2)(n+3)}$;
4. $D_{n,m}^{(1/n,1/m)}((s-y)^2;x,y)=\frac{y(1-y)(3m^2-5m-6)+2(m+1)}{(m+1)(m+2)(m+3)}$.

Remark 8.3 ([24]) From Remark 8.2, we get

1. $D_{n,m}^{(1/n,1/m)}((t-x)^2;x,y)\leq\frac{3}{n+1}\left(x(1-x)+\frac{1}{n+2}\right)$;
2. $D_{n,m}^{(1/n,1/m)}((s-y)^2;x,y)\leq\frac{3}{m+1}\left(y(1-y)+\frac{1}{m+2}\right)$.

Next, a Voronoskaja type asymptotic formula and the order of approximation in terms of Peetre's K-functional for the operators $D_{n,m}^{(1/n,1/m)}$ are given

Theorem 8.18 ([24]) *For $f\in C(I)$ we have*

$$\lim_{n\to\infty}n(D_{n,n}^{(1/n,1/n)}(f;x,y)-f(x,y))=(1-2x)\,f_x(x,y)+(1-2y)\,f_y(x,y)$$

$$+\frac{3}{2}\left\{x\,(1-x)\,f_{xx}(x,y)+y\,(1-y)\,f_{yy}(x,y)\right\}$$

uniformly on I.

Theorem 8.19 ([24]) *For the function $f\in C(I)$, we have*

$\left|D_{n,m}^{(1/n,1/m)}(f;x,y)-f(x,y)\right|$

$$\leq 4\mathcal{K}\left(f;\delta_{n,m}\left(x,y\right)\right)+\omega\left(f;\sqrt{\left(\frac{2x-1}{n+2}\right)^{2}+\left(\frac{2y-1}{m+2}\right)^{2}}\right)$$

$$\leq C\omega_{2}\left(f;\sqrt{\delta_{n,m}\left(x,y\right)}\right)+\omega\left(f;\sqrt{\left(\frac{2x-1}{n+2}\right)^{2}+\left(\frac{2y-1}{m+2}\right)^{2}}\right)$$

where

$$\delta_{n,m}\left(x,y\right)=\frac{1}{n+1}\delta_{n}^{2}\left(x\right)+\frac{1}{m+1}\delta_{m}^{2}\left(y\right)\ \textit{and}\ \delta_{k}^{2}\left(z\right)=z\left(1-z\right)+\frac{1}{k+2},\ k=n,m$$

and $z\in I$.

8.3.3 q-Durrmeyer Operators

Gupta [117] introduced the q-Durrmeyer operators as

$$D_{m,q}(f;x)=[m+1]_{q}\sum_{k=0}^{m}q^{-k}p_{m.k}(q;x)\int_{0}^{1}f(s)p_{m,k}(q;qs)d_{q}s \quad (8.3.8)$$

for all positive integer m and $f\in C[0,1]$. Bărbosu et al. [56] extended the operators (8.3.8) in two dimensions as follows:

For $f\in C(I)$ and $(x,y)\in I$, let $I=[0,1]\times[0,1]$, $q_{1},q_{2}\in(0,1)$ and $f\in C(I)$. The parametric extensions of (8.3.8) are the operators

$$D_{m,q_{1}}^{x},D_{n,q_{2}}^{y}:C(I)\rightarrow C(I),$$

defined for each positive integers m,n, $f\in C(I)$, $(x,y)\in I$ as follows:

$$D_{m,q_{1}}^{x}(f;x,y)=[m+1]_{q_{1}}\sum_{k=0}^{m}p_{m,k}(q_{1};x)\int_{0}^{1}f(s,y)p_{m,k}(q_{1};q_{1}s)d_{q_{1}}s, \quad (8.3.9)$$

$$D_{n,q_{2}}^{y}(f;x,y)=[n+1]_{q_{2}}\sum_{j=0}^{n}p_{n,j}(q_{2};y)\int_{0}^{1}f(x,t)p_{n,j}(q_{2};q_{2}t)d_{q_{2}}t, \quad (8.3.10)$$

The operators (8.3.9) and (8.3.10) are linear and positive. They commute on $C(I)$ and their product is the bivariate linear positive operator

$$D_{m,n,q_1,q_2} : C(I) \to C(I)$$

defined for each positive integers m, n, $f \in C(I)$, $(x, y) \in I$ as follows:

$$D_{m,n,q_1,q_2}(f; x, y) = [m+1]_{q_1}[n+1]_{q_2} \sum_{k=0}^{m} \sum_{j=0}^{n} q^{-k} q^{-j} p_{m,k}(q_1; x) p_{n,j}(q_2; y)$$
$$\times \int_0^1 \int_0^1 f(s,t) p_{m,k}(q_1; q_1 s) p_{n,j}(q_2; q_2 t) d_{q_1} s d_{q_2} t. \tag{8.3.11}$$

Lemma 8.16 ([56]) *For each positive integers m, n and $(x, y) \in I$ the operators D_{m,n,q_1,q_2} satisfy*

1. $D_{m,n,q_1,q_2}(1; x, y) = 1;$
2. $D_{m,n,q_1,q_2}(s; x, y) = \dfrac{1 + q_1 x[m]_{q_1}}{[m+2]_{q_1}};$
3. $D_{m,n,q_1,q_2}(t; x, y) = \dfrac{1 + q_2 y[n]_{q_2}}{[n+2]_{q_2}};$
4. $D_{m,n,q_1,q_2}(st; x, y) = \dfrac{1 + q_1 x[m]_{q_1}}{[m+2]_{q_1}} \dfrac{1 + q_2 y[n]_{q_2}}{[n+2]_{q_2}};$
5. $D_{m,n,q_1,q_2}(s^2; x, y) = \dfrac{q_1^3 x^2 [m]_{q_1}([m]_{q_1} - 1) + (1 + q_1^2) q_1 x [m]_{q_1} + 1 + q_1}{[m+3]_{q_1}[m+2]_{q_1}};$
6. $D_{m,n,q_1,q_2}(t^2; x, y) = \dfrac{q_2^3 y^2 [n]_{q_2}([n]_{q_2} - 1) + (1 + q_2^2) q_2 y [n]_{q_2} + 1 + q_2}{[n+3]_{q_2}[n+2]_{q_2}}.$

Corollary 8.1 ([56]) *The following identities hold true*

1. $D_{m,n,q_1,q_2}((s-x)^2; x, y) = \dfrac{1}{[m+3]_{q_1}[m+2]_{q_1}} x^2 \{[m]_{q_1}([m]_{q_1} - 1) q_1^3 - 2[m]_{q_1}[m+3]_{q_1} q_1$
   $+ [m+2]_{q_1}[m+3]_{q_1}\} + \dfrac{x\{[m]_{q_1} q_1 (1+q_1)^2 - 2[m+3]_{q_1}\} + 1 + q_1}{[m+3]_{q_1}[m+2]_{q_1}};$
2. $D_{m,n,q_1,q_2}((t-y)^2; x, y) = y^2 \left\{ \dfrac{[n]_{q_2}([n]_{q_2} - 1) q_2^3 - 2[n]_{q_2}[n+3]_{q_2} q_2}{[n+3]_{q_2}[n+2]_{q_2}} \right.$
   $\left. + \dfrac{[n+2]_{q_2}[n+3]_{q_2}}{[n+3]_{q_2}[n+2]_{q_2}} \right\} + \dfrac{y\{[n]_{q_2} q_2 (1+q_2)^2 - 2[n+3]_{q_2}\} + 1 + q_2}{[n+3]_{q_2}[n+2]_{q_2}}.$

Lemma 8.17 ([56]) *Let $m, n \in \mathbb{N}$ and $q_1, q_2 \in (0, 1)$. Then*

1. $D_{m,n,q_1,q_2}((s-x)^2; x, y) \le \dfrac{2}{[m+2]_{q_1}} \delta^2_{m,q_1}(x);$
2. $D_{m,n,q_1,q_2}((t-y)^2; x, y) \le \dfrac{2}{[n+2]_{q_2}} \delta^2_{n,q_2}(y).$

8.3.4 Bivariate q-Bernstein–Chlodowsky–Durrmeyer Operators

Büyükyazici and Sharma [67] introduced the Durrmeyer type generalization of the q-Bernstein–Chlodowsky operators on a rectangle domain as follows:

For any $\alpha_n > 0,\ \beta_m > 0$, if

$$\square_{\alpha_n,\beta_m} := \{(x,y) : 0 \le x \le \alpha_n,\ 0 \le y \le \beta_m\},$$

then the operator $\widetilde{D}_{n,m}^{q_n,q_m}$ is defined as

$$\widetilde{D}_{n,m}^{q_n,q_m}(f;x,y) = \frac{[n+1]_{q_n}}{\alpha_n}\frac{[m+1]_{q_m}}{\beta_m}\sum_{k=0}^{n}\sum_{j=0}^{m} q_n^{-k} q_m^{-j} \psi_{n,m}^{q_n,q_m}\left(\frac{x}{\alpha_n},\frac{y}{\beta_m}\right)$$

$$\times\left(\int_{t_1=0}^{\alpha_n}\int_{t_2=0}^{\beta_m} \psi_{n,m}^{q_n,q_m}\left(q_n\frac{t_1}{\alpha_n}, q_m\frac{t_2}{\beta_m} f(t_1,t_2) d_{q_n}t_1 d_{q_m}t_2\right)\right),$$

where

$$\psi_{n,m}^{q_n,q_m}(u,v) = \Omega_{k,n,q_n}(u)\Omega_{j,m,q_m}(v),$$

$$\Omega_{k,n,q_n}(u) = \begin{bmatrix} n \\ k \end{bmatrix}_{q_n} u^k \prod_{s=0}^{n-k-1}(1-q_n^s)\ (k=0,1,\ldots,n).$$

Lemma 8.18 ([67]) *We have the following identities:*

$$\sum_{k=0}^{n}\frac{[k]_{q_n}}{[n]_{q_n}}\alpha_n\Omega_{k,n,q_n}(u)\left(\frac{x}{\alpha_n}\right) = x,$$

$$\sum_{k=0}^{n}\frac{[k]_{q_n}^2}{[n]_{q_n}^2}\alpha_n^2\Omega_{k,n,q_n}(u)\left(\frac{x}{\alpha_n}\right) = x^2 + \frac{x(\alpha_n - x)}{[n]_{q_n}},$$

$$\sum_{k=0}^{n}\frac{[k]_{q_n}^3}{[n]_{q_n}^3}\alpha_n^3\Omega_{k,n,q_n}(u)\left(\frac{x}{\alpha_n}\right) = \left(1-\frac{3}{[n]_{q_n}}+\frac{2}{[n]_{q_n}^2}\right)x^3 + \left(\frac{3\alpha_n}{[n]_{q_n}}-\frac{3\alpha_n}{[n]_{q_n}^2}\right)x^2$$
$$+\frac{\alpha_n^2}{[n]_{q_n}^2}x,$$

$$\sum_{k=0}^{n}\frac{[k]_{q_n}^4}{[n]_{q_n}^4}\alpha_n^4\Omega_{k,n,q_n}(u)\left(\frac{x}{\alpha_n}\right) = \left(1-\frac{6}{[n]_{q_n}}+\frac{11}{[n]_{q_n}^2}-\frac{6}{[n]_{q_n}^3}\right)x^4 + \left(\frac{6\alpha_n}{[n]_{q_n}}-\frac{18\alpha_n}{[n]_{q_n}^2}\right.$$
$$\left.+\frac{12\alpha_n}{[n]_{q_n}^3}\right)x^3 + \left(\frac{7\alpha_n^2}{[n]_{q_n}^2}-\frac{7\alpha_n^2}{[n]_{q_n}^3}\right)x^2 + \frac{\alpha_n^3}{[n]_{q_n}^3}x,$$

$$\sum_{j=0}^{m} \frac{[j]_{q_m}}{[m]_{q_m}} \beta_m \Omega_{j,m,q_m}(v)\left(\frac{y}{\beta_m}\right) = y,$$

$$\sum_{j=0}^{m} \frac{[j]_{q_m}^2}{[m]_{q_m}^2} \beta_m^2 \Omega_{j,m,q_m}(v)\left(\frac{y}{\beta_m}\right) = y^2 + \frac{y(\beta_m - y)}{[m]_{q_m}},$$

$$\sum_{j=0}^{m} \frac{[j]_{q_m}^3}{[m]_{q_m}^3} \beta_m^3 \Omega_{j,m,q_m}(v)\left(\frac{y}{\beta_m}\right) = \left(1 - \frac{3}{[m]_{q_m}} + \frac{2}{[m]_{q_m}^2}\right) y^3 + \left(\frac{3\beta_m}{[m]_{q_m}} - \frac{3\beta_m}{[m]_{q_m}^2}\right) y^2 + \frac{\beta_m^2}{[m]_{q_m}^2} y,$$

$$\sum_{j=0}^{m} \frac{[j]_{q_m}^4}{[m]_{q_m}^4} \beta_m^4 \Omega_{j,m,q_m}(v)\left(\frac{y}{\beta_m}\right) = \left(1 - \frac{6}{[m]_{q_m}} + \frac{11}{[m]_{q_m}^2} - \frac{6}{[m]_{q_m}^3}\right) y^4 + \left(\frac{6\beta_m}{[m]_{q_m}} - \frac{18\beta_m}{[m]_{q_m}^2} + \frac{12\beta_m}{[m]_{q_n}^3}\right) y^3 + \left(\frac{7\beta_m^2}{[m]_{q_m}^2} - \frac{7\beta_y^2}{[m]_{q_m}^3}\right) y^2 + \frac{\beta_m^3}{[m]_{q_m}^3} y.$$

Lemma 8.19 ([67]) *For $r = 0, 1, 2, \ldots$, we have the following identities:*

$$\widetilde{D}_{n,m}^{q_n,q_m}(t_1^r; x, y) = \frac{[n+1]!_{q_n} \alpha_n^r}{[n+r+1]_{q_n}!} \sum_{k=0}^{n} \Omega_{k,n,q_n}\left(\frac{x}{\alpha_n}\right) \frac{[r+k]!_{q_n}}{[k]!_{q_n}},$$

$$\widetilde{D}_{n,m}^{q_n,q_m}(t_2^r; x, y) = \frac{[m+1]!_{q_m} \beta_m^r}{[m+r+1]_{q_m}!} \sum_{j=0}^{m} \Omega_{j,m,q_m}\left(\frac{y}{\beta_m}\right) \frac{[r+j]!_{q_m}}{[j]!_{q_m}}.$$

In [67] authors obtained the Korovkin type approximation properties and the rates of convergence using the modulus of continuity and Peetre's K-functional.

Theorem 8.20 ([67]) *If $f \in C(\square_{\alpha_n,\beta_m})$, then for any sufficiently large fixed positive real a and b, the equality*

$$\lim_{n,m\to\infty} ||\widetilde{D}_{n,m}^{q_n,q_m}(f; x, y) - f(x, y)||_{C(\square_{\alpha_n,\beta_m})} = 0,$$

holds.

Lemma 8.20 ([67]) *Let $n, m > 3$ be a given natural number and let $q_0 = q_0(n, m) \in (0, 1)$ be the least number such that*

$$q_n^{n+2} + q_n + 2 < q_n^{n+1} + 2q_n^n + 2q_n^{n-1} + \ldots + 2q_n^3 + q_n^2$$

and

$$q_m^{m+2} + q_m + 2 < q_m^{m+1} + 2q_m^m + 2q_m^{m-1} + \ldots + 2q_m^3 + q_m^2$$

for every $q_n, q_m \in (q_0, 1)$. *Then,*

$$\widetilde{D}_{n,m}^{q_n,q_m}((t_1 - x)^2; x, y) \le \frac{2}{[n+2]_{q_n}} \left(\phi_1^2(x) + \frac{\alpha_n^2}{[n+3]_{q_n}} \right),$$

$$\widetilde{D}_{n,m}^{q_n,q_m}((t_2 - y)^2; x, y) \le \frac{2}{[m+2]_{q_m}} \left(\phi_2^2(y) + \frac{\beta_m^2}{[m+3]_{q_m}} \right).$$

where $\phi_1^2(x) = x(\alpha_n - x)$ *and* $\phi_2^2(y) = y(\beta_m - y)$, $(x, y) \in \square_{\alpha_n,\beta_m}$.

Corollary 8.2 ([67]) *From Lemma (8.20), for* $(x, y) \in \square_{\alpha_n,\beta_m}$, *we have*

$$\widetilde{D}_{n,m}^{q_n,q_m}((t_1 - x)^2; x, y) \le \frac{2(a\alpha_n[n+3]_{q_n} + \alpha_n^2)}{[n+2]_{q_n}[n+3]_{q_n}},$$

$$\widetilde{D}_{n,m}^{q_n,q_m}((t_2 - y)^2; x, y) \le \frac{2(b\beta_m[m+3]_{q_m} + \beta_m^2)}{[m+2]_{q_m}[m+3]_{q_m}}.$$

For the next theorem, we use the above corollary

Theorem 8.21 ([67]) *For any* $f \in C(\square_{\alpha_n,\beta_m})$ *and* $q_n, q_m \to 1$ *as* $n, m \to \infty$ *the following inequalities hold*

$$|\widetilde{D}_{n,m}^{q_n,q_m}(f; x, y) - f(x, y)| \le 2\left[\omega^{(1)}\left(f; \sqrt{\frac{2(a\alpha_n[n+3]_{q_n} + \alpha_n^2)}{[n+2]_{q_n}[n+3]_{q_n}}}\right) + \omega^{(1)}\left(f; \sqrt{\frac{2(b\beta_m[m+3]_{q_m} + \beta_m^2)}{[m+2]_{q_m}[m+3]_{q_m}}}\right)\right],$$

$$|\widetilde{D}_{n,m}^{q_n,q_m}(f; x, y) - f(x, y)| \le 2\omega\left(f; \left\{\frac{2(a\alpha_n[n+3]_{q_n} + \alpha_n^2)}{[n+2]_{q_n}[n+3]_{q_n}} + \frac{2(b\beta_m[m+3]_{q_m} + \beta_m^2)}{[m+2]_{q_m}[m+3]_{q_m}}\right\}^{1/2}\right).$$

Theorem 8.22 ([67]) *If* $f \in C(\square_{\alpha_n,\beta_m})$, *then*

$$||\widetilde{D}_{n,m}^{q_n,q_m}(f; x, y) - f(x, y)||_{C(\square_{\alpha_n,\beta_m})} \le 2K\left(f; \frac{\delta_{n,m}}{2}\right),$$

where

$$\delta_{n,m} = \max\left\{\frac{\alpha_n}{[n+2]_{q_n}}, \frac{2(a\alpha_n[n+3]_{q_n}+\alpha_n^2)}{[n+2]_{q_n}[n+3]_{q_n}}, \frac{2(b\beta_m[m+3]_{q_m}+\beta_m^2)}{[m+2]_{q_m}[m+3]_{q_m}}\right\}$$

8.4 Bivariate Chlodowsky–Szász–Kantorovich–Charlier Type Operators

Agrawal and Ispir in [21] introduced the Szász variant based Charlier polynomials defined as

$$S_m(f;x,a) = \sum_{j=0}^{\infty} \Pi_{m,j}(b_m y, a) f\left(\frac{j}{c_m}\right); a > 1$$

where (b_m), (c_m) are increasing sequences of positive numbers such that

$$c_m \geq 1,\ b_m \geq 1 \text{ and } \lim_{n\to\infty}(1/c_m) = 0, b_m/c_m = 1 + O(1/c_m).$$

Also, Agrawal and Ispir [21] introduced bivariate operators by combining the Bernstein–Chlodowsky operators and Szász–Charlier type operators as follows:

$$S_{n,m}^a(f;x,y) = \sum_{k=0}^{n}\sum_{j=0}^{\infty} p_{n,k}\left(\frac{x}{a_n}\right)\Pi_{m,j}(b_m y, a) f\left(\frac{k}{n}a_n, \frac{j}{c_n}\right) \tag{8.4.1}$$

for all $n, m \in N,\ f \in C(I_{a_n})$ with

$$I_{a_n} = \left\{(x,y) : 0 \leq x \leq a_n, y \geq 0\right\}$$

and

$$C(I_{a_n}) = \{f : I_{a_n} \to R^+ \text{ is continuous}\}.$$

The weighted approximation properties of bivariate modified Szász operators are studied in [97, 143, 217]. Note that the operator $S_{n,m}^a$ is the tensorial product of ${}_xC_n$ and ${}_yS_m^a$, i.e., $S_{n,m}^a = {}_xC_n \times {}_yS_m^a$ where

$${}_xC_n(f;x,y) = \sum_{k=0}^{n} p_{n,k}\left(\frac{x}{a_n}\right) f\left(\frac{ka_n}{n}, y\right)$$

and

$$ {}_yS_m^a(f;x,y)=\sum_{j=0}^{\infty}\Pi_{m,j}(b_my,a)f\left(x,\frac{j}{c_m}\right). $$

Agrawal et al. [25] introduced a new bivariate operator associated with a combination of Kantorovich variant of the operators given by (8.4.1) as follows: For all $n,m\in N$ and $f\in C(I_{a_n})$, we define

$$ C_{n,m}^a(f;x,y)=\frac{n}{a_n}c_m\sum_{k=0}^{n}\sum_{j=0}^{\infty}p_{n,k}\left(\frac{x}{a_n}\right)\Pi_{m,j}(b_my,a)\int_{\frac{j}{c_m}}^{\frac{j+1}{c_m}}\int_{\frac{k}{n}a_n}^{\frac{k+1}{n}a_n}f(t,s)dtds \tag{8.4.2} $$

where $a>1$, and the sequences $(a_n),\ (b_m),(c_m)$ are defined as above and satisfy the following conditions

$$ \lim_{n\to\infty}(a_n/n)=0 \text{ and } \lim_{m\to\infty}(1/c_m)=0,\ b_m/c_m=1+O(1/c_m). \tag{8.4.3} $$

For operators defined by (8.4.2) we have

$$ C_{n,m}^a(f;x,y)={}_xC_n^*({}_y^*S_m^a(f;x,y))={}_y^*S_m^a({}_xC_n^*(f;x,y)) $$

where

$$ {}_xC_n^*(f;x,y)=\frac{n}{a_n}\sum_{k=0}^{n}p_{n,k}\left(\frac{x}{a_n}\right)\int_{\frac{k}{n}a_n}^{\frac{k+1}{n}a_n}f(t,y)dt $$

and

$$ {}_y^*S_m^a(f;x,y)=c_m\sum_{j=0}^{\infty}\Pi_{m,j}(b_my,a)\int_{\frac{j}{c_m}}^{\frac{j+1}{c_m}}f(x,s)ds. $$

Next, the degree of approximation of the operator $C_{n,m}^a$ given by (8.4.2) will be established in the space of continuous functions on compact set

$$ I_{de}=[0,d]\times[0,e]\subset I_{a_n}. $$

For $I_{de}=[0,d]\times[0,e]$, let $C(I_{de})$, denote the space of all real valued continuous functions on I_{de}, endowed with the norm

$$\| f \|_{C(I_{de})} = \sup_{(x,y)\in I_{de}} |f(x,y)|.$$

In what follows, let

$$e_{ij} : I_{a_n} \to R,\ e_{ij}(x,y) = x^i y^j,\ (x,y) \in I_{a_n},\ (i,j) \in N^0 \times N^0 \text{ with } i+j \le 4$$

be the two-dimensional test functions.

Lemma 8.21 ([25]) *For the operators ${}_xC_n^*$ and ${}_y^*S_m^a$, we have these inequalities:*

1. ${}_xC_n^*(e_{00}; x, y) = 1;$
2. ${}_xC_n^*(e_{10}; x, y) = x;$
3. ${}_xC_n^*(e_{20}; x, y) = \left(1 - \frac{1}{n}\right)x^2 + \frac{a_n x}{n};$
4. ${}_y^*S_m^a(e_{01}; x, y) = \frac{b_m y+1}{c_m};$
5. ${}_y^*S_m^a(e_{02}; x, y) = \frac{b_m^2 y^2}{c_m^2} + \frac{b_m}{c_m^2}\left(2 + \frac{1}{(a-1)}\right)y + \frac{1}{c_m^2};$
6. ${}_y^*S_m^a(e_{03}; x, y) = \frac{b_m^3 y^3}{c_m^3} + \frac{3b_m^2}{c_m^3}\left(1 + \frac{1}{(a-1)}\right)y^2 + \frac{b_m}{c_m^3}\left(3 + \frac{3}{(a-1)} + \frac{2}{(a-1)^2}\right)y + \frac{1}{c_m^3};$
7. ${}_y^*S_m^a(e_{04}; x, y) = \frac{b_m^4 y^4}{c_m^4} + \frac{2b_m^3}{c_m^3}\left(1 + \frac{1}{(a-1)}\right)y^3 + \frac{b_m^2 y^2}{c_m^4}\left(4 + \frac{6}{(a-1)} + \frac{11}{(a-1)^2}\right) + \frac{2b_m y}{c_m^4}\left(2 + \frac{2}{(a-1)} + \frac{2}{(a-1)^2} + \frac{3}{(a-1)^3}\right) + \frac{1}{c_m^4}.$

Lemma 8.22 ([25]) *The following statements hold:*

1. $C_{n,m}^a(e_{00}; x, y) = 1;$
2. $C_{n,m}^a(e_{10}; x, y) = x + \frac{a_n}{2n};$
3. $C_{n,m}^a(e_{01}; x, y) = \frac{b_m y}{c_m} + \frac{3}{2c_m};$
4. $C_{n,m}^a(e_{20}; x, y) = \left(1 - \frac{1}{n}\right)x^2 + 2\frac{a_n}{n}x + \frac{a_n^2}{3n^2};$
5. $C_{n,m}^a(e_{30}; x, y) = x^3 + \frac{3a_n x^2}{n}\left(1 - \frac{x}{a_n}\right) + \frac{a_n^2 x}{n^2}\left(1 - \frac{x}{a_n}\right)\left(1 - \frac{2x}{a_n}\right) + \frac{3a_n}{2n}\left(\frac{a_n^2}{3n^2} + \left(1 - \frac{1}{n}\right)x^2 + \frac{2a_n x}{n}\right) + \frac{a_n^2}{n^2}\left(x + \frac{a_n}{2n}\right) + \frac{1}{4a_n^3};$
6. $C_{n,m}^a(e_{40}; x, y) = x^4 + 6x^3\frac{a_n}{n}\left(1 - \frac{x}{a_n}\right) + \frac{a_n^2 x^2}{n^2}\left(6\left(1 - \frac{x}{a_n^2}\right)^2 - 3\frac{x}{a_n}\left(1 - \frac{x}{a_n}\right) + \left(1 - \frac{x}{a_n}\right)\left(1 - \frac{2x}{a_n}\right)\right) + \frac{x a_n^3}{n^3}\left(\left(1 - \frac{x}{a_n}\right)^2\left(1 - \frac{2x}{a_n}\right) - \frac{x}{a_n}\left(1 - \frac{x}{a_n}\right)\left(1 - \frac{2x}{a_n}\right) - \frac{2x}{a_n}\left(1 - \frac{x}{a_n}\right)^2\right) + \frac{2a_n}{n}\left(x^3 + \frac{3a_n x^2}{n}\left(1 - \frac{x}{a_n}\right) + \frac{a_n^2 x}{n^2}\left(1 - \frac{x}{a_n}\right)\left(1 - \frac{2x}{a_n}\right)\right.$

$$+\frac{3a_n}{2n}\left(\frac{a_n^2}{3n^2}+\left(1-\frac{1}{n}\right)x^2+\frac{2a_nx}{n}\right)+\frac{a_n^2}{n^2}\left(x+\frac{a_n}{2n}\right)+\frac{1}{4a_n^3}\right)$$

$$+\frac{2a_n^2}{n^2}\left(\frac{a_n^2}{3n^2}+\left(1-\frac{1}{n}\right)x^2+\frac{2a_nx}{n}\right)+\frac{a_n^3}{n^3}\left(x+\frac{a_n}{2n}\right)+\frac{a_n^4}{5n^4};$$

7. $C^a_{n,m}(e_{01};x,y)=\frac{b_my}{c_m}+\frac{3}{2c_m};$

8. $C^a_{n,m}(e_{02};x,y)=\frac{b_m^2y^2}{c_m^2}+\frac{b_my}{c_m^2}\left(3+\frac{1}{(a-1)}\right)+\frac{7}{3c_m^2};$

9. $C^a_{n,m}(e_{03};x,y)=\frac{b_m^3y^3}{c_m^3}+\frac{b_m^2y^2}{c_m^3}\left(\frac{9}{2}+\frac{3}{a-1}\right)+\frac{b_my}{c_m^3}\left(7+\frac{9}{2(a-1)}+\frac{2}{(a-1)^2}\right)+\frac{15}{4c_m^3};$

10. $C^a_{n,m}(e_{04};x,y)=\frac{b_m^4y^4}{c_m^4}+\frac{b_m^3y^3}{c_m^4}\left(4+\frac{6}{a-1}\right)+\frac{b_m^2y^2}{c_m^4}\left(12+\frac{12}{a-1}+\frac{11}{(a-1)^2}\right)$
$+\frac{b_my}{c_m^4}\left(15+\frac{12}{a-1}+\frac{8}{(a-1)^2}+\frac{6}{(a-1)^3}\right)+\frac{31}{5c_m^4}.$

Theorem 8.23 *If $f\in C(I_{de})$, then the operators $C^a_{n,m}$ given by (8.4.2) converge uniformly to f on the compact set I_{de}, as $n,m\to\infty$.*

Below we present the rate of convergence of the approximation of the bivariate operators (8.4.2) defined by means of modulus of continuity of the functions.

Theorem 8.24 ([25]) *For any $f\in C(I_{de})$, the following inequalities are satisfied*

$$\left|C^a_{n,m}(f;x,y)-f(x,y)\right|\leq 2\big(\omega^{(1)}(f;\delta_n)+\omega^{(2)}(f;\delta_m)\big)$$

$$\left|C^a_{n,m}(f;x,y)-f(x,y)\right|\leq 2\omega(f;\delta_{n,m})$$

where $\delta_n=\delta_n(x)$, $\delta_m=\delta_m(y)$ and $\delta_{n,m}=\delta_{n,m}(x,y)$.

Theorem 8.25 ([25]) *Suppose that $f\in Lip_L(f;\gamma_1,\gamma_2)$. Then, for every $(x,y)\in I_{de}$, we have*

$$\left|C^a_{n,m}(f;x,y)-f(x,y)\right|\leq L(\delta_n)^{\gamma_1/2}(\delta_m)^{\gamma_2/2},$$

where $\delta_n=\delta_n(x)$ and $\delta_m=\delta_m(y)$.

Theorem 8.26 ([25]) *Let $f\in C^1(I_{de})$. Then, for every $(x,y)\in I_{de}$, we have the following inequality*

$$\left|C^a_{n,m}(f;x,y)-f(x,y)\right|\leq\| f_x^{'}\|_{C(I_{de})}\sqrt{\delta_n(x)}+\| f_y^{'}\|_{C(I_{de})}\sqrt{\delta_m(y)}.$$

Theorem 8.27 ([25]) *Let $f\in C(I_{de})$. Consider the operators*

$$\hat{C}^a_{n,m}(f;x,y)=C^a_{n,m}(f;x,y)+f(x,y)-f\left(x+\frac{a_n}{2n},\frac{b_my}{c_m}+\frac{3}{2c_m}\right). \tag{8.4.4}$$

Then, for all $g\in C^2(I_{de})$, we have the estimate

$$C_{n,m}^{a}(f;x,y)-f(x,y)\leq L\Big\{\omega_2\left(f;\sqrt{\chi_{n,m}(x,y)}\right)$$

$$+\,min\left\{1,\chi_{n,m}(x,y)\right\}\|\,f\,\|_{C(I_{de})}\Big\}+\omega\left(f;\sqrt{(\tfrac{a_n}{2n})^2+(\tfrac{b_m y}{c_m}+\tfrac{3}{2c_m}-y)^2}\right),$$

where

$$\chi_{n,m}(x,y)=O\left(\frac{a_n}{n}\right)(x^2+x+1)+(\frac{a_n}{2n})^2+\frac{\tau(a)}{c_m}(y^2+y+1)+\frac{((b_m-c_m)y+3)^2}{c_m^2}.$$

Theorem 8.28 (From [91, 92].) *Let $K_{n,m}$ be a sequence of linear operators acting from $C_\rho(R_+^2)$ to $B_\rho(R_+^2)$, and let $\rho_1(x,y)\geq 1$ be a continuous function for which*

$$\lim_{|v|\to\infty}\frac{\rho(v)}{\rho_1(v)}=0,\ (where\ v=(x,y)). \tag{8.4.5}$$

If $K_{n,m}$ satisfies the conditions of Theorem 2.2, then

$$\lim_{n,m\to\infty}\|\,K_{n,m}f-f\,\|_{\rho_1}=0,$$

for all $f\in C_\rho(R_+^2)$.

Now, we consider the following positive linear operators $K_{n,m}$, defined by

$$K_{n,m}(f;x,y)=\begin{cases}C_{n,m}^{a}(f;x,y) & when\ (x,y)\in I_{a_nd_m}\\ \\ f(x,y) & when\ (x,y)\in R_+^2\backslash I_{a_nd_m}\end{cases} \tag{8.4.6}$$

where $I_{a_nd_m}=\{(x,y):0\leq x\leq a_n,0\leq y\leq d_m\}$, (d_m) be a sequences such that $\lim\limits_{m\to\infty}d_m=\infty$.

Theorem 8.29 ([25]) *Let $\rho(x,y)=1+x^2+y^2$ be a weight function and $K_{n,m}(f;x,y)$ be a sequence of linear positive operators defined by (8.4.6). Then, for all $f\in C_\rho(R_+^2)$, we have*

$$\lim_{n,m\to\infty}\|\,K_{n,m}f-f\,\|_{\rho_1}=0,$$

where $\rho_1(x,y)$ is a continuous function satisfying condition (8.4.5).

Theorem 8.30 ([25]) *Let $\{K_{n,m}\}$ be a sequence linear positive operators defined by (8.4.6). Then, for each function $f\in C_\rho(R_+^2)$, we have*

$$\lim_{n,m\to\infty}\|\,K_{n,m}f-f\,\|_{\rho}=0.$$

Now the order of approximation of the operators $C_{n,m}^{a}$ in the terms of the weighted modulus of continuity $\Omega(f;\delta_n,\delta_m)$ (see [142]) is defined by

$$\Omega(f;\delta_n,\delta_m)=\sup_{(x,y)\in R_+^2}\ \sup_{|h_1|\leq\delta_1,|h_2|\leq\delta_2}\frac{|f(x+h_1,y+h_2)-f(x,y)|}{\rho(x,y)\rho(h1,h2)},\ f\in C_\rho^*(R^+).$$

By the properties of weighted modulus of continuity $\Omega(f;\delta_n,\delta_m)$ (see [142], p. 577) the inequality:

$$|f(t,s)-f(x,y)|\leq 8\Omega(f;\delta_n,\delta_m)(1+x^2+y^2)g(t,x)g(s,y)$$

holds, where

$$g(t,x)=\left(\left(1+\frac{|t-x|}{\delta_n}\right)\left(1+(t-x)^2\right)\right),$$

$$g(s,y)=\left(\left(1+\frac{|s-y|}{\delta_m}\right)\left(1+(s-y)^2\right)\right).$$

Theorem 8.31 ([25]) *For each $f\in C_\rho^*(R^+)$, there exists a positive constant M independent of n,m such that the inequality*

$$\| C_{n,m}^a(f;x,y)-f(x,y)\|_{\rho^3}\ \leq M\Omega(f;\delta_n,\delta_m),$$

is satisfied for a sufficiently large n,m, where $\delta_n=\frac{a_n}{n}$ and $\delta_m=\frac{\upsilon(a)}{n}$.

8.5 Bivariate q-Dunkl Analogue of the Szász–Mirakjan–Kantorovich Operator

In 2016, Mursaleen et al. [179] defined the q-Dunkl analogue of Szász–Mirakjan–Kantorovich operator as:

$$T_{n,q}^*(f,x)=\frac{[n]_q}{e_{\mu,q}([n]_q\frac{x}{q})}\sum_{k=0}^{\infty}\frac{([n]_qx)^k}{\gamma_{\mu,q}(k)(q)^k}\int_{\frac{q[k+2\mu\theta_k]_q}{[n]_q}}^{\frac{[k+1+2\mu\theta_k]_q}{[n]_q}}f(t)d_qt,\qquad(8.5.1)$$

where $\mu>\frac{1}{2}, n\in\mathbb{N}, 0<q<1, 0\leq x<\frac{1}{1-q^n}$, and f is a continuous nondecreasing function on the interval $[0,\infty)$. Let $0<q_{n_i}<1$ be sequences in $(0,1)$ such that $q_{n_i}\to 1$ and $q_{n_i}^{n_i}\to a_i,\quad(0\leq a_i<1)$, as $n_i\to\infty$ for $i=1,2$. For $f\in C_B(I)$, where $I=[0,\infty)\times[0,\infty)$, the class of bounded and uniformly continuous functions on I endowed with the norm $\|f\|_{C_B(I)}=\sup_{x,y\in I}|f(x,y)|$, the bivariate generalization of the operator, given by (8.5.1) is introduced by Agrawal et al. [26] which is defined as follows:
$T_{n_1,n_2,q_{n_1},q_{n_2}}^*(f;x,y)$

$$= \frac{[n_1]_{q_{n_1}}}{e_\mu([n_1]_{q_{n_1}}\frac{x}{q_{n_1}})}\frac{[n_2]_{q_{n_2}}}{e_\mu([n_2]_{q_{n_2}}\frac{y}{q_{n_2}})}\sum_{k_2=0}^{\infty}\sum_{k_1=0}^{\infty}\frac{([n_1]_{q_{n_1}}x)^{k_1}}{\gamma_\mu(k_1)q_{n_1}^{k_1}}\frac{([n_2]_{q_{n_2}}x)^{k_2}}{\gamma_\mu(k_2)q_{n_2}^{k_2}}$$

$$\times \int_{\frac{q_{n_2}[k_2+2\mu\theta_{k_2}]_{q_{n_2}}}{[n_2]_{q_{n_2}}}}^{\frac{[k_2+1+2\mu\theta_{k_2}]_{q_{n_2}}}{[n_2]_{q_{n_2}}}}\int_{\frac{q_{n_1}[k_1+2\mu\theta_{k_1}]_{q_{n_1}}}{[n_1]_{q_{n_1}}}}^{\frac{[k_1+1+2\mu\theta_{k_1}]_{q_{n_1}}}{[n_1]_{q_{n_1}}}} f(t,s)d_{q_1}(t)\, d_{q_2}(s). \tag{8.5.2}$$

Let $e_{ij}(t,s)=t^is^j$, $(t,s)\in I$, i, j be the nonnegative integers with $i+j\leq 2$.

Lemma 8.23 ([26]) *For the operators $T^*_{n_1,n_2,q_{n_1},q_{n_2}}(f;x,y)$ there hold:*

1. $T^*_{n_1,n_2,q_{n_1},q_{n_2}}(e_{00};x,y)=1$;
2. $T^*_{n_1,n_2,q_{n_1},q_{n_2}}(e_{10};x,y)=\frac{1}{[2]_{q_{n_1}}[n_1]_{q_{n_1}}}+\frac{2x}{[2]_{q_{n_1}}}$,
3. $T^*_{n_1,n_2,q_{n_1},q_{n_2}}(e_{01};x,y)=\frac{1}{[2]_{q_{n_2}}[n_2]_{q_{n_2}}}+\frac{2y}{[2]_{q_{n_2}}}$,
4. $\frac{1}{[3]_{q_{n_1}}[n]^2_{q_{n_1}}}+\frac{3x}{[3]_{q_{n_1}}[n]_{q_{n_1}}}\left(1+q_{n_1}^{2\mu+1}[1-2\mu]_{q_{n_1}}\frac{e_\mu([n]_{q_{n_1}}x)}{e_\mu([n]_{q_{n_1}}\frac{x}{q_{n_1}})}\right)+\frac{3x^2}{[3]_{q_{n_1}}}$
$\leq T^*_{n_1,n_2,q_{n_1},q_{n_2}}(e_{20};x,y)$
$\leq \frac{1}{[3]_{q_{n_1}}[n]^2_{q_{n_1}}}+\frac{3x}{[3]_{q_{n_1}}[n]_{q_{n_1}}}(1+[1+2\mu]_{q_{n_1}})+\frac{3x^2}{[3]_{q_{n_1}}}$,
5. $\frac{1}{[3]_{q_{n_2}}[n]^2_{q_{n_2}}}+\frac{3y}{[3]_{q_{n_2}}[n]_{q_{n_2}}}\left(1+q_{n_2}^{2\mu+1}[1-2\mu]_{q_{n_2}}\frac{e_\mu([n]_{q_{n_2}}y)}{e_\mu([n]_{q_{n_2}}\frac{y}{q_{n_2}})}\right)+\frac{3y^2}{[3]_{q_{n_2}}}$
$\leq T^*_{n_1,n_2,q_{n_1},q_{n_2}}(e_{02};x,y)$
$\leq \frac{1}{[3]_{q_{n_2}}[n]^2_{q_{n_2}}}+\frac{3y}{[3]_{q_{n_2}}[n]_{q_{n_2}}}(1+[1+2\mu]_{q_{n_2}})+\frac{3y^2}{[3]_{q_{n_2}}}$.

For $g\in C_B(I)$, $I=[0,\infty)\times[0,\infty)$, the total modulus of continuity is defined as follows:

$$\omega(g;\delta_1,\delta_2)=sup\left\{|g(t,s)-g(x_0,y_0)|:\ |t-x_0|\leq\delta_1, |s-y_0|\leq\delta_2\right\},$$

where $\delta_1,\delta_2>0$. In what follows, let $\delta_{n_1}(x)=T^*_{n_1,n_2,q_{n_1},q_{n_2}}((t-x)^2;x,y)$ and $\delta_{n_2}(y)=T^*_{n_1,n_2,q_{n_1},q_{n_2}}((s-y)^2;x,y)$.

First we estimate the rate of convergence of the operators $T^*_{n_1,n_2,q_{n_1},q_{n_2}}$ for functions in $C_B(I)$.

Theorem 8.32 ([26]) *For $f\in C_B(I)$ and $(x,y)\in I$, we have*

$$\left|T^*_{n_1,n_2,q_{n_1},q_{n_2}}(f;x,y)-f(x,y)\right|\leq 4\omega(f;\sqrt{\delta_{n_1}(x)},\sqrt{\delta_{n_2}(y)}).$$

Theorem 8.33 ([26]) *If $f(x,y)\in C_B(I)$ accepts the partial derivatives $\frac{\partial f}{\partial x}$ and $\frac{\partial f}{\partial y}$ in I, then the following inequality holds*

$$\left|T^*_{n_1,n_2,q_{n_1},q_{n_2}}(f;x,y)-f(x,y)\right| \leq M_1\lambda_{n_1}(x)+\omega(f_x^{'},\delta_{n_1}(x))(1+\sqrt{\delta_{n_1}(x)})$$
$$+M_2\lambda_{n_2}(y)+\omega(f_y^{'},\delta_{n_2}(y))(1+\sqrt{\delta_{n_2}(y)}).$$

where M_1 and M_2 are positive constants such that

$$\left|\frac{\partial f}{\partial x}\right| \leq M_1 \text{ and } \left|\frac{\partial f}{\partial y}\right| \leq M_2,$$

and

$$\lambda_{n_1}(x)=\left|\frac{1}{[2]_{q_{n_1}}[n_1]_{q_{n_1}}}+x\left(\frac{2}{[2]_{q_{n_1}}}-1\right)\right|,\ \lambda_{n_2}(y)=\left|\frac{1}{[2]_{q_{n_2}}[n_2]_{q_{n_2}}}+y\left(\frac{2}{[2]_{q_{n_2}}}-1\right)\right|.$$

Theorem 8.34 ([26]) *Let $f \in Lip_M(\xi_1,\xi_2)$. Then for all $(x,y) \in I$, we have*

$$\left|T^*_{n_1,n_2,q_{n_1},q_{n_2}}(f;x,y)-f(x,y)\right| \leq M(\delta_{n_1}(x))^{\frac{\xi_1}{2}}(\delta_{n_2}(y))^{\frac{\xi_2}{2}}.$$

Let $C_B^1(I)$ denote the space of all continuous functions on I such that their first partial derivatives are continuous in I.

Theorem 8.35 ([26]) *For $f \in C_B^1(I)$ and $(x,y) \in I$, we have*

$$\left|T^*_{n_1,n_2,q_{n_1},q_{n_2}}(f;x,y)-f(x,y)\right| \leq \|f_x^{'}\|_{C_B(I)}\sqrt{\delta_{n_1}(x)}+\|f_y^{'}\|_{C_B(I)}\sqrt{\delta_{n_2}(y)}.$$

Theorem 8.36 ([26]) *For $f \in C_B(I)$ and $(x,y) \in I$, we have*

$$\left|T^*_{n_1,n_2,q_{n_1},q_{n_2}}(f;x,y)-f(x,y)\right| \leq 2\{\bar{\omega}_1(f;\sqrt{\delta_{n_1}(x)})+\bar{\omega}_2(f;\sqrt{\delta_{n_2}(y)})\}.$$

Chapter 9
Convergence of GBS Operators

In [59, 60], Bögel introduced a new concept of Bögel-continuous and Bögel-differentiable functions and also established some important theorems using these concepts. Dobrescu and Matei [80] showed the convergence of the Boolean sum of bivariate generalization of Bernstein polynomials to the B-continuous function on a bounded interval. Subsequently, Badea and Cottin [46] obtained Korovkin theorems for GBS operators.

9.1 General Definitions of GBS Operators

Now, we give some basic definitions and notations.

A function f on J^2 where $J = [0, 1]$ is called a B-continuous (Bögel continuous) function if for every $(x_0, y_0) \in J^2$ we have

$$\lim_{(x,y)\to(x_0,y_0)} \Delta f[(x_0, y_0); (x, y)] = 0,$$

where $\Delta f[(x_0, y_0); (x, y)]$ denotes the mixed difference defined by

$$\Delta f[(x_0, y_0); (x, y)] = f(x, y) - f(x, y_0) - f(x_0, y) + f(x_0, y_0).$$

By $B_b\left(J^2\right)$, we denote the space of all B-bounded functions $f : J^2 \to J$, equipped with the norm

$$\|f\|_B = \sup_{(x,y),(t,s)\in A} |\Delta f\,[(t, s); (x, y)]|\,.$$

V. Gupta et al., *Recent Advances in Constructive Approximation Theory*,
Springer Optimization and Its Applications 138,
https://doi.org/10.1007/978-3-319-92165-5_9

We denote by $C_b\left(J^2\right)$, the space of all B-continuous functions on J^2. $B\left(J^2\right)$, $C\left(J^2\right)$ denote the space of all bounded functions and the space of all continuous (in the usual sense) functions on J^2 endowed with the sup-norm $\|.\|_\infty$, respectively. It is known that $C\left(J^2\right) \subset C_b\left(J^2\right)$ [61, p. 52].

A function $f : J^2 \longrightarrow J$ is called a B-differentiable (Bögel differentiable) function at $(x_0, y_0) \in J^2$ if the limit

$$\lim_{(x,y)\to(x_0,y_0)} \frac{\Delta f[(x_0, y_0); (x, y)]}{(x - x_0)(y - y_0)}$$

exists and is finite.

The limit is said to be the B-differential of f at the point (x_0, y_0) and is denoted by $D_B(f; x_0, y_0)$ and the space of all B-differentiable functions is denoted by $D_b(J^2)$.

Let $C(J^2)$ denote the space of all real valued continuous functions on J, endowed with the norm $\|f\|_{C(J^2)} = \sup\limits_{(x,y)\in J^2} |f(x, y)|$.

The Peetre's K-functional of the function $f \in C(J^2)$ is defined by

$$\mathcal{K}(f; \delta) = \inf_{g\in C^2(J^2)} \{||f - g||_{C(J^2)} + \delta||g||_{C(J^2)}\}, \delta > 0.$$

Also, from [66, p. 192] it is known that

$$\mathcal{K}(f; \delta) \le M \left\{\bar{\omega}_2(f; \sqrt{\delta}) + min(1, \delta)||f||_{C(J^2)}\right\},$$

holds for all $\delta > 0$. The constant M in the above inequality is independent of δ and f and $\bar{\omega}_2(f; \sqrt{\delta})$ is the second order complete modulus of continuity which is defined as

$$\bar{\omega}_2(f; \sqrt{\delta}) = \sup_{|h|\le\delta,|k|\le\delta} \left\{\left|\sum_{\nu=0}^{2}(-1)^{2-\nu} f(x + \nu h, y + \nu k)\right| : (x, y), (x + 2h, y + 2k) \in J^2\right\}.$$

The mixed modulus of smoothness of $f \in C_b(J^2)$ is defined as

$$\omega_{mixed}\,(f; \delta_1, \delta_2) := \sup\{|\Delta f\,[(t, s); (x, y)]| : |x - t| < \delta_1, |y - s| < \delta_2\},$$

for all $(x, y), (t, s) \in J^2$ and for any $(\delta_1, \delta_2) \in (0, \infty) \times (0, \infty)$ with

$$\omega_{mixed} : [0, \infty) \times [0, \infty) \to \mathbb{R}.$$

The basic properties of ω_{mixed} were obtained by Badea et al. in [46, 48] which are similar to the properties of the usual modulus of continuity.

Badea et al. [47] proved the following Korovkin-type theorem in order to approximate B-continuous functions by using GBS-operators.

Lemma 9.1 *Let* (L_{n_1,n_2}), $L_{n_1,n_2} : C_b(J^2) \to B(J^2)$, $n_1, n_2 \in \mathbb{N}$ *be a sequence of bivariate linear positive operators,* G_{n_1,n_2} *be the GBS-operators associated with* L_{n_1,n_2} *and the following conditions are satisfied*

1. $L_{n_1,n_2}(e_{00}; x, y) = 1$
2. $L_{n_1,n_2}(e_{10}; x, y) = x + u_{n_1,n_2}(x, y)$
3. $L_{n_1,n_2}(e_{01}; x, y) = y + v_{n_1,n_2}(x, y)$
4. $L_{n_1,n_2}(e_{20} + e_{02}; x, y) = x^2 + y^2 + w_{n_1,n_2}(x, y)$

for all $(x, y) \in A$. *If the sequences* (u_{n_1,n_2}), (v_{n_1,n_2}) *and* (w_{n_1,n_2}) *converge to zero uniformly on* J^2, *then the sequence* $(G_{n_1,n_2} f)$ *converges to* f *uniformly on* J^2 *for all* $f \in C_b(J^2)$.

Badea et al. [48] established the following Shisha–Mond type theorem to obtain the degree of approximation for B-continuous functions using GBS operators:

Theorem 9.1 *Let* $L : C_b(J^2) \to C_b(J^2)$ *be a bivariate linear positive operator and* $G : C_b(J^2) \to C_b(J^2)$ *be the associated GBS-operator. The following inequality*

$$
\begin{aligned}
|G(f; x, y) - f(x, y)| \leq |f(x, y)||L(1; x, y) - 1| + \Big\{ & L(1; x, y) \\
& + \delta_1^{-1}\sqrt{L((t-x)^2; x, y)} + \delta_2^{-1}\sqrt{L((s-y)^2; x, y)} \\
& + \delta_1^{-1}\sqrt{L((t-x)^2; x, y)}\delta_2^{-1}\sqrt{L((s-y)^2; x, y)} \Big\} \\
& \omega_{mixed}(f; \delta_1, \delta_2),
\end{aligned}
$$

holds for all $f \in C_b(J^2)$, $(x, y) \in J^2$ *and* $\delta_1, \delta_2 > 0$.

9.2 GBS Operators of Discrete Type

9.2.1 *q-Bernstein–Schurer–Stancu Type*

For any $(x, y) \in J^2$, the q-GBS operator of Bernstein–Schurer–Stancu type $U_{m,n,p_1,p_2}^{(\alpha_1,\beta_1,\alpha_2,\beta_2)} : C_b(J^2) \to C_b(J^2)$, associated with $S_{m,n,p_1,p_2}^{(\alpha_1,\beta_1,\alpha_2,\beta_2)}$ is defined as:

$$\begin{aligned} & U_{m,n,p_1,p_2}^{(\alpha_1,\beta_1,\alpha_2,\beta_2)}(f(t,s);q_m,q_n,x,y) \\ &= S_{m,n,p_1,p_2}^{(\alpha_1,\beta_1,\alpha_2,\beta_2)}(f(t,y)+f(x,s)-f(t,s);q_m,q_n,x,y) \\ &= \sum_{k_1=0}^{m+p_1}\sum_{k_2=0}^{n+p_2}\begin{bmatrix} m+p_1 \\ k_1 \end{bmatrix}_{q_m}\begin{bmatrix} n+p_2 \\ k_2 \end{bmatrix}_{q_n}\prod_{s=0}^{m+p_1-k_1-1}(1-q_m^s x) \\ &\times \prod_{r=0}^{n+p_2-k_2-1}(1-q_n^r y)x^{k_1}y^{k_2}\{f_{k_1}+f_{k_2}-f_{k_1,k_2}\}, \end{aligned} \tag{9.2.1}$$

where

$$f_{k_1}(y)=f\left(\frac{[k_1]_{q_m}+\alpha_1}{[m]_{q_m}+\beta_1},y\right),\quad f_{k_2}(x)=f\left(x,\frac{[k_2]_{q_n}+\alpha_2}{[n]_{q_n}+\beta_2}\right),$$

$$f_{k_1,k_2}(x,y)=f\left(\frac{[k_1]_{q_m}+\alpha_1}{[m]_{q_m}+\beta_1},\frac{[k_2]_{q_n}+\alpha_2}{[n]_{q_n}+\beta_2}\right).$$

The Lipschitz class $Lip_M(\xi,\gamma)$ with $\xi,\gamma\in(0,1]$, for B-continuous functions is defined by

$$Lip_M(\xi,\gamma)=\left\{f\in C_b(I):\left|\Delta_{(x,y)}f[t,s;x,y]\right|\le M|t-x|^{\xi}|s-y|^{\gamma},\text{for }(t,s),(x,y)\in J\right\},$$

where $f\in C_b(J^2)$.

Theorem 9.2 ([216]) *Let $f\in Lip_M(\xi,\gamma)$ then we have*

$$\left|U_{m,n,p_1,p_2}^{(\alpha_1,\beta_1,\alpha_2,\beta_2)}(f;q_m,q_n,x,y)-f(x,y)\right|\le M\delta_m^{\xi/2}\delta_n^{\gamma/2}.$$

for $M>0$, $\xi,\gamma\in(0,1]$.

Proof By the definition of the operator $U_{m,n,p_1,p_2}^{(\alpha_1,\beta_1,\alpha_2,\beta_2)}$ and by linearity of the operator $S_{m,n,p_1,p_2}^{(\alpha_1,\beta_1,\alpha_2,\beta_2)}$ given by (9.2.1), we can write

$$\begin{aligned} & U_{m,n,p_1,p_2}^{(\alpha_1,\beta_1,\alpha_2,\beta_2)}(f;q_m,q_n,x,y) \\ &= S_{m,n,p_1,p_2}^{(\alpha_1,\beta_1,\alpha_2,\beta_2)}(f(x,s)+f(t,y)-f(t,s);q_m,q_n,x,y) \\ &= S_{m,n,p_1,p_2}^{(\alpha_1,\beta_1,\alpha_2,\beta_2)}\left(f(x,y)-\Delta_{(x,y)}f[t,s;x,y];q_m,q_n,x,y\right) \\ &= f(x,y)\,S_{m,n,p_1,p_2}^{(\alpha_1,\beta_1,\alpha_2,\beta_2)}(e_{00};q_m,q_n,x,y) \\ &\quad -S_{m,n,p_1,p_2}^{(\alpha_1,\beta_1,\alpha_2,\beta_2)}\left(\Delta_{(x,y)}f[t,s;x,y];q_m,q_n,x,y\right). \end{aligned}$$

By the hypothesis, we get

$$\left|U_{m,n,p_1,p_2}^{(\alpha_1,\beta_1,\alpha_2,\beta_2)}(f;q_m,q_n,x,y)-f(x,y)\right|$$

$$\leq S_{m,n,p_1,p_2}^{(\alpha_1,\beta_1,\alpha_2,\beta_2)}\left(\left|\Delta_{(x,y)}f[t,s;x,y]\right|;q_m,q_n,x,y\right)$$
$$\leq MS_{m,n,p_1,p_2}^{(\alpha_1,\beta_1,\alpha_2,\beta_2)}\left(|t-x|^{\xi}|s-y|^{\gamma};q_m,q_n,x,y\right)$$
$$=MS_{m,n,p_1,p_2}^{(\alpha_1,\beta_1,\alpha_2,\beta_2)}\left(|t-x|^{\xi};q_m,x\right)S_{m,n,p_1,p_2}^{(\alpha_1,\beta_1,\alpha_2,\beta_2)}\left(|s-y|^{\gamma};q_n,y\right).$$

Now, using the Hölder's inequality with

$$u_1=2/\xi,\, v_1=2/(2-\xi)\quad\text{and}\quad u_2=2/\gamma,\, v_2=2/(2-\gamma),$$

we have

$$\left|U_{m,n,p_1,p_2}^{(\alpha_1,\beta_1,\alpha_2,\beta_2)}(f;q_m,q_n,x,y)-f(x,y)\right|$$

$$\leq M\left(S_{m,n,p_1,p_2}^{(\alpha_1,\beta_1,\alpha_2,\beta_2)}(t-x)^2;q_m,x\right)^{\xi/2}S_{m,n,p_1,p_2}^{(\alpha_1,\beta_1,\alpha_2,\beta_2)}(e_0;q_m,x)^{(2-\xi)/2}$$
$$\times S_{m,n,p_1,p_2}^{(\alpha_1,\beta_1,\alpha_2,\beta_2)}\left((s-y)^2;y\right)^{\gamma/2}S_{m,n,p_1,p_2}^{(\alpha_1,\beta_1,\alpha_2,\beta_2)}(e_0;q_n,y)^{(2-\gamma)/2}.$$

Considering Lemma 9.1, we obtain the degree of local approximation for B-continuous functions belonging to $Lip_M(\xi,\gamma)$. ■

Theorem 9.3 ([216]) *Let the function $f\in D_b(J^2)$ with $D_Bf\in B(J^2)$. Then, for each $(x,y)\in J^2$, we have*

$$|U_{m,n,p_1,p_2}^{(\alpha_1,\beta_1,\alpha_2,\beta_2)}(f;q_m,q_n,x,y)-f(x,y)|$$

$$\leq\frac{C}{[m]_{q_m}^{1/2}[n]_{q_n}^{1/2}}\left\{||D_Bf||_\infty+\omega_{mixed}(D_Bf;[m]_{q_m}^{-1/2}[n]_{q_n}^{-1/2})\right\}.$$

Proof Since $f\in D_b(J^2)$, we have the identity

$$\Delta_{x,y}f[t,s;x,y]=(t-x)(s-y)D_Bf(\xi,\eta),\ \ with\ x<\xi<t;\ \ y<\eta<s.$$

It is clear that

$$D_Bf(\xi,\eta)=\Delta_{(x,y)}D_Bf(\xi,\eta)+D_Bf(\xi,y)+D_Bf(x,\eta)-D_Bf(x,y).$$

Since $D_Bf\in B(J^2)$, by the above relations, we can write

$$
\begin{aligned}
&|S_{m,n,p_1,p_2}^{(\alpha_1,\beta_1,\alpha_2,\beta_2)}(\Delta_{(x,y)} f[t,s;x,y];q_m,q_n,x,y)|\\
&= |S_{m,n,p_1,p_2}^{(\alpha_1,\beta_1,\alpha_2,\beta_2)}((t-x)(s-y)D_B f(\xi,\eta);q_m,q_n,x,y)|\\
&\leq S_{m,n,p_1,p_2}^{(\alpha_1,\beta_1,\alpha_2,\beta_2)}(|t-x||s-y||\Delta_{(x,y)} D_B f(\xi,\eta)|;q_m,q_n,x,y)\\
&\quad + S_{m,n,p_1,p_2}^{(\alpha_1,\beta_1,\alpha_2,\beta_2)}(|t-x||s-y|(|D_B f(\xi,y)|\\
&\quad + |D_B f(x,\eta)| + |D_B f(x,y)|);q_m,q_n,x,y)\\
&\leq S_{m,n,p_1,p_2}^{(\alpha_1,\beta_1,\alpha_2,\beta_2)}(|t-x||s-y|\omega_{mixed}(D_B f;|\xi-x|,|\eta-y|);q_m,q_n,x,y)\\
&\quad +3\,||D_B f||_\infty\; S_{m,n,p_1,p_2}^{(\alpha_1,\beta_1,\alpha_2,\beta_2)}(|t-x||s-y|;q_m,q_n,x,y).
\end{aligned}
$$

Since the mixed modulus of smoothness ω_{mixed} is nondecreasing, we have

$$
\begin{aligned}
&\omega_{mixed}(D_B f;|\xi-x|,|\eta-y|)\\
&\qquad\leq \omega_{mixed}(D_B f;|t-x|,|s-y|)\\
&\qquad\leq (1+\delta_m^{-1}|t-x|)(1+\delta_n^{-1}|s-y|)\;\omega_{mixed}(D_B f;\delta_m,\delta_n).
\end{aligned}
$$

Substituting in the above inequality, using the linearity of $S_{m,n,p_1,p_2}^{(\alpha_1,\beta_1,\alpha_2,\beta_2)}$, and applying the Cauchy–Schwarz inequality we obtain

$$
\begin{aligned}
&|U_{m,n,p_1,p_2}^{(\alpha_1,\beta_1,\alpha_2,\beta_2)}(f;q_m,q_n,x,y)-f(x,y)|\\
&= |S_{m,n,p_1,p_2}^{(\alpha_1,\beta_1,\alpha_2,\beta_2)}\Delta_{(x,y)} f[t,s;x,y];q_m,q_n,x,y|\\
&\leq 3||D_B f||_\infty \sqrt{S_{m,n,p_1,p_2}^{(\alpha_1,\beta_1,\alpha_2,\beta_2)}((t-x)^2(s-y)^2;q_m,q_n,x,y)}\\
&\quad +\Bigg(S_{m,n,p_1,p_2}^{(\alpha_1,\beta_1,\alpha_2,\beta_2)}(|t-x||s-y|;q_m,q_n,x,y)\\
&\quad +\delta_m^{-1} S_{m,n,p_1,p_2}^{(\alpha_1,\beta_1,\alpha_2,\beta_2)}((t-x)^2|s-y|;q_m,q_n,x,y)\\
&\quad +\delta_n^{-1} S_{m,n,p_1,p_2}^{(\alpha_1,\beta_1,\alpha_2,\beta_2)}(|t-x|(s-y)^2;q_m,q_n,x,y)\\
&\quad +\delta_m^{-1}\delta_n^{-1} S_{m,n,p_1,p_2}^{(\alpha_1,\beta_1,\alpha_2,\beta_2)}((t-x)^2(s-y)^2;q_m,q_n,x,y)\Bigg)\omega_{mixed}(D_B f;\delta_m,\delta_n)\\
&\leq 3||D_B f||_\infty \sqrt{S_{m,n,p_1,p_2}^{(\alpha_1,\beta_1,\alpha_2,\beta_2)}((t-x)^2(s-y)^2;q_m,q_n,x,y)}\\
&\quad +\Bigg(\sqrt{S_{m,n,p_1,p_2}^{(\alpha_1,\beta_1,\alpha_2,\beta_2)}((t-x)^2(s-y)^2;q_m,q_n,x,y)}\\
&\quad +\delta_m^{-1}\sqrt{S_{m,n,p_1,p_2}^{(\alpha_1,\beta_1,\alpha_2,\beta_2)}((t-x)^4(s-y)^2;q_m,q_n,x,y)}
\end{aligned}
$$

$$+\delta_n^{-1}\sqrt{S_{m,n,p_1,p_2}^{(\alpha_1,\beta_1,\alpha_2,\beta_2)}((t-x)^2(s-y)^4;q_m,q_n,x,y)}$$

$$\left.+\delta_m^{-1}\delta_n^{-1}S_{m,n,p_1,p_2}^{(\alpha_1,\beta_1,\alpha_2,\beta_2)}((t-x)^2(s-y)^2;q_m,q_n,x,y)\right)\omega_{mixed}(D_B f;\delta_m,\delta_n).$$

It is observed that for $(x, y), (t, s) \in J^2$ and $i, j \in \{1, 2\}$

$$S_{m,n,p_1,p_2}^{(\alpha_1,\beta_1,\alpha_2,\beta_2)}((t-x)^{2i}(s-y)^{2j};q_m,q_n,x,y) = S_{m,p_1}^{(\alpha_1,\beta_1)}((t-x)^{2i};q_m,x,y)$$
$$\times S_{n,p_2}^{(\alpha_2,\beta_2)}((s-y)^{2j};q_n,x,y).$$

Hence choosing

$$\delta_m = \frac{1}{[m]_{q_m}^{1/2}},\quad \delta_n = \frac{1}{[n]_{q_n}^{1/2}},$$

we get the required result. ■

Example In Figures 9.1 and 9.2, respectively, for $m, n = 10, \alpha_1, \alpha_2 = 1, \beta_1, \beta_2 = 2, p_1, p_2 = 1$ and for $m, n = 5, \alpha_1 = 0.4, \beta_1 = 0.7, \alpha_2 = 0.5, \beta_2 = 0.9, p_1, p_2 = 2$, the comparison of convergence of the operators

$$S_{m,n,p_1,p_2}^{(\alpha_1,\beta_1,\alpha_2,\beta_2)}(f;q_m,q_n,x,y) \text{ (green)}$$

Fig. 9.1

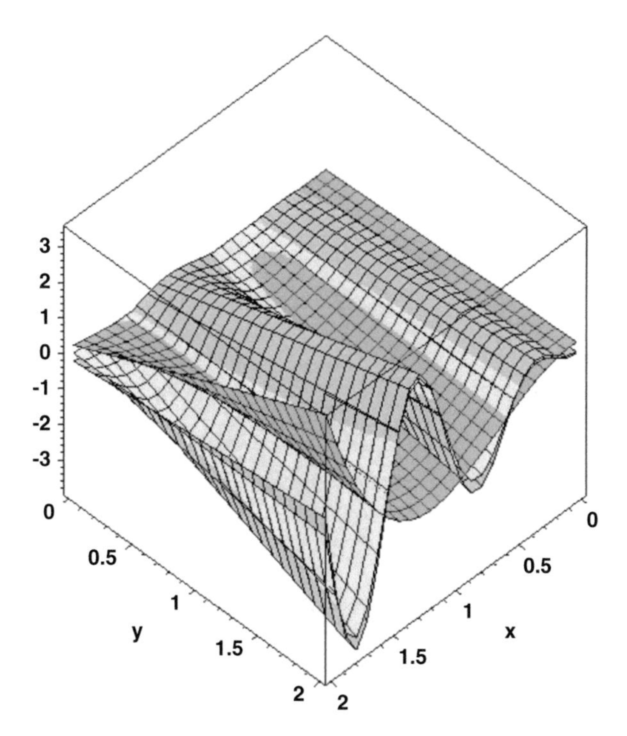

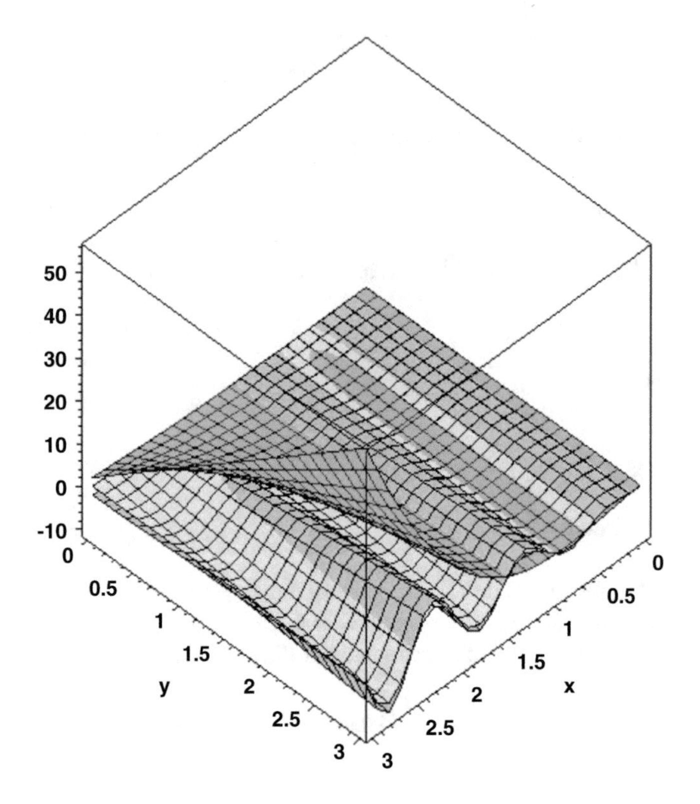

Fig. 9.2

and its GBS type operators

$$U_{m,n,p_1,p_2}^{(\alpha_1,\beta_1,\alpha_2,\beta_2)}(f;q_m,q_n,x,y) \text{ (pink)}$$

to

$$f(x,y) = x\sin(\pi x)\,y; \quad \text{(yellow)}$$

with

$$q_m = m/(m+1), q_n = 1 - 1/\sqrt{n}$$

is illustrated. It is clearly seen that the operator

$$U_{m,n,p_1,p_2}^{(\alpha_1,\beta_1,\alpha_2,\beta_2)}$$

provides a better approximation than the operator

$$S_{m,n,p_1,p_2}^{(\alpha_1,\beta_1,\alpha_2,\beta_2)}.$$

9.2.2 *Bivariate Chlodowsky–Szász–Charlier Type Operators*

Agrawal and Ispir [21] introduced an operator associated with a combination of Chlodowsky and generalized Szász–Charlier type operators as follows:

$$T_{n,m}(f;x,y,a)=\sum_{k=0}^{n}\sum_{j=0}^{\infty}p_{n,k}\left(\frac{x}{\alpha_n}\right)s_{m,j}(\beta_m y,a)f\left(\frac{k}{n}\alpha_n,\frac{j}{\gamma_m}\right), \tag{9.2.2}$$

for all $n,m\in\mathbb{N}$, $f\in C(A_{\alpha_n})$ with $A_{\alpha_n}=\{(x,y):0\leq x\leq\alpha_n,\ 0\leq y<\infty\}$ and $C(A_{\alpha_n})=\{f:A_{\alpha_n}\to\mathbb{R}\text{ is continuous}\}$. Here (α_n) is an unbounded sequence of positive numbers such that $\lim_{n\to\infty}(\alpha_n/n)=0$ and also (γ_m), (β_m) denote the unbounded sequences of positive numbers such that

$$\frac{\beta_m}{\gamma_m}=1+O(1/\gamma_m).$$

Also the basis elements are

$$p_{n,k}\left(\frac{x}{\alpha_n}\right)=\binom{n}{k}\left(\frac{x}{\alpha_n}\right)^k\left(1-\frac{x}{\alpha_n}\right)^{n-k},$$

$$s_{m,j}(\beta_m y,a)=e^{-1}\left(1-\frac{1}{a}\right)^{(a-1)\beta_m y}\frac{C_j^{(a)}(-(a-1)\beta_m y)}{j!}$$

with $C_j^{(a)}(u)$ being the Charlier polynomial and $a>1$.

For $I_{cd}=[0,c]\times[0,d]$, $C_b(I_{cd})$ denotes the space of all B-continuous function on I_{cd} and let $C(I_{cd})$ be the space of all ordinary continuous function on I_{cd}. Then the GBS operators associated with the operator $T_{n,m}(f;x,y,a)$ (9.2.2) is defined as follows:

$$GT_{n,m}(f;x,y,a):=G^*_{n,m}(f;x,y,a)=\sum_{k=0}^{n}\sum_{j=0}^{m}p_{n,k}\left(\frac{x}{\alpha_n}\right)s_{m,j}(\beta_m y,a)\times\left[f\left(\frac{k}{n}\alpha_n,y\right)+f\left(x,\frac{j}{\gamma_m}\right)-f\left(\frac{k}{n}\alpha_n,\frac{j}{\gamma_m}\right)\right] \tag{9.2.3}$$

Theorem 9.4 *For every $f \in C_b(I_{cd})$, in each point $(x, y) \in I_{cd}$, the operator (9.2.3) verifies the following inequality*

$$|G^*_{n,m}(f; x, y, a) - f(x, y)| \leq 4\omega_B(f; \delta_n(a), \delta_m(a)),$$

where $\delta_n(a) = (\upsilon(a)\alpha_n/n)^{1/2}$, $\delta_m(a) = (\eta(a)/\gamma_m)^{1/2}$ and $\upsilon(a)$, $\eta(a)$ are constants depending on a.

For $f \in C_b(I_{cd})$, the Lipschitz class $Lip_M(\lambda, \mu)$ with $\lambda, \mu \in (0, 1]$ is defined by

$$Lip_M(\lambda, \mu) = \left\{ f \in C_b(I_{cd}) : |\Delta_{(x,y)} f[t, s; x, y]| \leq M|t - x|^{\lambda}|s - y|^{\mu} \right\}$$

for $(t, s), (x, y) \in I_{cd}$, $M > 0$.

Theorem 9.5 *Let $f \in Lip_M(\lambda, \mu)$, then we have*

$$|G^*_{n,m}(f; x, y, a) - f(x, y)| \leq M\delta_n^{\lambda/2}, \delta_m^{\mu/2},$$

where

$$\delta_n = \|_x B_n((t - x)^2; ., a)\|_{\infty},$$

$$\delta_m = \|_y S_m((s - y)^2; ., a)\|_{\infty}$$

and

$$\lambda, \mu \in (0, 1], \ (x, y) \in I_{cd}.$$

9.2.3 Bivariate Chlodowsky–Szász–Appell Type Operators

Appell in [32] introduced a sequence of polynomials $P_n(x)$ of degree n which satisfies the differential equation

$$D\ P_n(x) = nP_{n-1}(x), \quad D \equiv \frac{d}{dx},$$

known as Appell polynomials. These polynomials have been studied widely because of their remarkable applications not only in mathematics [43] but also in physics and in chemistry. In [218], Sheffer extended the class of Appell polynomials and called these polynomials as zero type polynomials. For an extensive study of a broad variety of polynomial inequalities and zeros of polynomials the reader is referred to [94, 173, 174].

Using Appell polynomials, Jakimovski and Leviatan [146] introduced a generalization of the Favard–Szász operators as

$$P_n(f;x) = \frac{e^{-nx}}{g(1)} \sum_{k=0}^{\infty} p_k(nx) f\left(\frac{k}{n}\right), \tag{9.2.4}$$

where,

$$g(u)e^{ux} = \sum_{k=0}^{\infty} p_k(x)u^k$$

is the generating function for the Appell polynomials $p_k(x) \geq 0$, with

$$g(z) = \sum_{n=0}^{\infty} a_n z^n, \quad |z| < R, \;\; R > 1$$

and $g(1) \neq 0$.

Subsequently, the Stancu type generalization of the operators (9.2.4) was introduced by Atakut and Büyükyazici [42], wherein the authors established some approximation properties. A generalization of the operators given by (9.2.4) is defined as

$$P_n^*(f;x) = \frac{e^{-b_n x}}{g(1)} \sum_{k=0}^{\infty} p_k(b_n x) f\left(\frac{k}{c_n}\right), \tag{9.2.5}$$

where (b_n), (c_n) denote the unbounded and increasing sequences of positive real numbers such that

$$b_n \geq 1, \; c_n \geq 1, \text{ and } \lim_{n\to\infty} \frac{1}{c_n} = 0, \;\; \frac{b_n}{c_n} = 1 + O\left(\frac{1}{c_n}\right), \text{ as } n \to \infty.$$

In the special case $g(z) = 1$, these operators reduce to the modified Szász operators studied by Walczak [226]. Also, for $b_n = n = c_n$, these operators coincide with the operators (9.2.4).

On the interval $[0, a_n]$ with $a_n \to \infty$, as $n \to \infty$, the Bernstein–Chlodowsky polynomials are defined by

$$B_n(f;x) = \sum_{k=0}^{n} \binom{n}{k} \left(\frac{x}{a_n}\right)^k \left(1 - \frac{x}{a_n}\right)^{n-k} f\left(k\frac{a_n}{n}\right), \tag{9.2.6}$$

where $x \in [0, a_n]$ and $\lim_{n\to\infty} \frac{a_n}{n} = 0$.

By combining the Bernstein–Chlodowsky operators (9.2.6) and the operators (9.2.5), we introduce the bivariate operators as follows:

$$T_{n,m}(f;x,y)=\sum_{k=0}^{n}\sum_{j=0}^{\infty}\binom{n}{k}\left(\frac{x}{a_n}\right)^k\left(1-\frac{x}{a_n}\right)^{n-k}\frac{e^{-b_m y}}{g(1)}p_j(b_m y)f\left(k\frac{a_n}{n},\frac{j}{c_m}\right), \tag{9.2.7}$$

for all $n,m\in\mathbb{N},\ f\in C(A)_{a_n}$ with

$$A_{a_n}=\{(x,y):0\le x\le a_n, 0\le y<\infty\},$$

and

$$C(A_{a_n}):=\{f:A_{a_n}\to\mathbb{R}\text{ is continuous}\}.$$

Note that the operator (9.2.7) is the tensorial product of ${}_xB_n$ and ${}_yP_m^*$, i.e., $T_{n,m}={}_xB_n\circ{}_yP_m^*$, where

$${}_xB_n(f;x,y)=\sum_{k=0}^{n}\binom{n}{k}\left(\frac{x}{a_n}\right)^k\left(1-\frac{x}{a_n}\right)^{n-k}f\left(k\frac{a_n}{n},y\right),$$

and

$${}_yP_m^*(f;x,y)=\frac{e^{-b_m y}}{g(1)}\sum_{k=0}^{\infty}p_k(b_m y)f\left(x,\frac{k}{c_m}\right).$$

The basic results of the operators (9.2.7) using the test functions

$$e_{i,j}=t^i s^j\ (i,j=0,1,2)$$

are given as:

Lemma 9.2 ([217]) *The following hold true*

1. $T_{n,m}(e_{0,0};x,y)=1,$
2. $T_{n,m}(e_{1,0};x,y)=x,$
3. $T_{n,m}(e_{0,1};x,y)=\frac{b_m}{c_m}y+\frac{1}{c_m}\frac{g'(1)}{g(1)},$
4. $T_{n,m}(e_{2,0};x,y)=x^2+\frac{x}{n}(a_n-x),$
5. $T_{n,m}(e_{0,2};x,y)=\frac{b_m^2}{c_m^2}y^2+\frac{b_m}{c_m^2}\left(2\frac{g'(1)}{g(1)}+1\right)y+\frac{1}{c_m^2}\left(\frac{g''(1)}{g(1)}+\frac{g'(1)}{g(1)}\right),$
6. $T_{n,m}(e_{3,0};x,y)=x^3+\frac{x(a_n-x)}{n^2}(x(3n-2)+a_n),$

7. $T_{n,m}(e_{0,3};x,y)=\frac{b_m^3}{c_m^3}y^3+\frac{b_m^2}{c_m^3}\left(3\frac{g'(1)}{g(1)}+4\right)y^2+\frac{b_m}{c_m^3}\left(3\frac{g''(1)}{g(1)}+8\frac{g'(1)}{g(1)}+1\right)y+\frac{1}{c_m^3}\left(\frac{g'''(1)}{g(1)}+4\frac{g''(1)}{g(1)}+\frac{g'(1)}{g(1)}\right)$,
8. $T_{n,m}(e_{4,0};x,y)=x^4+\frac{x^3(a_n-x)}{n^3}(6n^2-5n+2)+\frac{x(a_n-x)^2}{n^3}(2x(3n-2)+a_n)+\frac{x^2a_n(a_n-x)(n-1)}{n^3}$,
9. $T_{n,m}(e_{0,4};x,y)=\frac{b_m^4}{c_m^4}y^4+\frac{b_m^3}{c_m^4}y^3\left(4\frac{g'(1)}{g(1)}+10\right)+\frac{b_m^2}{c_m^4}y^2\left(6\frac{g''(1)}{g(1)}+30\frac{g'(1)}{g(1)}+14\right)+\frac{b_m}{c_m^4}y\left(4\frac{g'''(1)}{g(1)}+30\frac{g''(1)}{g(1)}+28\frac{g'(1)}{g(1)}+1\right)+\frac{1}{c_m^4}\left(\frac{g^{(4)}(1)}{g(1)}+10\frac{g'''(1)}{g(1)}+14\frac{g''(1)}{g(1)}+\frac{g'(1)}{g(1)}\right)$.

Lemma 9.3 ([217]) *For the operator (9.2.7), we have the following results:*

1. $T_{n,m}((e_{1,0}-x)^2;x,y)=\frac{x}{n}(a_n-x)$,
2. $T_{n,m}((e_{0,1}-y)^2;x,y)=\left(\frac{b_m}{c_m}-1\right)^2y^2+\left(2\frac{b_m}{c_m^2}\frac{g'(1)}{g(1)}-\frac{2}{c_m}\frac{g'(1)}{g(1)}+\frac{b_m}{c_m^2}\right)y+\frac{1}{c_m^2}\left(\frac{g'(1)}{g(1)}+\frac{g''(1)}{g(1)}\right)$,
3. $T_{n,m}((e_{1,0}-x)^4;x,y)=\left(\frac{3}{n^2}-\frac{6}{n^3}\right)x^4-\frac{6a_n(n-2)}{n^3}x^3+a_n^2\left(\frac{3}{n^2}-\frac{7}{n^3}\right)x^2+\frac{a_n^3}{n^3}x$,
4. $T_{n,m}((e_{0,1}-y)^4;x,y)=\left(\frac{b_m}{c_m}-1\right)^4y^4+\left\{\frac{b_m^3}{c_m^4}\left(4\frac{g'(1)}{g(1)}+10\right)+6\frac{b_m}{c_m^2}\left(2\frac{g'(1)}{g(1)}+1\right)-4\frac{b_m^2}{c_m^3}\left(3\frac{g'(1)}{g(1)}+4\right)-\frac{4}{c_m}\frac{g'(1)}{g(1)}\right\}y^3+\left\{\frac{b_m^2}{c_m^4}\left(6\frac{g''(1)}{g(1)}+30\frac{g'(1)}{g(1)}+14\right)-4\frac{b_m}{c_m^3}\left(3\frac{g''(1)}{g(1)}+8\frac{g'(1)}{g(1)}+1\right)+\frac{6}{c_m^2}\left(\frac{g''(1)}{g(1)}+\frac{g'(1)}{g(1)}\right)\right\}y^2+\left\{\frac{b_m}{c_m^4}\left(4\frac{g'''(1)}{g(1)}+30\frac{g''(1)}{g(1)}+28\frac{g'(1)}{g(1)}+1\right)-\frac{4}{c_m^3}\left(\frac{g'''(1)}{g(1)}+4\frac{g''(1)}{g(1)}+\frac{g'(1)}{g(1)}\right)\right\}y+\frac{1}{c_m^4}\left(\frac{g^4(1)}{g(1)}+10\frac{g'''(1)}{g(1)}+14\frac{g''(1)}{g(1)}+\frac{g'(1)}{g(1)}\right)$.

Lemma 9.4 ([217]) *Taking into account the conditions on $(a_n),(b_n),(c_n)$ and using Lemma 9.2 and Lemma 9.3, we may write*

1. $T_{n,m}((e_{1,0}-x)^2;x,y)=O\left(\frac{a_n}{n}\right)(x^2+x)$, *as* $n\to\infty$,
2. $T_{n,m}((e_{0,1}-y)^2;x,y)\le\frac{\eta(g)}{c_m}(y^2+y+1)$,

3. $T_{n,m}((e_{1,0}-x)^4;x,y)=O\left(\frac{a_n}{n}\right)(x^4+x^3+x^2+x),$ *as* $n\to\infty$,
4. $T_{n,m}((e_{0,1}-y)^4;x,y)\leq\frac{\mu(g)}{c_m}(y^4+y^3+y^2+y+1)$,

where $\eta(g)$ *and* $\mu(g)$ *are certain constants depending on* g.

For every $f\in C_b(A_{\alpha_n})$, the GBS operator associated with the operator $T_{n,m}(f;x,y)$ is defined as follows:

$$U_{n,m}(f;x,y)=\sum_{k=0}^{n}\sum_{j=0}^{\infty}\binom{n}{k}\left(\frac{x}{a_n}\right)^k\left(1-\frac{x}{a_n}\right)^{n-k}\frac{e^{-b_my}}{g(1)}p_j(b_my)$$
$$\times\left[f\left(k\frac{a_n}{n},y\right)+f\left(x,\frac{j}{c_m}\right)-f\left(k\frac{a_n}{n},\frac{j}{c_m}\right)\right]. \tag{9.2.8}$$

Let $I_{cd}:=[0,c]\times[0,d]\subset A_{a_n}$.

Theorem 9.6 *For every* $f\in C_b(I_{cd})$ *and for all* $(x,y)\in I_{cd}$, *we have the following inequality for the operator defined in (9.2.8):*

$$|U_{n,m}(f;x,y)-f(x,y)|\leq 4\omega_{mixed}(f;\delta_n,\delta_m),$$

where

$$\delta_n=\left(\frac{a_n}{n}(c^2+c)\right)^{1/2},\ \delta_m=\delta_m(g):=\left(\frac{\rho(g)}{c_m}\right)^{1/2}$$

and $\rho(g)$ *is a constant depending on* g.

If $f\in C_b(I_{cd})$, for two parameters $0<\xi_1\leq 1$ and $0<\xi_2\leq 1$, $Lip_M(\xi_1,\xi_2)$ is defined by

$$Lip_M(\xi_1,\xi_2)=\left\{f\in C_b(I_{cd}):\left|\Delta_{(x,y)}f(t,s)\right|\leq M|t-x|^{\xi_1}|s-y|^{\xi_2},\right\},$$

for $(t,s),(x,y)\in I_{cd}$.

Theorem 9.7 *For* $f\in Lip_M(\xi_1,\xi_2)$ *and* $(x,y)\in I_{cd}$, *we have*

$$\left|U_{n,m}(f;x,y)-f(x,y)\right|\leq M\delta_n^{\xi_1/2}\delta_m^{\xi_2/2},$$

where $\delta_n=||_xB_n((t-x)^2;\cdot)||_\infty$, $\delta_m=||_yP_m^*((s-y)^2;\cdot)||_\infty$ *and* M *is a certain positive constant.*

Theorem 9.8 *If $f \in D_b(I_{cd})$ and $D_B f \in B(I_{cd})$, then for each $(x, y) \in I_{cd}$, we get*

$$\begin{aligned}
&|U_{n,m}(f; x, y) - f(x, y)| \\
&\leq C\left\{3||D_B f||_\infty + 2\omega_{mixed}(f; \delta_n, \delta_m)\sqrt{x^2 + x}\sqrt{y^2 + y + 1}\right\}\delta_n\delta_m \\
&+ \left\{\omega_{mixed}(f; \delta_n, \delta_m)\left(\delta_m\sqrt{x^4 + x^3 + x^2 + x}\sqrt{y^2 + y + 1}\right.\right. \\
&\left.\left.+ \delta_n\sqrt{y^4 + y^3 + y^2 + y + 1}\sqrt{x^2 + x}\right)\right\},
\end{aligned}$$

where $\delta_n = \sqrt{\frac{a_n}{n}}$, $\delta_m = \sqrt{\frac{\sigma(g)}{c_m}}$, $\sigma(g) = \max\{\eta(g), \mu(g)\}$ and C is a constant depending on n, m only.

9.2.4 Numerical Examples

In this section we give some numerical results regarding the approximation properties of Chlodowsky–Szász–Appell operators defined in (9.2.7).

Example 9.1 Let us consider the function $f(x, y) = e^{-y}\cos(\pi x)$, $g(u) = u$ and $a_n = \sqrt{n}$, $b_n = n$, $c_n = n + \frac{1}{\sqrt{n}}$. For $n = m = 5$ and $n = m = 40$ the convergence of $T_{n,m}(f; x, y)$ to $f(x, y)$ is illustrated in Figure 9.3.

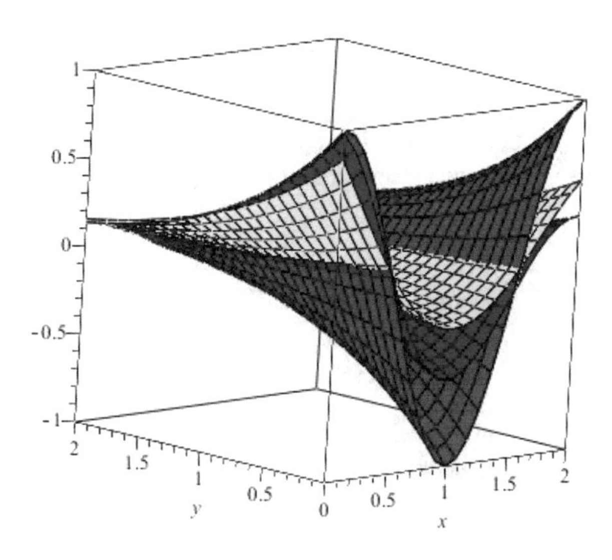

Fig. 9.3 The convergence of $T_{n,m}(f; x, y)$ to $f(x, y)$ (red f, blue $T_{40,40}$, yellow $T_{5,5}$)

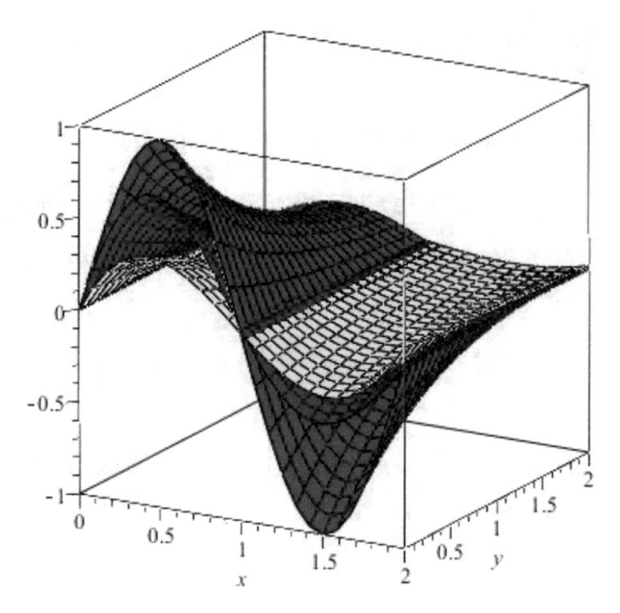

Fig. 9.4 The convergence of $T_{n,m}(f; x, y)$ to $f(x, y)$ (red f, blue $T_{40,40}$, yellow $T_{5,5}$)

Example 9.2 Let us consider the function $f(x, y) = e^{-y}\sin(\pi x)$, $g(u) = u$ and $a_n = \sqrt{n}$, $b_n = n$, $c_n = n + e^{-n}$. For $n = m = 5$ and $n = m = 40$ the convergence of $T_{n,m}(f; x, y)$ to $f(x, y)$ is illustrated in Figure 9.4.

We notice from the above examples that for $n = m = 40$ the approximation of the operator $T_{n,m}$ to the function f is better than $n = m = 5$.

Example 9.3 If $f \in C(I_{ab})$, then

$$|T_{n,m}(f; x, y) - f(x, y)| \le 2\left(\omega_1(f; \delta_n) + \omega_2(f; \delta_m)\right)$$
$$\le 2\left(\|f^{(1,0)}\|_\infty \delta_n + \|f^{(0,1)}\|_\infty \delta_m\right),$$

where δ_n and δ_m are defined in Theorem 9.2.7.

In Table 9.1 we compute the error of approximation of $f(x, y) = xye^{-y}$ by using the above relation for $I_{ab} = [0, 4] \times [0, 4]$.

9.2.5 Bernstein Type in Three Variables

In this section, we recall some results which we will use in this section. In the following, let X, Y, and Z be compact real intervals and $D = X \times Y \times Z$. A function $f : D \to \mathbb{R}$ is called a B-continuous function at $(x_0, y_0, z_0) \in D$ iff for any $\epsilon > 0$, there exists a $\delta > 0$ such that

$$|\Delta f[(x, y, z), (x_0, y_0, z_0)]| < \epsilon$$

Table 9.1 Error of approximation for $T_{n,m}$

$n=m$	$a_n=\sqrt{n},\ b_n=n,\ c_n=n+\frac{1}{\sqrt{n}}$	$a_n=\sqrt{n},\ b_n=n,\ c_n=n+e^{-n}$
20	3.9062769320	3.9678794400
50	2.6253029800	2.6362709420
100	1.9595674840	1.9624455190
500	0.9975826189	0.9977048873
1000	0.7505644223	0.7505954100
1500	0.6370476284	0.6370614836
2000	0.5677575415	0.5677653628
2500	0.5196173630	0.5196223806
3000	0.4835609538	0.4835644444

for any $(x, y, z) \in D$, with $|x - x_0| < \delta$, $|y - y_0| < \delta$ and $|z - z_0| < \delta$. Here

$$\begin{aligned}\Delta f[(x, y, z), (x_0, y_0, z_0)] &= f(x, y, z) - f(x, y, z_0) - f(x, y_0, z) - f(x_0, y, z) \\ &\quad + f(x, y_0, z_0) + f(x_0, y, z_0) + f(x_0, y_0, z) \\ &\quad - f(x_0, y_0, z_0)\end{aligned}$$

denote a mixed difference of f. The function f is B-continuous (Bögel continuous) on D iff f is B-continuous at each point $(x_0, y_0, z_0) \in D$. These notions were introduced by Bögel in [59–61].

The function $f : D \to \mathbb{R}$ is B-bounded on D iff there exists $K > 0$ such that

$$|\Delta f[(x, y, z), (x_0, y_0, z_0)]| \leq K$$

for any $(x, y, z), (x_0, y_0, z_0) \in D$. Let $B_b(D)$ denote the space of all B-bounded functions on D, equipped with the norm

$$\|f\|_B = \sup_{(x,y),(t,s)\in A} |\Delta f\,[(t, s); (x, y)]|\,.$$

We denote by $C_b(D)$, the space of all B-continuous functions on D. $B(D), C(D)$ denote the space of all bounded functions and the space of all continuous (in the usual sense) functions on D endowed with the sup-norm $\|.\|_\infty$ respectively. Let $f \in B_b(D)$, then the function

$$\omega_{mixed}(f; \delta_1, \delta_2, \delta_3) = \sup\Big\{|f[(x, y, z), (x_0, y_0, z_0)]| : |x - x_0| \leq \delta_1, |y - y_0| \leq \delta_2, \\ |z - z_0| \leq \delta_3\Big\}$$

for any $\delta_1, \delta_2, \delta_3 \in [0, \infty)$ is called the mixed modulus of smoothness for three variables.

Lemma 9.5 ([85]) *Let $f \in B_b(D)$, then*

$$\omega_{mixed}(f; \delta_1, \delta_2, \delta_3) \leq \omega_{mixed}(f; \delta_1', \delta_2', \delta_3')$$

for any $\delta_1, \delta_2, \delta_3, \delta_1', \delta_2', \delta_3' \in [0, \infty)$ such that $\delta_1 \leq \delta_1'$, $\delta_2 \leq \delta_2'$ and $\delta_3 \leq \delta_3'$;

$$\Delta f[(x, y, z), (u, v, w)] \leq \omega_{mixed}(f; |x-u|, |y-v|, |z-w|);$$

and

$$\Delta f[(x, y, z), (u, v, w)] \leq \left(1 + \frac{|x-u|}{\delta_1}\right)\left(1 + \frac{|y-v|}{\delta_2}\right)\left(1 + \frac{|z-w|}{\delta_3}\right) \omega_{mixed}(f; \delta_1, \delta_2, \delta_3);$$

for $\delta_1, \delta_2, \delta_3 > 0$ and $\lambda_1, \lambda_2, \lambda_3 > 0$

$$\omega_{mixed}(f; \lambda_1\delta_1, \lambda_2\delta_2, \lambda_3\delta_3) \leq (1+\lambda_1)(1+\lambda_2)(1+\lambda_3)\omega_{mixed}(f; \delta_1, \delta_2, \delta_3).$$

Lemma 9.6 ([85]) *Let $f \in C_b(D)$, then*

$$\lim_{\delta_1, \delta_2, \delta_3 \to 0} \omega_{mixed}(f; \delta_1, \delta_2, \delta_3) = 0.$$

Let L be a positive operator of three variables, which maps the space $\mathbb{R}^{[a,b]\times[a',b']\times[a'',b'']}$ into itself. The operator

$$UL : \mathbb{R}^{[a,b]\times[a',b']\times[a'',b'']} \to \mathbb{R}^{[a,b]\times[a',b']\times[a'',b'']}$$

defined by

$$(ULf)(x, y, z) = L[f(\bullet, y, z) + f(x, *, z) + f(x, y, \circ) - f(\bullet, *, z) - f(\bullet, y, \circ) - f(x, *, \circ); x, y, z]$$

is called GBS (generalized Boolean sum) operator associated with L, where $\bullet, *, \circ$ stand for the first, the second, and the third variable. The term of GBS operator was introduced by Badea and Cottin.

Lemma 9.7 ([85]) *For any $f \in C_b(D)$, and any $\delta_1, \delta_2, \delta_3 > 0$ the following inequality holds*

$$|(ULf)(x, y, z) - f(x, y, z)|$$
$$\leq \left(1 + \frac{\sqrt{(L(\bullet - x)^2)(x, y, z)}}{\delta_1} + \frac{\sqrt{(L(* - y)^2)(x, y, z)}}{\delta_2}\right.$$

$$+\frac{\sqrt{(L(\circ - z)^2)(x,y,z)}}{\delta_3} + \frac{\sqrt{(L(\bullet - x)^2(* - y)^2)(x,y,z)}}{\delta_1\delta_2}$$
$$+\frac{\sqrt{(L(* - y)^2(\circ - z)^2)(x,y,z)}}{\delta_2\delta_3} + \frac{\sqrt{(L(\circ - z)^2(\bullet - x)^2)(x,y,z)}}{\delta_3\delta_1}$$
$$+\left.\frac{\sqrt{(L(\bullet - x)^2(* - y)^2)(\circ - z)^2(x,y,z)}}{\delta_1\delta_2\delta_3}\right)\omega_{mixed}(f;\delta_1,\delta_2,\delta_3),$$

where $L : B(D) \to B(D)$ *is a positive linear operator which reproduces the constants.*

Let

$$\Delta_3 = \{(x,y,z) \in \mathbb{R}\times\mathbb{R}\times\mathbb{R} : x,y,z \geq 0, x+y+z \leq 1\}$$

and

$$F(\Delta_3) = \{f/f : \Delta_3 \to \mathbb{R}\}.$$

For m a nonnegative integer, let the operator $B_m : F(\Delta_3) \to (\Delta_3)$ be defined by

$$(B_m f)(x,y,z) = \sum_{i,j,k=0, i+j+k\leq m} p_{m,i,j,k}(x,y,z) f\left(\frac{i}{m},\frac{j}{m},\frac{k}{m}\right)$$

for any $(x,y,z) \in \Delta_3$ where

$$p_{m,i,j,k}(x,y,z) = \frac{m!}{i!j!k!(m-i-j-k)!} x^i y^j z^k (1-x-y-z)^{m-i-j-k}.$$

The operators are named as Bernstein polynomial of three variables and they satisfy linearity and positivity on $F(\Delta_3)$.

Lemma 9.8 *If* m, p_1, p_2, p_3, k *are positive given integers, then*

$$(B_m e_{p_1,p_2,p_3})(x,y,z) = \frac{1}{m^{p_1+p_2+p_3}} \sum_{n_1=1}^{p_1}\sum_{n_2=1}^{p_2}\sum_{n_3=1}^{p_3} m^{[n_1+n_2+n_3]}$$
$$\times S(p_1,n_1)S(p_2,n_2)S(p_3,n_3)x^{n_1}y^{n_2}z^{n_3},$$

where

$$e_{p_1,p_2,p_3}(x,y,z) = x^{p_1}y^{p_2}z^{p_3}, (x,y,z) \in \Delta_3, m^{[k]} = m(m-1)\ldots(m-k+1)$$

and $S(m,k)$ *denoted the Stirling numbers of second kind.*

Lemma 9.9 *If* $(x, y, z) \in \Delta_3$, *then*

$$xy \leq \frac{1}{4}, 3x^2y^2 - xy(x+y) + xy \leq \frac{3}{16};$$

$$xyz \leq \frac{xy + yz + zx}{9} \leq \frac{x+y+z}{27} \leq \frac{1}{27};$$

$$xy + yz + zx - \frac{9}{4}xyz \leq \frac{1}{4};$$

$$17xyz - 5xyz(x+y+z) \leq \frac{4}{9};$$

$$-2xyz(x+y+z) + \frac{xy+yz+zx}{9} + 2xyz \leq \frac{1}{27};$$

$$5x^2y^2z^2 - xyz(xy+yz+zx) + \frac{xyz(x+y+z)}{3} - \frac{xyz}{3} + \frac{4}{729} \geq 0;$$

$$\frac{-5m+6}{3}xyz(x+y+z) + \frac{17m-18}{3}xyz - \frac{4m-4}{27} \leq 0, \ \ m \geq 1.$$

Let for any $(x, y, z) \in \Delta_3$ and m a nonnegative integer, the GBS operator of Bernstein type [44] $UB_m : C_b(\Delta_3) \to B(\Delta_3)$ be defined by

$$\begin{aligned}(UB_m f)(x, y, z) &= (B_m(f(\bullet, y, z) + f(x, *, z) + f(x, y, \circ) - f(\bullet, *, z) \\ &\quad - f(\bullet, y, \circ) - f(x, *, \circ))(x, y, z) \\ &= \sum_{i,j,k=0, i+j+k \leq m} p_{m,i,j,k}\left(f\left(\frac{i}{m}, y, z\right) + f\left(x, \frac{j}{m}, z\right)\right. \\ &\quad + f\left(x, y, \frac{k}{m}\right) - f\left(\frac{i}{m}, \frac{j}{m}, z\right) - f\left(\frac{i}{m}, y, \frac{k}{m}\right) \\ &\quad \left. - f\left(x, \frac{j}{m}, \frac{k}{m}\right) + f\left(\frac{i}{m}, \frac{j}{m}, \frac{k}{m}\right)\right).\end{aligned}$$

Lemma 9.10 *The operators* $(B_m)_{m \geq 1}$ *satisfy for any* $(x, y, z) \in \Delta_3$ *the following*

1. $(B_m e_{00})(x, y, z) = 1$;
2. $m(B_m(\bullet - x)^2)(x, y, z) = x(1-x) \leq \frac{1}{4}, m \geq 1$;
3. $m^3(B_m(\bullet - x)^2(* - y)^2)(x, y, z) = 3(m-2)x^2y^2 - (m-2)xy(x+y) + (m-1)xy \leq \dfrac{3m-2}{16}, \ m \geq 2$;

4. $m^5(B_m(\bullet - x)^2(* - y)^2(\circ - z)^2)(x, y, z) = (-15m^2 + 130m - 120)x^2y^2z^2 + (3m^2 - 26m + 24)xy(xy + yz + zx) + (-m^2 + 7m - 6)xyz(x + y + z) + (m^2 - 3m + 2)xyz \leq \dfrac{12m^2 + 4m - 12}{729}$, $m \geq 8$.

Theorem 9.9 ([85]) *If $f \in C_b(\Delta_3)$, then*

$$|(UB_m f)(x, y, z) - f(x, y, z)| \leq 4\omega_{mixed}\left(f; \frac{1}{\sqrt{m}}, \frac{1}{\sqrt{m}}, \frac{1}{\sqrt{m}}\right), \quad m \geq 8.$$

9.3 GBS Operators of Continuous Type

9.3.1 *q*-Durrmeyer–Pólya Type

The q-Durrmeyer type operator based on Pólya distribution was defined by

$$P_{n_1,q_{n_1}}^{\frac{1}{[n_1]_{q_{n_1}}}}(f; x) = [n_1 + 1]_{q_{n_1}} \sum_{k_1=0}^{n_1} q_{n_1}^{-k_1} p_{n_1,k_1}^{\frac{1}{[n_1]_{q_{n_1}}}}(x) \int_0^1 p_{n_1,k_1}(q_{n_1}t_1) f(t_1) d_{q_{n_1}} t_1, \quad (9.3.1)$$

where

$$p_{n_1,k_1}^{\frac{1}{[n_1]_{q_{n_1}}}}(u) = \begin{bmatrix} n \\ k \end{bmatrix}_{q_{n_1}} \frac{\prod_{j=0}^{k-1}\left(u + \frac{[j]_{q_{n_1}}}{[n_1]_{q_{n_1}}}\right) \prod_{j=0}^{n_1-k_1-1}\left(1 - u + \frac{[j]_{q_{n_1}}}{[n_1]_{q_{n_1}}}\right)}{\prod_{j=0}^{n_1-1}\left(u + \frac{[j]_{q_{n_1}}}{[n_1]_{q_{n_1}}}\right)},$$

and

$$p_{n_1,k_1}(q_{n_1}t) = \begin{bmatrix} n_1 \\ k_1 \end{bmatrix}_{q_{n_1}} q_{n_1}^{k_1} t^{k_1} \prod_{j=0}^{n_1-k_1-1}\left(1 - q_{n_1}^{j+1} t\right).$$

Let

$$P_{n_1,n_2,n_3,q_{n_1},q_{n_2},q_{n_3}}^{\frac{1}{[n_1]_{q_{n_1}}} \frac{1}{[n_2]_{q_{n_2}}} \frac{1}{[n_3]_{q_{n_3}}}} : C(I^3) \to C(I^3),$$

be the q- Durrmeyer–Pólya operators for the functions with three variables be defined by

$$P_{n_1,n_2,n_3,q_{n_1},q_{n_2},q_{n_3}}^{\frac{1}{[n_1]_{q_{n_1}}}\frac{1}{[n_2]_{q_{n_2}}}\frac{1}{[n_3]_{q_{n_3}}}}(f;x,y,z)$$

$$=\prod_{i=1}^{3}[n_i+1]_{q_{n_i}}\sum_{k_1=0}^{n_1}\sum_{k_2=0}^{n_2}\sum_{k_3=0}^{n_3}\prod_{i=1}^{3}q_{n_i}^{-k_i}p_{n_1,k_1}^{\frac{1}{[n_1]_{q_{n_1}}}}(x)p_{n_2,k_2}^{\frac{1}{[n_2]_{q_{n_2}}}}(y)p_{n_3,k_3}^{\frac{1}{[n_3]_{q_{n_3}}}}(z)$$

$$\times\int_0^1\int_0^1\int_0^1\prod_{i=1}^{3}p_{n_i,k_i}(q_{n_i}t_i)f(t_1,t_2,t_3)d_{q_{n_1}}t_1d_{q_{n_2}}t_2d_{q_{n_3}}t_3,$$

and let

$$P_{n_1,q_{n_1}}^{x,\frac{1}{[n_1]_{q_{n_1}}}},\ P_{n_2,q_{n_2}}^{y,\frac{1}{[n_2]_{q_{n_2}}}}\text{ and }P_{n_3,q_{n_3}}^{z,\frac{1}{[n_3]_{q_{n_3}}}}$$

be the parametric extensions of the operators (9.3.1) defined for any function $f\in C(I^3)$ by

$$P_{n_1,q_{n_1}}^{x,\frac{1}{[n_1]_{q_{n_1}}}}(f;x)=[n_1+1]_{q_{n_1}}\sum_{k_1=0}^{n_1}q_{n_1}^{-k_1}p_{n_1,k_1}^{\frac{1}{[n_1]_{q_{n_1}}}}(x)\int_0^1p_{n_1,k_1}(q_{n_1}t_1)f(t_1,v,w)d_{q_{n_1}}t_1,$$

$$P_{n_2,q_{n_2}}^{y,\frac{1}{[n_2]_{q_{n_2}}}}(f;y)=[n_2+1]_{q_{n_2}}\sum_{k_2=0}^{n_2}q_{n_2}^{-k_2}p_{n_2,k_2}^{\frac{1}{[n_2]_{q_{n_2}}}}(y)\int_0^1p_{n_2,k_2}(q_{n_2}t_2)f(u,t_2,w)d_{q_{n_2}}t_2,$$

$$P_{n_3,q_{n_3}}^{z,\frac{1}{[n_3]_{q_{n_3}}}}(f;z)=[n_3+1]_{q_{n_3}}\sum_{k_3=0}^{n_3}q_{n_3}^{-k_3}p_{n_3,k_3}^{\frac{1}{[n_3]_{q_{n_3}}}}(z)\int_0^1p_{n_3,k_3}(q_{n_3}t_3)f(u,v,t_3)d_{q_{n_3}}t_3.$$

For any $(x,y,z)\in I^3$, the GBS operators of q-Durrmeyer–Pólya type

$$Q_{n_1,n_2,n_3,q_{n_1},q_{n_2},q_{n_3}}^{\frac{1}{[n_1]_{q_{n_1}}}\frac{1}{[n_2]_{q_{n_2}}}\frac{1}{[n_3]_{q_{n_3}}}}:C_b(I^3)\to C(I^3)$$

associated with

$$P_{n_1,n_2,n_3,q_{n_1},q_{n_2},q_{n_3}}^{\frac{1}{[n_1]_{q_{n_1}}}\frac{1}{[n_2]_{q_{n_2}}}\frac{1}{[n_3]_{q_{n_3}}}}$$

is defined by

$$Q_{n_1,n_2,n_3,q_{n_1},q_{n_2},q_{n_3}}^{\frac{1}{[n_1]_{q_{n_1}}}\frac{1}{[n_2]_{q_{n_2}}}\frac{1}{[n_3]_{q_{n_3}}}}(f;x,y,z) = P_{n_1,n_2,n_3,q_{n_1},q_{n_2},q_{n_3}}^{\frac{1}{[n_1]_{q_{n_1}}}\frac{1}{[n_2]_{q_{n_2}}}\frac{1}{[n_3]_{q_{n_3}}}}\Big(f(u,y,z)+f(x,v,z)$$
$$+f(x,y,w)-f(u,v,z)-f(x,v,w)$$
$$-f(u,y,w)+f(u,v,w);x,y,z\Big)$$

It can be easily observed that the GBS q-Durrmeyer–Pólya operator is the Boolean sum of parametric extensions

$$Q_{n_1,n_2,n_3,q_{n_1},q_{n_2},q_{n_3}}^{\frac{1}{[n_1]_{q_{n_1}}}\frac{1}{[n_2]_{q_{n_2}}}\frac{1}{[n_3]_{q_{n_3}}}} = P_{n_1,q_{n_1}}^{x,\frac{1}{[n_1]_{q_{n_1}}}} + P_{n_2,q_{n_2}}^{y,\frac{1}{[n_2]_{q_{n_2}}}} + P_{n_3,q_{n_3}}^{z,\frac{1}{[n_3]_{q_{n_3}}}} - P_{n_1,q_{n_1}}^{x,\frac{1}{[n_1]_{q_{n_1}}}}P_{n_2,q_{n_2}}^{y,\frac{1}{[n_2]_{q_{n_2}}}}$$
$$-P_{n_1,q_{n_1}}^{x,\frac{1}{[n_1]_{q_{n_1}}}}P_{n_3,q_{n_3}}^{z,\frac{1}{[n_3]_{q_{n_3}}}} - P_{n_2,q_{n_2}}^{y,\frac{1}{[n_2]_{q_{n_2}}}}P_{n_3,q_{n_3}}^{z,\frac{1}{[n_3]_{q_{n_3}}}}$$
$$+P_{n_1,q_{n_1}}^{x,\frac{1}{[n_1]_{q_{n_1}}}}P_{n_2,q_{n_2}}^{y,\frac{1}{[n_2]_{q_{n_2}}}}P_{n_3,q_{n_3}}^{z,\frac{1}{[n_3]_{q_{n_3}}}}.$$

The approximation properties of the triple Bernstein type operators and corresponding generalized Boolean sum operators were investigated in [84, 85]. For B-continuous functions we have the following

Theorem 9.10 *Suppose $q_{n_i} \in (0,1)$ such that $\lim_{n_i\to\infty} q_{n_i} = 1, i = 1,2,3$. Let for each $f \in C_b(I^3)$, $(x,y,z) \in I^3$ the following inequality*

$$|Q_{n_1,n_2,n_3,q_{n_1},q_{n_2},q_{n_3}}^{\frac{1}{[n_1]_{q_{n_1}}}\frac{1}{[n_2]_{q_{n_2}}}\frac{1}{[n_3]_{q_{n_3}}}}(f(u,v,w);x,y,z) - f(x,y,z)| \leq 8\omega_B(f;\delta_{n_1},\delta_{n_2},\delta_{n_3})$$

holds, where

$$\delta_{n_i} = \frac{3}{[n_i+2]_{q_{n_i}}}\left(\frac{1}{4}+\frac{3}{[n_i+2]_{q_{n_i}}}\right) \text{ and } i = 1,2,3.$$

Now, we study the degree of approximation for the operators

$$Q_{n_1,n_2,n_3,q_{n_1},q_{n_2},q_{n_3}}^{\frac{1}{[n_1]_{q_{n_1}}}\frac{1}{[n_2]_{q_{n_2}}}\frac{1}{[n_3]_{q_{n_3}}}}(f;x,y,z)$$

by means of the Lipschitz class for B-continuous functions.

Let $Lip_L(\gamma)$ be the Lipschitz class of B-continuous functions, for $0<\gamma\leq 1$, be defined by

$$Lip_L(\gamma) = \left\{f \in C(I^3) : |\Delta_{(x,y,z)} f[u,v,w;x,y,z]| \leq L\|r-s\|^{\gamma}\right\},$$

where $r = (u, v, w)$, $s = (x, y, z) \in (I^3)$ and

$$\|r - s\| = \left\{(u - x)^2 + (v - y)^2 + (w - z)^2\right\}^{1/2}$$

is the Euclidean norm.

Theorem 9.11 *If $f \in Lip_L(\gamma)$, then for $L > 0$ and $\gamma \in (0, 1]$, we have*

$$\left| Q_{n_1,n_2,n_3,q_{n_1},q_{n_2},q_{n_3}}^{\frac{1}{[n_1]_{q_{n_1}}} \frac{1}{[n_2]_{q_{n_2}}} \frac{1}{[n_3]_{q_{n_3}}}}\left(f; x, y, z\right) - f(x, y, z)\right|$$
$$\leq \frac{L}{(x + y)^{\gamma/2}} \left\{\delta_{n_1}(x) + \delta_{n_2}(y) + \delta_{n_3}(z)\right\}^{\gamma/2}.$$

9.3.2 q-Bernstein–Schurer–Kantorovich Type

We define the GBS operator of the operator $K_{n_1,n_2,p}$ given by (8.2.1), for any $f \in C\left(I^2\right)$ and $m, n \in \mathbb{N}$, by

$$T_{n_1,n_2,p}(f(t,s); q_{n_1}, q_{n_2}, x, y) := K_{n_1,n_2,p}\left(f(t, y) + f(x, s) - f(t, s); q_{n_1}, q_{n_2}, x, y\right),$$

for all $(x, y) \in I^2$.
More precisely for any $f \in C_b(I^2)$, the GBS operator of q-Bernstein–Schurer–Kantorovich type is given by

$$T_{n_1,n_2,p}(f; q_{n_1}, q_{n_2}, x, y)$$
$$= [n_1 + 1]_{q_{n_1}} [n_2 + 1]_{q_{n_2}} \sum_{k_1=0}^{n_1+p} \sum_{k_2=0}^{n_2+p} b_{n_1+p,n_2+p,k_1,k_2}^{q_{n_1},q_{n_2}}(x, y) q_{n_1}^{-k_1} q_{n_2}^{-k_2}$$
$$\times \int_{[k_2]_{q_{n_2}}/[n_2+1]_{q_{n_2}}}^{[k_2+1]_{q_{n_2}}/[n_2+1]_{q_{n_2}}} \int_{[k_1]_{q_{n_1}}/[n_1+1]_{q_{n_1}}}^{[k_1+1]_{q_{n_1}}/[n_1+1]_{q_{n_1}}} [f(x, s) + f(t, y)$$
$$-f(t, s)]\, d_{q_{n_1}}^R(t) d_{q_{n_2}}^R(s), \tag{9.3.2}$$

where

$$b_{n_1+p,n_2+p,k_1,k_2}^{q_{n_1},q_{n_2}}(x, y) = \binom{n_1 + p}{k_1}_{q_{n_1}} \binom{n_2 + p}{k_2}_{q_{n_2}} x^{k_1} y^{k_2}$$
$$\times (1 - x)_{q_{n_1}}^{n_1+p-k_1} (1 - y)_{q_{n_2}}^{n_2+p-k_2}.$$

Here the operator $T_{n_1,n_2,p}$ is a linear positive operator and is well defined from the space $C_b(I^2)$ on itself.

Theorem 9.12 ([215]) *For every $f \in C_b(I^2)$ and at each point $(x, y) \in I^2$, the operator (9.3.3) satisfies the following inequality*

$$|T_{n_1,n_2,p}(f; q_{n_1}, q_{n_2}, x, y) - f(x, y)| \leq 4\,\omega_{mixed}(f; \delta_{n_1}, \delta_{n_2}).$$

Proof By the definition of $\omega_{mixed}(f; \delta_{n_1}, \delta_{n_2})$ and using the elementary inequality

$$\omega_{mixed}(f; \lambda_1\delta_{n_1}, \lambda_{n_2}\delta_{n_2}) \leq (1+\lambda_1)(1+\lambda_2)\ \omega_{mixed}(f; \delta_{n_1}, \delta_{n_2});\ \ \lambda_1, \lambda_2 > 0,$$

we may write,

$$\begin{aligned} &|\Delta_{(x,y)} f[t, s; x, y]| \\ &\leq \omega_{mixed}(f; |t-x|, |s-y|) \\ &\leq \left(1 + \frac{|t-x|}{\delta_1}\right)\left(1 + \frac{|s-y|}{\delta_2}\right)\omega_{mixed}(f; \delta_1, \delta_2), \end{aligned} \tag{9.3.3}$$

for every $(x, y), (t, s) \in I^2$ and for any $\delta_1, \delta_2 > 0$. From the definition of $\Delta_{(x,y)} f[t, s; x, y]$, we get

$$f(x, s) + f(t, y) - f(t, s) = f(x, y) - \Delta_{(x,y)} f[t, s; x, y].$$

By applying the linear positive operator $K_{n_1,n_2,p}$ to this equality and by the definition of operator $T_{n_1,n_2,p}$, we can write

$$\begin{aligned} T_{n_1,n_2,p}(f; q_{n_1}, q_{n_2}, x, y) &= f(x, y)\, K_{n_1,n_2,p}(e_{00}; q_{n_1}, q_{n_2}, x, y) \\ &\quad - K_{n_1,n_2,p}\left(\Delta_{(x,y)} f[t, s; x, y]; q_{n_1}, q_{n_2}, x, y\right). \end{aligned}$$

Since $K_{n_1,n_2,p}(e_{00}; q_{n_1}, q_{n_2}, x, y) = 1$, considering the inequality (9.3.3) and applying the Cauchy–Schwarz inequality we obtain,

$$\begin{aligned} &|T_{n_1,n_2,p}(f; q_{n_1}, q_{n_2}, x, y) - f(x, y)| \\ &\leq K_{n_1,n_2,p}\left(|\Delta_{(x,y)} f[t, s; x, y]|; q_{n_1}, q_{n_2}, x, y\right) \\ &\leq \Bigg(K_{n_1,n_2,p}(e_{00}; q_{n_1}, q_{n_2}, x, y) + \delta_{n_1}^{-1}\sqrt{K_{n_1,n_2,p}((t-x)^2; q_{n_1}, q_{n_2}, x, y)} \\ &\quad + \delta_{n_2}^{-1}\sqrt{K_{n_1,n_2,p}((s-y)^2; q_{n_1}, q_{n_2}, x, y)} + \delta_{n_1}^{-1}\delta_{n_2}^{-1}\sqrt{K_{n_1,n_2,p}((t-x)^2; q_{n_1}, q_{n_2}, x, y)} \\ &\quad \times \sqrt{K_{n_1,n_2,p}((s-y)^2; q_{n_1}, q_{n_2}, x, y)}\Bigg)\ \omega_{mixed}(f; \delta_{n_1}, \delta_{n_2})) \\ &\leq 4\ \omega_{mixed}(f; \delta_{n_1}, \delta_{n_2}), \end{aligned}$$

from which the desired result is immediate on choosing δ_{n_1} and δ_{n_2}, where

$$\delta_{n_1} = \{K_{n_1,n_2,p}((u-x)^2;\ q_{n_1}, q_{n_2}, x, y)\}^{1/2}$$

and

$$\delta_{n_2} = \{K_{n_1,n_2,p}((v-y)^2;\ q_{n_1}, q_{n_2}, x, y)\}^{1/2}$$

■

The Lipschitz class for B-continuous functions $Lip_M(\alpha, \beta)$ with $\alpha, \beta \in (0, 1]$ is defined by

$$Lip_M(\alpha, \beta) = \left\{f \in C_b\left(I^2\right) : \left|\Delta_{(x,y)} f[t, s; x, y]\right| \le M|t-x|^{\alpha}|s-y|^{\beta}\right\},$$

where for $(t, s), (x, y) \in I^2$.

The next theorem gives the degree of approximation for the operators $T_{n_1,n_2,p}$ by means of the Lipschitz class of Bögel continuous functions.

Theorem 9.13 ([215]) *Let $f \in Lip_M(\alpha, \beta)$, then we have*

$$\left|T_{n_1,n_2,p}\left(f; q_{n_1}, q_{n_2}, x, y\right) - f(x, y)\right| \le M\delta_{n_1}^{\alpha/2}\delta_{n_2}^{\beta/2}$$

for $M > 0$, $\alpha, \beta \in (0, 1]$.

Proof By the definition of the operator $T_{n_1,n_2,p}\left(f; q_{n_1}, q_{n_2}, x, y\right)$ and by linearity of the operator $K_{n_1,n_2,p}$ given by (8.2.1), we can write

$$\begin{aligned}
T_{n_1,n_2,p}\left(f; q_{n_1}, q_{n_2}, x, y\right) &= K_{n_1,n_2,p}\left(f(x,s) + f(t,y) - f(t,s)\right)\\
&= K_{n_1,n_2,p}\left(f(x,y) - \Delta_{(x,y)} f[t, s; x, y]; x, y\right)\\
&= f(x,y)\, K_{n_1,n_2,p}\left(e_{00}; x, y\right)\\
&\quad - K_{n_1,n_2,p}\left(\Delta_{(x,y)} f[t, s; x, y]; x, y\right).
\end{aligned}$$

By the hypothesis, we get

$$\begin{aligned}
&\left|T_{n_1,n_2,p}\left(f; q_{n_1}, q_{n_2}, x, y\right) - f(x, y)\right|\\
&\qquad \le K_{n_1,n_2,p}\left(\left|\Delta_{(x,y)} f[t, s; x, y]\right|; x, y\right)\\
&\qquad \le M K_{n_1,n_2,p}\left(|t-x|^{\alpha}|s-y|^{\beta}; x, y\right)\\
&\qquad = M K_{n_1,n_2,p}\left(|t-x|^{\alpha}; x\right) \times K_{n_1,n_2,p}\left(|s-y|^{\beta}; y\right).
\end{aligned}$$

Now, using Hölder's inequality with

$$p_1 = 2/\alpha, q_1 = 2/(2-\alpha) \text{ and } p_2 = 2/\beta, q_2 = 2/(2-\beta),$$

we have

$$\begin{aligned}
&\left|T_{n_1,n_2,p}\left(f; q_{n_1}, q_{n_2}, x, y\right) - f(x,y)\right| \\
&\qquad \le M\left(K_{n_1,n_2,p}(t-x)^2; x\right)^{\alpha/2} K_{n_1,n_2,p}(e_0; x)^{(2-\alpha)/2} \\
&\qquad\quad \times K_{n_1,n_2,p}\left((s-y)^2; y\right)^{\beta/2} K_{n_1,n_2,p}(e_0; y)^{(2-\beta)/2}.
\end{aligned}$$

Considering Lemma 8.9 and choosing

$$\delta_{n_1} = K_{n_1,n_2,p}\left((t-x)^2; x\right), \ \delta_{n_2} = K_{n_1,n_2,p}\left((s-y)^2; y\right)$$

we obtain the degree of local approximation for B-continuous functions belonging to $Lip_M(\alpha,\beta)$. ■

Theorem 9.14 ([215]) *Let the function* $f \in D_b(I^2)$ *with* $D_B f \in B(I^2)$. *Then, for each* $(x,y) \in I^2$, *we have*

$$\begin{aligned}
|T_{n_1,n_2,p}(f; q_{n_1}, q_{n_2}, x, y) - f(x,y)| \le \frac{M}{[n_1]_{q_{n_1}}^{1/2}[n_2]_{q_{n_2}}^{1/2}}\Big(||D_B f||_\infty \\
+\omega_{mixed}(D_B f; [n_1]_{q_{n_1}}^{-1/2}, [n_2]_{q_{n_2}}^{-1/2})\Big).
\end{aligned}$$

Proof Since $f \in D_b(I^2)$, we have the identity

$$\Delta_{x,y} f[t,s;x,y] = (t-x)(s-y)D_B f(\xi,\eta), \ \ with\ x < \xi < t;\ y < \eta < s.$$

It is clear that

$$D_B f(\xi,\eta) = \Delta_{(x,y)} D_B f(\xi,\eta) + D_B f(\xi,y) + D_B f(x,\eta) - D_B f(x,y).$$

Since $D_B f \in B(I^2)$, *by the above relations, we can write* $|K_{n_1,n_2,p}(\Delta_{(x,y)} f[t,s;x,y]; q_{n_1}, q_{n_2}, x, y)|$

$$\begin{aligned}
&= |K_{n_1,n_2,p}((t-x)(s-y)D_B f(\xi,\eta); q_{n_1}, q_{n_2}, x, y)| \\
&\le K_{n_1,n_2,p}(|t-x||s-y||\Delta_{(x,y)} D_B f(\xi,\eta)|; x, y) \\
&\quad + K_{n_1,n_2,p}(|t-x||s-y|(|D_B f(\xi,y)|
\end{aligned}$$

$$+|D_B f(x,\eta)| + |D_B f(x,y)|); q_{n_1}, q_{n_2}, x, y)$$
$$\leq K_{n_1,n_2,p}(|t-x||s-y|\omega_{mixed}(D_B f; |\xi - x|, |\eta - y|); x, y)$$
$$+3\ ||D_B f||_\infty\ K_{n_1,n_2,p}(|t-x||s-y|; q_{n_1}, q_{n_2}, x, y).$$

Since the mixed modulus of smoothness ω_{mixed} is nondecreasing, we have

$$\omega_{mixed}(D_B f; |\xi - x|, |\eta - y|) \leq \omega_{mixed}(D_B f; |t-x|, |s-y|)$$
$$\leq (1 + \delta_{n_1}^{-1}|t-x|)(1 + \delta_{n_2}^{-1}|s-y|)\ \omega_{mixed}(D_B f; \delta_{n_1}, \delta_{n_2}).$$

Substituting in the above inequality, using the linearity of $K_{n_1,n_2,p}$ and applying the Cauchy–Schwarz inequality, we obtain

$$|T_{n_1,n_2,p}(f; q_{n_1}, q_{n_2}, x, y) - f(x,y)|$$
$$= |K_{n_1,n_2,p}\Delta_{(x,y)} f[t,s;x,y]; x, y|$$
$$\leq 3||D_B f||_\infty \sqrt{K_{n_1,n_2,p}((t-x)^2(s-y)^2; q_{n_1}, q_{n_2}, x, y)}$$
$$+\Big(K_{n_1,n_2,p}(|t-x||s-y|; q_{n_1}, q_{n_2}, x, y)$$
$$+\delta_{n_1}^{-1} K_{n_1,n_2,p}((t-x)^2|s-y|; q_{n_1}, q_{n_2}, x, y)$$
$$+\delta_{n_2}^{-1} K_{n_1,n_2,p}(|t-x|(s-y)^2; q_{n_1}, q_{n_2}, x, y)$$
$$+\delta_{n_1}^{-1}\delta_{n_2}^{-1} K_{n_1,n_2,p}((t-x)^2(s-y)^2; q_{n_1}, q_{n_2}, x, y) \Big)\omega_{mixed}(D_B f; \delta_{n_1}, \delta_{n_2})$$
$$\leq 3||D_B f||_\infty \sqrt{K_{n_1,n_2,p}((t-x)^2(s-y)^2; q_{n_1}, q_{n_2}, x, y)}$$
$$+\Big(\sqrt{K_{n_1,n_2,p}((t-x)^2(s-y)^2; q_{n_1}, q_{n_2}, x, y)}$$
$$+\delta_{n_1}^{-1} \sqrt{K_{n_1,n_2,p}((t-x)^4(s-y)^2; q_{n_1}, q_{n_2}, x, y)}$$
$$+\delta_{n_2}^{-1} \sqrt{K_{n_1,n_2,p}((t-x)^2(s-y)^4; q_{n_1}, q_{n_2}, x, y)}$$
$$+\delta_{n_1}^{-1}\delta_{n_2}^{-1} K_{n_1,n_2,p}((t-x)^2(s-y)^2; q_{n_1}, q_{n_2}, x, y) \Big)\omega_{mixed}(D_B f; \delta_{n_1}, \delta_{n_2}).$$
$$(9.3.4)$$

We observe that for $(x, y), (t, s) \in I^2$ and $i, j \in \{1, 2\}$ the following holds

$$K_{n_1,n_2,p}((t-x)^{2i}(s-y)^{2j}; q_{n_1}, q_{n_2}, x, y) = K_{n_1,p}((t-x)^{2i}; q_{n_1}, x, y)$$
$$\times K_{n_2,p}((s-y)^{2j}; q_{n_2}, x, y). \quad (9.3.5)$$

If $\{q_n\}$ is a sequence in $(0, 1)$ such that $q_n \to 1$ and $q_n^n \to 0$ as $n \to \infty$, then it is known from [20, Theorem 4.5] that

$$K_{n,p}((t-x)^2; q_n, x) \leq \frac{M_1}{[n]_{q_n}} \tag{9.3.6}$$

$$K_{n,p}((t-x)^4; q_n, x) \leq \frac{M_2}{[n]_{q_n}^2} \tag{9.3.7}$$

for some constants $M_1, M_2 > 0$.

Let $\delta_{n_1} = \frac{1}{[n_1]_{q_{n_1}}^{1/2}}$, and $\delta_{n_2} = \frac{1}{[n_2]_{q_{n_2}}^{1/2}}$. Thus, (9.3.4) becomes

$$\begin{aligned}
&|T_{n_1,n_2,p}(f; q_{n_1}, q_{n_2}, x, y) - f(x, y)| \\
&\quad = 3||D_B||_\infty O\left(\frac{1}{[n_1]_{q_{n_1}}^{1/2}}\right) O\left(\frac{1}{[n_2]_{q_{n_2}}^{1/2}}\right) \\
&\quad\quad + O\left(\frac{1}{[n_1]_{q_{n_1}}^{1/2}}\right) O\left(\frac{1}{[n_2]_{q_{n_2}}^{1/2}}\right) \omega_{mixed}(D_B f; [n_1]_{q_{n_1}}^{-1/2}, [n_2]_{q_{n_2}}^{-1/2})
\end{aligned} \tag{9.3.8}$$

Hence, combining (9.3.4)–(9.3.8), we obtain the desired result. ■

9.3.3 q-Stancu–Durrmeyer Type

The GBS operator of $D_{n_1,n_2}^{\alpha_{n_1},\alpha_{n_2}}$ given by (8.3.3) for any $f \in C_b(I^2)$ and $n_1, n_2 \in \mathbb{N}$ is defined by

$$G_{n_1,n_2}^{\alpha_{n_1},\alpha_{n_2}}(f; q_{n_1}, q_{n_2}, x, y) = D_{n_1,n_2}^{\alpha_{n_1},\alpha_{n_2}}(f(t, y) + f(x, s) - f(t, s); q_{n_1}, q_{n_2}, x, y),$$

for all $(x, y) \in I^2$. More precisely for any $f \in C_b(I^2)$, the GBS operator of q-Stancu–Durrmeyer type operator is given by

$$\begin{aligned}
G_{n_1,n_2}^{\alpha_{n_1},\alpha_{n_2}}(f; q_{n_1}, q_{n_2}, x, y) &= [n_1+1]_{q_{n_1}} [n_2+1]_{q_{n_2}} \sum_{k_1=0}^{n_1} \sum_{k_2=0}^{n_2} p_{n_1,k_1}^{q_{n_1},\alpha_{n_1}}(x) p_{n_2,k_2}^{q_{n_2},\alpha_{n_2}}(y) \\
&\int_0^1 \int_0^1 p_{n_1,k_1}^{q_{n_1}}(t) p_{n_2,k_2}^{q_{n_2}}(s)[f(t, y) + f(x, s) - f(t, s)] d_{q_{n_1}} t d_{q_{n_2}} s.
\end{aligned}$$

Evidently, the operator $G_{n_1,n_2}^{\alpha_{n_1},\alpha_{n_2}}$ is a linear operator.

Theorem 9.15 *For every $f \in C_b(I^2)$ and $(x, y) \in I^2$, we have*

$$|G_{n_1,n_2}^{\alpha_{n_1},\alpha_{n_2}}(f; q_{n_1}, q_{n_2}, x, y) - f(x, y)| \leq 4\omega_{mixed}(f, \delta_{n_1,q_{n_1}}^{\alpha_{n_1}}(x), \delta_{n_2,q_{n_2}}^{\alpha_{n_2}}(y)).$$

Let for $(t, s), (x, y) \in I^2$, the Lipschitz class for B-continuous functions be defined as

$$Lip_M(\beta, \gamma) = \left\{ f \in C_b\left(I^2\right) : |\Delta f\,[t, s; x, y]| \leq M\,|t - x|^{\beta}\,|s - y|^{\gamma}, \beta, \gamma \in (0, 1] \right\},$$

Our next theorem gives the rate of convergence for the operators $G_{n_1,n_2}^{\alpha_{n_1},\alpha_{n_2}}$ by means of the class $Lip_M(\beta, \gamma)$.

Theorem 9.16 ([182]) *Let $f \in Lip_M(\beta, \gamma)$ and $(x, y) \in I^2$. Then for $M > 0$, $\beta, \gamma \in (0, 1]$, we have*

$$|G_{n_1,n_2}^{\alpha_{n_1},\alpha_{n_2}}(f; q_{n_1}, q_{n_2}, x, y) - f(x, y)| \leq M\left(\delta_{n_1,q_{n_1}}^{\alpha_{n_1}}(x)\right)^{\beta}\left(\delta_{n_2,q_{n_2}}^{\alpha_{n_2}}(y)\right)^{\gamma}.$$

Theorem 9.17 ([182]) *Let $f \in D_b(I^2)$ with $D_B f \in B(I^2)$. Then, for each $(x, y) \in I$, we have*

$$|G_{n_1,n_2}^{\alpha_{n_1},\alpha_{n_2}} f; q_{n_1}, q_{n_2}, x, y - f(x, y)| \leq \frac{M}{[n_1]_{q_{n_1}}^{1/2}[n_2]_{q_{n_2}}^{1/2}}\Bigg(||D_B f||_\infty$$

$$+\omega_{mixed}(D_B f; [n_1]_{q_{n_1}}^{-1/2}, [n_2]_{q_{n_2}}^{-1/2})\Bigg).$$

9.3.4 Bernstein–Schurer–Stancu–Kantorovich Type Based on q-Integers

Let $I = [0, 1 + p]$ and $p \in \mathbb{N} \cup \{0\}$. For $f \in C(I)$, the space of all continuous functions on I endowed with the norm

$$\| f \| = \sup_{x \in [0,1+p]} |f(x)|$$

and $0 < q < 1$. Ren and Zeng [205] defined the following new version of q-Bernstein–Schurer operator which preserves the linear functions:

$$\overline{B}_n^p(f(t); q, x) = \sum_{k=0}^{n+p} \tilde{p}_{n,k}^*(q, x) f\left(\frac{[k]_q}{[n]_q}\right), \tag{9.3.9}$$

where,

$$\tilde{p}^{*}_{n,k}(q,x)=\frac{[n]_q^{n+p}}{[n+p]_q^{n+p}}\binom{n+p}{k}_q x^k\left(\frac{[n+p]_q}{[n]_q}-x\right)_q^{n+p-k}.$$

Later, Acu [11] proposed a q-Durrmeyer modification of the operators (9.3.9) as follows

$$D_{n,p}(f;q,x)=\frac{[n+p+1]_q[n]_q}{[n+p]_q}\sum_{k=0}^{n+p}\tilde{p}^{*}_{n,k}(q,x)\int_0^{\frac{[n+p]_q}{[n]_q}} f(t)\tilde{b}^{p}_{n,k}(q,qt)d_qt,$$

and discussed the rate of convergence in terms of the modulus of continuity, a Lipschitz class function, and a Voronovskaja type result. Subsequently, for $\alpha,\beta\in\mathbb{R}$ such that $0\le\alpha\le\beta$ and $f\in C(I)$, Agrawal et al. [22] introduced a Stancu type Kantorovich modification of the operators (9.3.9), defined as

$$\mathcal{K}^{(\alpha,\beta)}_{n,p}(f;q,x)=\sum_{k=0}^{n+p}\tilde{p}^{*}_{n,k}(q,x)\int_0^1 f\left(\frac{[k]_q+q^kt+\alpha}{[n+1]_q+\beta}\right)d_qt \tag{9.3.10}$$

and discussed the basic convergence theorem, the rate of convergence involving modulus of continuity, and Lipschitz function.

Lemma 9.11 ([22]) *For the operators given by (9.3.10), the following equalities hold:*

1. $\mathcal{K}^{(\alpha,\beta)}_{n,p}(1;q,x)=1;$
2. $\mathcal{K}^{(\alpha,\beta)}_{n,p}(t;q,x)=\dfrac{\alpha}{[n+1]_q+\beta}+\dfrac{2q[n]_qx+1}{[2]_q([n+1]_q+\beta)};$
3. $\mathcal{K}^{(\alpha,\beta)}_{n,p}(t^2;q,x)=\dfrac{1}{[2]_q[3]_q([n+1]_q+\beta)^2}\Big\{\dfrac{[n]_q^2[n+p-1]_q}{[n+p]_q}([3]_qq^2+3q^4)x^2+\{(4\alpha+3)q[3]_q+q^2(1+[2]_q)\}[n]_qx+[4]_q\alpha^2+2\alpha[3]_q+(1+q\alpha^2)[2]_q\Big\}.$

Lemma 9.12 ([22]) *For $m\in\mathbb{N}\cup\{0\}$, the mth order central moment of $\mathcal{K}^{(\alpha,\beta)}_{n,p}(f;q,x)$ defined as*

$$\mu^{*}_{n,m,q}(x)=\mathcal{K}^{\alpha,\beta}_{n,p}((t-x)^m;q,x),$$

we have

1. $\mu^{*}_{n,1,q}(x)=\dfrac{(2-[2]_q)q[n]_qx-(\beta+1)[2]_qx+1}{[2]_q([n+1]_q+\beta)}+\dfrac{\alpha}{[n+1]_q+\beta};$

2. $\mu^*_{n,2,q}(x) = \left\{\frac{[n]_q^2[n+p-1]_q([3]_q q^2+3q^4)}{[n+p]_q([n+1]_q+\beta)^2[2]_q[3]_q} - \frac{4q[n]_q}{[2]_q[n+1]_q+\beta}+1\right\}x^2$
$+\left\{\frac{\{(4\alpha+3)[3]_q q+q^2(1+[2]_q)\}[n]_q}{([n+1]_q+\beta)^2[2]_q[3]_q} - \frac{2\alpha}{[n+1]_q+\beta} - \frac{2}{[2]_q[n+1]_q+\beta}\right\}x +$
$\frac{[4]_q\alpha^2+2\alpha[3]_q+(1+q\alpha^2)[2]_q}{([n+1]_q+\beta)^2[2]_q[3]_q}.$

By the definition of the Jackson integral, the inequality

$$(a+b)^4 \leq 8(a^4+b^4), \text{ where } a>0, b>0$$

and by Lemma 2.4 [205], the fourth order central moment of the operators (9.3.10), obtained by Chauhan et al. [70] satisfy:

$$\mathcal{K}_{n,p}^{(\alpha,\beta)}((t-x)^4;q_n,x) \leq 64\frac{1}{[n]_{q_n}^2} + 64M_2\frac{1/4}{[n]_{q_n}^2} + \frac{8}{[n]_{q_n}^2} = \frac{64+16M_2+8}{[n]_{q_n}^2}.$$

For any $f \in C_b(I_1 \times I_2)$, the GBS operator

$$T_{n_1,n_2,p_1,p_2}^{(\alpha_1,\alpha_2,\beta_1,\beta_2)} : C_b(I_1\times I_2) \longrightarrow C(I_1\times I_2)$$

of q-Bernstein–Schurer–Kantorovich type is given by

$$T_{n_1,n_2,p_1,p_2}^{(\alpha_1,\alpha_2,\beta_1,\beta_2)}(f;q_{n_1},q_{n_2},x,y)$$
$$= \sum_{k_1=0}^{n_1+p_1}\sum_{k_2=0}^{n_2+p_2} \tilde{p}^*_{n_1,n_2,k_1,k_2}(q_{n_1},q_{n_2};x,y)\int_0^1\int_0^1 f\left\{\left(\frac{[k_1]_{q_{n_1}}+q_{n_1}^{k_1}t+\alpha_1}{[n_1+1]_{q_{n_1}}+\beta_1},y\right)\right.$$
$$+f\left(x,\frac{[k_2]_{q_{n_2}}+q_{n_2}^{k_2}s+\alpha_2}{[n_2+1]_{q_{n_2}}+\beta_2}\right)$$
$$\left.-f\left(\frac{[k_1]_{q_{n_1}}+q_{n_1}^{k_1}t+\alpha_1}{[n_1+1]_{q_{n_1}}+\beta_1},\frac{[k_2]_{q_{n_2}}+q_{n_2}^{k_2}s+\alpha_2}{[n_2+1]_{q_{n_2}}+\beta_2}\right)\right\}d_{q_{n_1}}t\ d_{q_{n_2}}s, \tag{9.3.11}$$

where,

$$\tilde{p}^*_{n_1,n_2,k_1,k_2}(q_{n_1},q_{n_2},x,y)$$
$$= \frac{[n_1]_{q_{n_1}}^{n_1+p_1}}{[n_1+p_1]_{q_{n_1}}^{n_1+p_1}}\begin{bmatrix} n_1+p_1 \\ k_1 \end{bmatrix}_{q_{n_1}} x^{k_1}\left(\frac{[n_1+p_1]_{q_{n_1}}}{[n_1]_{q_{n_1}}}-x\right)_{q_{n_1}}^{n_1+p_1-k_1}$$

$$\times \frac{[n_2]_{q_{n_2}}^{n_2+p_2}}{[n_2+p_2]_{q_{n_2}}^{n_2+p_2}} \begin{bmatrix} n_2+p_2 \\ k_2 \end{bmatrix}_{q_{n_2}} y^{k_2} \left(\frac{[n_2+p_2]_{q_{n_2}}}{[n_2]_{q_{n_2}}} - y \right)_{q_{n_2}}^{n_2+p_2-k_2}.$$

Clearly, the operator $T_{n_1,n_2,p_1,p_2}^{(\alpha_1,\alpha_2,\beta_1,\beta_2)}$ is linear and preserves linear functions.

Theorem 9.18 ([70]) *For every $f \in C_b(I_1 \times I_2)$, at each point $(x, y) \in J^2$, the operator (9.3.11) satisfies the following inequality*

$$|T_{n_1,n_2,p_1,p_2}^{(\alpha_1,\alpha_2,\beta_1,\beta_2)}(f; q_{n_1}, q_{n_2}, x, y) - f(x, y)| \le 4\,\omega_{mixed}(f; \sqrt{\delta_{n_1}(x)}, \sqrt{\delta_{n_2}(y)}).$$

Theorem 9.19 ([70]) *For $f \in Lip_M(\xi, \eta)$, we have*

$$|T_{n_1,n_2,p_1,p_2}^{(\alpha_1,\alpha_2,\beta_1,\beta_2)}(f; q_{n_1}, q_{n_2}, x, y) - f(x, y)| \le M(\delta_{n_1}(x))^{\frac{\xi}{2}}(\delta_{n_2}(y))^{\frac{\eta}{2}},$$

for $M > 0$, $\xi, \eta \in (0, 1]$.

Theorem 9.20 ([70]) *For $f \in D_b(I_1 \times I_2)$ with $D_B f \in B(I_1 \times I_2)$ and each $(x, y) \in J^2$, we have*

$$|T_{n_1,n_2,p_1,p_2}^{(\alpha_1,\beta_1,\alpha_2,\beta_2)}(f; q_{n_1}, q_{n_2}, x, y) - f(x, y)|$$

$$\le \frac{M}{[n_1]_{q_{n_1}}^{1/2}[n_2]_{q_{n_2}}^{1/2}} \left(||D_B f||_\infty + \omega_{mixed}(D_B f; [n_1]_{q_{n_1}}^{-1/2}, [n_2]_{q_{n_2}}^{-1/2}) \right).$$

9.3.5 Durrmeyer–Lupaş Type

We define the GBS operator of the operator $D_{n,m}^{(1/n,1/m)}$ given by (8.3.7), for any $f \in C(I^2)$ and $m, n \in \mathbb{N}$, by

$$T_{n,m}^{(1/n,1/m)}(f(t,s); x, y) := D_{n,m}^{(1/n,1/m)}(f(t,y) + f(x,s) - f(t,s); x, y)$$

for all $(x, y) \in I^2$. More precisely, the Lupaş–Durrmeyer type GBS operator is defined as follows:

$$T_{n,m}^{(1/n,1/m)}(f(t,s); x, y) := (n+1)(m+1) \sum_{k=0}^{n} \sum_{j=0}^{m} p_{n,m,k,j}^{(1/n,1/m)}(x, y)$$

$$\times \int_0^1 \int_0^1 p_{n,m,k,j}(t,s)\,[f(x,s) + f(t,y) - f(t,s)]\,dtds. \tag{9.3.12}$$

where the operator $T_{n,m}^{(1/n,1/m)}$ is well defined from the space $C_b\left(I^2\right)$ on itself and $f \in C_b\left(I^2\right)$. It is clear that $T_{n,m}^{(1/n,1/m)}$ is a positive linear operator.

We shall estimate the rate of convergence of the sequences of the operators (9.3.12) to $f \in C_b(I^2)$ using the mixed modulus of smoothness. For this, we use the well-known Shisha–Mond theorem for B-continuous functions established by Gonska [98] and Badea and Cottin [46].

Theorem 9.21 ([24]) *For every $f \in C_b\left(I^2\right)$, at each point $(x, y) \in I^2$, the operator (9.3.12) satisfies the following inequality*

$$\left|T_{n,m}^{(1/n,1/m)}(f; x, y) - f(x, y)\right| \leq 9\ \omega_{mixed}(f; (n+1)^{-1/2}, (m+1)^{-1/2}). \tag{9.3.13}$$

Theorem 9.22 ([24]) *Let the function $f \in D_b\left(I^2\right)$ with $D_B f \in B(I^2)$. Then, for each $(x, y) \in I^2$, we have*

$$\begin{aligned} &\mid T_{n,m}^{(1/n,1/m)}(f; x, y) - f(x, y) \mid \\ &\quad \leq 4\left[\|D_B f\|_\infty + 9\ \omega_{mixed}\left(D_B f; (n+1)^{-1/2}, (m+1)^{-1/2}\right)\right] \\ &\qquad ((n+1)(m+1))^{-1/2}. \end{aligned}$$

9.3.6 GBS Operators of q-Durrmeyer Type

For all positive integers m, n, $f \in R^I$ and $(x, y) \in I$, the GBS case of the operator (8.3.11) is defined by

$$\begin{aligned} U_{m,n,q_1,q_2}(f; x, y) &= [m+1]_{q_1}[n+1]_{q_2} \sum_{k=0}^{m}\sum_{j=0}^{n} q^{-k} q^{-j} p_{m,k}(q_1; x) p_{n,j}(q_2; y) \\ &\int_0^1\int_0^1 \{f(s, y) + f(x, t) - f(s, t)\} p_{m,k}(q_1; q_1 s) p_{n,j}(q_2; q_2 t) d_{q_1} s d_{q_2} t. \end{aligned}$$

It can easily be shown that the GBS q-Durrmeyer operator is the Boolean sum of parametric extension D_{m,q_1}^x, D_{n,q_2}^y, i.e.,

$$U_{m,n,q_1,q_2} = D_{m,q_1}^x \oplus D_{n,q_2}^y = D_{m,q_1}^x + D_{n,q_2}^y - D_{m,n,q_1,q_2}.$$

Theorem 9.23 ([56]) *If* $q_{1m}, q_{2n} \in (0, 1)$ *such that*

$$\lim_{m\to\infty} q_{1m} = 1, \lim_{n\to\infty} q_{2n} = 1,$$

then the sequence U_{m,n,q_1,q_2} *converges uniformly to* f *on* I^2, *for each* $f \in C_b(I^2)$.

Theorem 9.24 ([56]) *Suppose* $q_{1m}, q_{2n} \in (0, 1)$ *be such that*

$$\lim_{m\to\infty} q_{1m} = 1, \lim_{n\to\infty} q_{2n} = 1,$$

then the following inequality

$$|f(x, y) - U_{m,n,q_1,q_2}(f; x, y)| \leq 4\omega_{mixed}(f; \delta_{1m}^{1/2}(x), \delta_{2n}^{1/2}(y))$$

holds for each $f \in C_b(I^2)$, $(x, y) \in I^2$, *where*

$$\delta_{1m} = \frac{2}{[m+2]_{q_1 m}}\left(x(1-x) + \frac{3}{2[m+2]_{q_1 m}}\right)$$

and

$$\delta_{2n} = \frac{2}{[n+2]_{q_2 n}}\left(y(1-y) + \frac{3}{2[n+2]_{q_2 n}}\right).$$

For $f \in C_b(I_1 \times I_2)$, the Lipschitz class $Lip_M(\alpha, \beta)$ with $\alpha, \beta \in (0, 1]$ is defined by

$$Lip_M(\alpha, \beta) = \Big\{ f \in C_b(I^2) : |\Delta f[s, t; x, y]| \leq M|t-x|^{\alpha}|s-y|^{\beta}, \text{ for } (s, t), (x, y) \in I^2 \Big\},$$

which is the Lipschitz class for B-continuous functions.

Theorem 9.25 ([56]) *Let* $q_{1m}, q_{2n} \in (0, 1)$ *such that*

$$\lim_{m\to\infty} q_{1m} = 1, \lim_{n\to\infty} q_{2n} = 1.$$

If $f \in Lip_M(\alpha, \beta)$, *we have*

$$|U_{m,n,q_1,q_2}(f; x, y) - f(x, y)| \leq M(\delta_{1m}(x))^{\frac{\alpha}{2}}(\delta_{2n}(y))^{\frac{\beta}{2}},$$

for $M > 0$, $\alpha, \beta \in (0, 1]$ *and* $\delta_{1m}(x), \delta_{2n}(y)$ *are defined in Theorem 9.24.*

Theorem 9.26 ([56]) *Suppose* $q_{1m}, q_{2n} \in (0, 1)$ *such that*

$$\lim_{m\to\infty} q_{1m} = 1, \lim_{n\to\infty} q_{2n} = 1.$$

Let $f \in D_b(I^2)$ *with* $D_B f \in B(I^2)$. *There exists a constant* $M > 0$ *such that for each* $(x, y) \in I^2$, *we have*

$$|U_{m,n,q_1,q_2}(f; x, y) - f(x, y)| \leq \frac{M}{[m]_{q_{1m}}^{1/2}[n]_{q_{2n}}^{1/2}}\left(||D_B f||_\infty + \omega_{mixed}(D_B f; [m]_{q_{1m}}^{-1/2}, [n]_{q_{2n}}^{-1/2})\right).$$

9.4 GBS Operator of Chlodowsky–Szász–Kantorovich–Charlier Operators

For $I_{de} = [0, d] \times [0, e]$, let $C_b(I_{de})$ denote the space of all B-continuous functions on I_{de} and let $C(I_{de})$ be the space of all ordinary continuous functions on I_{de}.

Agrawal et al. [25] define the GBS operators of the $C_{n,m}^a$ given by (8.4.2), for any $f \in C(I_{de})$ and $n, m \in N$, by

$$S_{n,m}^a(f(t, s); x, y) = C_{n,m}^a(f(t, y) + f(x, s) - f(t, s); x, y),$$

for all $(x, y) \in I_{de}$.

More precisely, for any $f \in C(I_{de})$, the GBS operator of Chlodowsky–Szász–Kantorovich–Charlier operators is given by

$$S_{n,m}^a(f; x, y) = \frac{n}{a_n} c_m \sum_{k=0}^{n} \sum_{j=0}^{\infty} p_{n,k}\left(\frac{x}{a_n}\right) \Pi_{m,j}(b_m y, a)$$

$$\times \int_{\frac{j}{c_m}}^{\frac{j+1}{c_m}} \int_{\frac{k}{n}a_n}^{\frac{k+1}{n}a_n} (f(t, y) + f(x, s) - f(t, s); x, y) dt ds.$$

Theorem 9.27 ([25]) *If* $f \in C_b(I_{de})$, *then for any* $(x, y) \in I_{de}$, *and any* $m, n \in N$, *we have*

$$|S_{m,n}^a(f(t, s); x, y) - f(x, y)| \leq 4\omega_B(f; \delta_n, \delta_m),$$

where

$$\delta_n = \left(\rho(a)\frac{a_n}{n}\right)^{1/2} \text{ and } \delta_m = \left(\frac{\varsigma(a)}{c_m}\right)^{1/2}.$$

For $0 < \gamma \leq 1$, let

$$Lip_L\gamma = \left\{ f \in C(I_{a_n}) : |\Delta_{(x,y)} f[t, s; x, y]| \leq L||r - s||^\gamma \right\},$$

be the Lipschitz class of B-continuous functions, where

$$r = (u, v), s = (x, y) \in I_{a_n} \text{ and } \| r - s \| = \left\{ (u - x)^2 + (v - y)^2 \right\}^{1/2}$$

is the Euclidean norm.

The following results provide the rate of convergence of the operator $S^a_{n,m}(f(t, s); x, y)$ in terms of the Lipschitz class.

Theorem 9.28 ([25]) *If $f \in Lip_L\gamma$, then for every $(x, y) \in I_{de}$, we have*

$$\left|S^a_{n,m}(f(t, s); x, y) - f(x, y)\right| \leq L \left\{\delta_n(x) + \delta_m(y)\right\}^{\gamma/2},$$

for $L > 0$, and $\gamma \in (0, 1]$.

Theorem 9.29 ([25]) *If $f \in D_b(I_{de})$ and $D_B f \in B(I_{de})$, then for each $(x, y) \in I_{de}$, we get*

$$|S^a_{n,m}(f; x, y) - f(x, y)| \leq C\left\{3||D_B f||_\infty + 2\omega_{mixed}(f; \delta_n, \delta_m)\sqrt{x^2 + x}\sqrt{y^2 + y + 1}\right\}\delta_n\delta_m$$
$$+ \left\{\omega_{mixed}(f; \delta_n, \delta_m)\left(\delta_m\sqrt{x^4 + x^3 + x^2 + x}\sqrt{y^2 + y + 1}\right.\right.$$
$$\left.\left. + \delta_n\sqrt{y^4 + y^3 + y^2 + y + 1}\sqrt{x^2 + x}\right)\right\},$$

where $\delta_n = \sqrt{\frac{a_n}{n}}$, $\delta_m = \sqrt{\frac{\eta(a)}{c_m}}$, $\eta(a) = \max\{\tau(a), \omega(a)\}$ and C is a constant depending on n, m only.

9.4.1 q-Dunkl Analogue of the Szász–Mirakjan–Kantorovich Type

For any $f \in C_b(I_{cd})$ the GBS operator associated with the operator $T^*_{n_1,n_2,q_{n_1},q_{n_2}}(f; x, y)$ given by (8.5.2), is defined as:

$$\mathcal{K}^*_{n_1,n_2,q_{n_1},q_{n_2}}(f(t, s); x, y) := T^*_{n_1,n_2,q_{n_1},q_{n_2}}(f(t, y) + f(x, s) - f(t, s); x, y), \tag{9.4.1}$$

for all $(x, y) \in I_{cd}$.

Hence for any $f \in C_b(I_{cd})$, the GBS operator of Dunkl generalization of q-Szász–Mirakjan–Katorovich operator [26] is $\mathcal{K}^*_{n_1,n_2,q_{n_1},q_{n_2}} : C_b(I_{cd}) \longrightarrow C(I_{cd})$ given by

$$\mathcal{K}^*_{n_1,n_2,q_{n_1},q_{n_2}}(f; x, y)$$

$$= \frac{[n_1]_{q_{n_1}}}{e_\mu([n_1]_{q_{n_1}}\frac{x}{q_{n_1}})} \frac{[n_2]_{q_{n_2}}}{e_\mu([n_2]_{q_{n_2}}\frac{y}{q_{n_2}})} \sum_{k_2=0}^{\infty}\sum_{k_1=0}^{\infty} \frac{([n_1]_{q_{n_1}}x)^{k_1}}{\gamma_\mu(k_1)q_{n_1}^{k_1}} \frac{([n_2]_{q_{n_2}}x)^{k_2}}{\gamma_\mu(k_2)q_{n_2}^{k_2}}$$

$$\times \int_{\frac{q_{n_2}[k_2+2\mu\theta_{k_2}]_{q_{n_2}}}{[n_2]_{q_{n_2}}}}^{\frac{[k_2+1+2\mu\theta_{k_2}]_{q_{n_2}}}{[n_2]_{q_{n_2}}}} \int_{\frac{q_{n_1}[k_1+2\mu\theta_{k_1}]_{q_{n_1}}}{[n_1]_{q_{n_1}}}}^{\frac{[k_1+1+2\mu\theta_{k_1}]_{q_{n_1}}}{[n_1]_{q_{n_1}}}} \Big\{ f(t,s) + f(x,s) - f(t,s) \Big\} d_{q_1}t \, d_{q_2}s.$$

Theorem 9.30 ([26]) *For every $f \in C_b(I_{cd})$, the operator (9.4.1) satisfies the following inequality*

$$|\mathcal{K}^*_{n_1,n_2,q_{n_1},q_{n_2}}(f; x, y) - f(x, y)| \leq 4\,\omega_{mixed}(f; \sqrt{\delta_{n_1}(x)}, \sqrt{\delta_{n_2}(y)}).$$

In the following, we estimate the degree of approximation for the operators $\mathcal{K}^*_{n_1,n_2,q_{n_1},q_{n_2}}$ by means of the class $Lip_M(\xi, \eta)$.

Theorem 9.31 ([26]) *If $f \in Lip_M(\xi, \eta)$, we have*

$$|\mathcal{K}^*_{n_1,n_2,q_{n_1},q_{n_2}}(f; x, y) - f(x, y)| \leq M(\delta_{n_1}(x))^{\frac{\xi}{2}}(\delta_{n_2}(y))^{\frac{\eta}{2}},$$

for $M > 0$, $\xi, \eta \in (0, 1]$.

Theorem 9.32 ([26]) *For $f \in D_b(I_{cd})$ with $D_B f \in B(I_{cd})$ and each $(x, y) \in I_{cd}$, we have*

$$|\mathcal{K}^*_{n_1,n_2,q_{n_1},q_{n_2}}(f; x, y) - f(x, y)|$$

$$\leq M\Bigg\{ \frac{1}{[n_1]_{q_{n_1}}^{1/2}[n_2]_{q_{n_2}}^{1/2}} ||D_B f||_\infty + \omega_{mixed}(D_B f; [n_1]_{q_{n_1}}^{-1/2}, [n_2]_{q_{n_2}}^{-1/2})$$

$$\times \left(\frac{1}{[n_1]_{q_{n_1}}^{1/2}[n_2]_{q_{n_2}}^{1/2}} + \frac{1}{[n_1]_{q_{n_1}}^{1/2}} + \frac{1}{[n_2]_{q_{n_2}}^{1/2}} \right) \Bigg\}.$$

where M is some constant.

Bibliography

1. U. Abel, V. Gupta, An estimate of the rate of convergence of a Bézier variant of the Baskakov-Kantorovich operators for bounded variation functions. Demonstr. Math. **36**(1), 123–136 (2003)
2. U. Abel, M. Ivan, On a generalization of an approximation operator defined by A. Lupaş. General Math. **15**(1), 21–34 (2007)
3. T. Acar, (p, q)-Generalization of Szász-Mirakyan operators. Math. Methods Appl. Sci. **39**(10), 2685–2695 (2016)
4. T. Acar, A. Aral, I. Raşa, Modified Bernstein-Durrmeyer operators. General Math. **22**(1), 27–41 (2014)
5. T. Acar, A. Aral, S.A. Mohiuddine, On Kantorovich modification of (p, q)-Baskakov operators. J. Inequal. Appl **2016**, 98 (2016)
6. T. Acar, P.N. Agrawal, T. Neer, Bézier variant of the Bernstein-Durrmeyer type operators. Results Math. **72**(3), 1341–1358 (2017)
7. T. Acar, A. Aral, S.A. Mohiuddine, On Kantorovich modification of (p, q)-Bernstein operators. Iran. J. Sci. Technol. Trans. Sci., 1–6 (2017). https://doi.org/10.1007/s40995-017-0154-8
8. T. Acar, A. Aral, H. Gonska, On Szász-Mirakyan operators preserving e^{2ax}, $a > 0$. Mediterr. J. Math. **14**(6) (2017). https://doi.org/10.1007/s00009-016-0804-7
9. T. Acar, A. Aral, D.C. Moreles, P. Garrancho, Szász-Mirakyan type operators which fix exponentials. Results Math. **72**(3), 1393–1404 (2017)
10. T. Acar, P.N. Agrawal, A.S. Kumar, On a modification of (p, q)-Szász-Mirakyan operators. Compl. Anal. Oper. Theory **12**(1), 155–167 (2018)
11. A.M. Acu, Stancu-Schurer-Kantorovich operators based on q-integers. Appl. Math. Comput. **259**, 896–907 (2015)
12. A.M. Acu, H. Gonska, Ostrowski-type inequalities and moduli of smoothness. Results Math. **53**(3–4), 217–228 (2009)
13. A.M. Acu, M.D. Rusu, New results concerning Chebyshev-Gruss-type inequalities via discrete oscillations. Appl. Math. Comput. **243**, 585–593 (2014)
14. A.M. Acu, H. Gonska, Composite Bernstein cubature. Banach J. Math. Anal. **10**(2), 235–250 (2016)
15. A.M. Acu, I. Rasa, New estimates for the differences of positive linear operators. Numer. Algorithms **73**(3), 775–789 (2016)
16. A.M. Acu, H. Gonska, I. Raşa, Grüss and Ostrowski-type in approximation theory. Ukr. Math. J. **63**(6), 843–864 (2011)

V. Gupta et al., *Recent Advances in Constructive Approximation Theory*,
Springer Optimization and Its Applications 138,
https://doi.org/10.1007/978-3-319-92165-5

17. A.M. Acu, P.N. Agrawal, T. Neer, Approximation properties of the modified Stancu operators. Numer. Funct. Anal. Optim. **38**(3), 279–292 (2017)
18. O. Agratini, On a sequence of linear positive operators. Facta Univ. (Nis). Ser. Math. Inform. **14**, 41–48 (1999)
19. P.N. Agrawal, A.J. Mohammad, On L_p-inverse theorem for a linear combination of a new sequence of linear positive operators. Soochow J. Math. **32**(3), 399–411 (2006)
20. P.N. Agrawal, Z. Finta, A.S. Kumar, Bivariate q-Bernstein-Schurer-Kantorovich operators. Results Math. **67**(3-4), 365–380 (2015)
21. P.N. Agrawal, N. Ispir, Degree of approximation for bivariate Chlodowsky-Szasz-Charlier type operators. Results Math. **69**(3), 369–385 (2016)
22. P.N. Agrawal, M. Goyal, A. Kajla, On q-Bernstein-Schurer-Kantorovich type operators. Boll. dell'Unione Mat. Ital. **8**, 169–180 (2015)
23. P.N. Agrawal, N. Ispir, A. Kajla, Rate of convergence of Lupaş Kantorovich operators based on Polya distribution. Appl. Math. Comput. **261**, 323–329 (2015)
24. P.N. Agrawal, N. Ispir, A. Kajla, GBS operators of Lupas-Durrmeyer type based on Polya distribution. Results Math. **69**(3), 397–418 (2016)
25. P.N. Agrawal, B. Baxhaku, R. Chauhan, The approximation of bivariate Chlodowsky-Szász-Kantorovich-Charlier type operators. J. Inequal. Appl. **2017**, 195 (2017). https://doi.org/10.1186/s13660-017-1465-1
26. P.N. Agrawal, A. Yadav, S. Araci, Dunkl generalization of q-Szász-Mirakyan-Kantorovich operators for functions of two variables, Communicated
27. F. Altomare, Korovkin-type theorems and approximation by positive linear operators. Surv. Approx. Theory **5**, 92–164 (2010)
28. F. Altomare, M. Campiti, *Korovkin-type Approximation Theory and its Applications* (de Gruyter, New York, 1994)
29. G.A. Anastassiou, Ostrowski type inequalities. Proc. Am. Math. Soc. **123**, 3775–3781 (1995)
30. G.A. Anastassiou, S.G. Gal, *Approximation Theory: Moduli of Continuity and Global Smoothness Preservation* (Birkhäuser, Boston, 2000)
31. D. Andrica, C. Badea, Grüss' inequality for positive linear functionals. Period. Math. Hung. **19**, 155–167 (1988)
32. P.E. Appell, Sur une classe de polynomes. Ann. Sci. l'E.N.S **2**(9), 119–144 (1880)
33. A. Aral, V. Gupta, On the Durrmeyer type modification of the q Baskakov type operators. Nonlinear Anal. Theory Methods Appl. **72**(3-4), 1171–1180 (2010)
34. A. Aral, V. Gupta, Generalized q Baskakov operators. Math. Slovaca **61**(4), 619–634 (2011)
35. A. Aral, T. Acar, On approximation properties of generalized Durrmeyer operators, in *Modern Mathematical Methods and High Performance Computing in Science and Technology, M3HPCST, Ghaziabad, India, December 2015*, ed. by V.K. Singh, H.M. Srivastava, E. Venturino, M. Resch, V. Gupta. Springer Proceedings in Mathematics & Statistics (2016), pp. 1–15
36. A. Aral, V. Gupta, (p, q)-type beta functions of second kind. Adv. Oper. Theory **1**(1), 134–146 (2016)
37. A. Aral, V. Gupta, Applications of (p, q)-gamma function to Szász Durrmeyer operators. Publ. Inst. Math. **102**(116), 211–220 (2017)
38. A. Aral, V. Gupta, (p, q) variant of Szász-beta operators. Revista de la Real Academia de Ciencias Exactas, Físicas y Naturales. Serie A. Matemáticas **111**, 719–733 (2017)
39. A. Aral, V. Gupta, R.P. Agarwal, *Applications of q Calculus in Operator Theory* (Springer, Cham, 2013)
40. A. Aral, D. Inoan, I. Raşa, On the generalized Szász-Mirakyan operators. Results Math. **65**(3–4), 441–452 (2014)
41. A. Aral, G. Ulusoy, E. Deniz, A new construction of Szász-Mirakyan operators. Numer. Algorithms (2017). https://doi.org/10.1007/s11075-017-0317-x
42. Ç. Atakut, I. Büyükyazici, Stancu type generalization of the Favard-Szàsz operators. Appl. Math. Lett. **23**, 1479–1482 (2010)

43. F. Avram, M.S. Taqqu, Noncentral limit theorems and Appell polynomials. Ann. Probab. **15**(2), 767–775 (1987)
44. I. Badea, Modulus of continuity in Bögel sense and its applications. Stud. Univ. Babeş-Bolyai, Ser. Math. Mech. **2**(4), 69–78 (1973, in Romanian)
45. C. Badea, K-functionals and moduli of smoothness of functions defined on compact metric spaces. Comput. Math. Appl. **30**, 23–31 (1995)
46. C. Badea, C. Cottin, Korovkin-type theorems for generalized Boolean sum operators. Colloq. Math. Soc. János Bolyai, Approx. Theory, Kecskemét (Hungary) **58**, 51–67 (1990)
47. C. Badea, I. Badea, H.H. Gonska, A test function theorem and approximation by pseudopolynomials. Bull. Aust. Math. Soc. **34**, 53–64 (1986)
48. C. Badea, I. Badea, H.H. Gonska, Notes on the degree of approximation of B-continuous and B-differentiable functions. J. Approx. Theory Appl. **4**, 95–108 (1988)
49. D. Bărbosu, GBS operators of Schurer-Stancu type. Ann. Univ. Craiova Math. Comput. Sci. Ser. **30**, 34–39 (2003)
50. D. Bărbosu, O. Pop, On the Bernstein bivariate approximation formula. Carpathian J. Math. **24**(3), 293–298 (2008)
51. D. Bărbosu, O. Pop, A note on the Bernstein's cubature formula. Gen. Math. **17**(3), 161–172 (2009)
52. D. Bărbosu, O.T. Pop, Bivariate Schurer-Stancu operators revisited. Carpathian J. Math. **26**(1), 24–35 (2010)
53. D. Bărbosu, D. Miclăuş, On the composite Bernstein type cubature formula. Gen. Math. **18**(3), 73–81 (2010)
54. D. Bărbosu, D. Miclăuş, On the composite Bernstein type quadrature formula. Rev. Anal. Numér. Théory Approx. **39**(1), 3–7 (2010)
55. D. Bărbosu, C.V. Muraru, Approximating B-continuous functions using GBS operators of Bernstein-Schurer-Stancu type based on q-integers. Appl. Math. Comput. **259**, 80–87 (2015)
56. D. Bărbosu, A.M. Acu, C.V. Muraru, On certain GBS-Durrmeyer operators based on q-integers. Turk. J. Math. **41**(2), 368–380 (2017)
57. L. Beutel, H. Gonska, D. Kacsó, G. Tachev, Variation-diminishing splines revised, in *Proceedings of the International Symposium on Numerical Analysis and Approximation Theory*, ed. by R. Trâmbiţaş (Presa Universitară Clujeană, Cluj-Napoca, 2002), pp. 54–75
58. M. Birou, Discrete operators associated with the Durrmeyer operator. Stud. Univ. Babeş-Bolyai Math. **60**, 295–302 (2015)
59. K. Bögel, Mehrdimensionale differentiation von Funktionen mehrerer Veränderlicher. J. Reine Angew. Math. **170**, 197–217 (1934)
60. K. Bögel, Über die mehrdimensionale differentiation, integration und beschränkte variation. J. Reine Angew. Math. **173**, 5–29 (1935)
61. K. Bögel, Über die mehrdimensionale differentiation. Jahresber. Deutsch. Math. Verein. **65**, 45–71 (1962)
62. H. Bohman, On approximation of continuous and analytic functions. Ark. Mat. **2**, 43–56 (1952)
63. B.D. Boyanov, V.M. Veselinov, A note on the approximation of functions in an infinite interval by linear positive operators. Bull. Math. Soc. Sci. Math. Roum. **14**(62), 9–13 (1970)
64. J. Bustamante, J.M. Quesada, L.M. Cruz, Direct estimate for positive linear operators in polynomial weighted spaces. J. Approx. Theory **162**, 1495–1508 (2010)
65. P.L. Butzer, On two dimensional Bernstein polynomials. Can. J. Math. **5**, 107–113 (1953)
66. P.L. Butzer, H. Berens, *Semi-Groups of Operators and Approximation* (Springer, New York, 1967)
67. I. Büyükyazici, H. Sharma, Approximation properties of two dimensional q-Bernstein-Chlodowsky-Durrmeyer operators. Numer. Funct. Anal. Optim. **33**(12), 1351–1371 (2012)
68. J.D. Cao, H. Gonska, Approximation by Boolean sums of positive linear operators III: estimates for some numerical approximation schemes. Numer. Funct. Anal. Optim. **10**(7), 643–672 (1989)

69. D. Cárdenas-Morales, P. Garrancho, I. Raşa, Bernstein-type operators which preserve polynomials. Comput. Math. Appl. **62**, 158–163 (2011)
70. R. Chauhan, N. Ispir, P.N. Agrawal, A new kind of Bernstein-Schurer-Stancu-Kantorovich type operators based on q-integers. J. Inequal. Appl. **2017**(50) (2017). https://doi.org/10.1186/s13660-017-1298-y
71. P.L. Chebyshev, Sur les expressions approximatives des intégrales définies par les autres prises entre les mêmes limites. Proc. Math. Soc. Kharkov **2**, 93–98 (1882, in Russian), translated in Oeuvres **2**, 716–719 (1907)
72. W.Z. Chen, *Approximation Theory of Operators* (Xiamen University Publishing House, Xiamen, 1989)
73. V.A. Cleciu, Approximation properties of a class of Bernstein-Stancu type operators, in *Proceedings of the International Conference NAAT*, ed. by O. Agratini (2006), pp. 171–178
74. V.A. Cleciu, Bernstein-Stancu operators. Stud. Univ. Babes-Bolyai Math. **52**(4), 53–65 (2007)
75. C. Cottin, Mixed K-functionals: a measure of smoothness for blending-type approximation. Math. Z. **204**, 69–83 (1990)
76. E. Deniz, A. Aral, V. Gupta, Note on Szász-Mirakyan-Durrmeyer operators preserving e^{2ax}, $a > 0$. Numer. Funct. Anal. Optim. **39**(2), 201–207 (2018)
77. M.-M. Derriennic, Modified Bernstein polynomials and Jacobi polynomials in q-calculus. Rendiconti Del Circolo Matematico Di Palermo, Serie II, Suppl. **76**, 269–290 (2005)
78. R.A. DeVore, G.G. Lorentz, *Constructive Approximation* (Springer, Berlin, 1993)
79. Z. Ditzian, V. Totik, *Moduli of Smoothness* (Springer, New York, 1987)
80. E. Dobrescu, I. Matei, The approximation by Bernstein type polynomials of bidimensionally continuous functions (Romanian). Ann. Univ. Timişoara Ser. Sti. Mat.-Fiz. **4**, 85–90 (1966)
81. S.S. Dragomir, Th.M. Rassias, *Ostrowski Type Inequalities and Applications in Numerical Integration* (Kluwer Academic, Dordrecht, 2002)
82. A. Erençin, F. Taşdelen, On certain Kantorovich type operators. Fasc. Math. **41**, 65–71 (2009)
83. T. Ernst, The history of q-calculus and a new method, *U.U.D.M Report 2000*, vol. 16. ISSN 1101-3591 (Department of Mathematics, Upsala University, 2000)
84. M.D. Farcaş, About approximation of B-continuous and B-differentiable functions of three variables by GBS operators of Bernstein type. Creat. Math. Inform. **17**, 20–27 (2008)
85. M.D. Farcaş, About approximation of B-continuous functions of three variables by GBS operators of Bernstein type on a tetrahedron. Acta Univ. Apulensis Math. Inform. **16**, 93–102 (2008)
86. J. Favard, Sur les multiplicateurs d'interpolation. J. Math. Pures Appl. **23**(9), 219–247 (1944)
87. Z. Finta, Korovkin type theorem for sequences of operators depending on a parameter. Demonstr. Math. **48**(3), 391–403 (2015)
88. Z. Finta, Approximation properties of (p,q)-Bernstein type operators. Acta Univ. Sapientiae Math. **8**(2), 222–232 (2016)
89. Z. Finta, V. Gupta, Direct and inverse estimates for Phillips type operators. J. Math. Anal. Appl. **303**(2), 627–642 (2005)
90. Z. Finta, V. Gupta, Approximation theorems for limit (p,q) Bernstein Durrmeyer operators. Facta Univ. (NIS) **32**(2), 195–207 (2017)
91. A. Gadjiev, Positive linear operators in weighted spaces of functions of several variables (Russian), izv. akad. nauk Azerbaijan ssr ser. fiz. Tehn. Mat. Nauk **1**, 32–37 (1980)
92. A. Gadjiev, H. Hacısalihoglu, *Convergence of the Sequences of Linear Positive Operators* (Ankara University, Ankara, 1995)
93. H. Gauchman, Integral inequalities in q-calculus. Comput. Math. Appl. **47**, 281–300 (2004)
94. W. Gautschi, G. Mastroianni, Th.M. Rassias, *Approximation and Computation - In Honor of Gradimir V. Milovanović* (Springer, New York, 2011)
95. B. Gavrea, Improvement of some inequalities of Chebysev-Grüss type. Comput. Math. Appl. **64**, 2003–2010 (2012)
96. B. Gavrea, I. Gavrea, Ostrowski type inequalities from a linear functional point of view. JIPAM. J. Inequal. Pure Appl. Math. **1**, article 11 (2000)

97. A.K. Gazanfer, I. Büyükyazıcı, Approximation by certain linear positive operators of two variables. Abst. Appl. Anal. **2014**, 6 (2014)
98. H. Gonska, *Quantitative Approximation in* $C(X)$, Habilitationsschrift (Universita at Duisburg, Duisburg, 1985)
99. H. Gonska, Degree of approximation by lacunary interpolators: (0,..., R-2,R) interpolation. Rocky Mt. J. Math. **19**, 157–171 (1989)
100. H. Gonska, R. Kovacheva, The second order modulus revised: remarks, applications, problems. Conf. Sem. Mat. Univ. Bari **257**, 1–32 (1994)
101. H. Gonska, P. Piţul, Remarks on an article of J.P. King. Comment. Math. Univ. Carol. **46**, 645–652 (2005)
102. H. Gonska, I. Raşa, Differences of positive linear operators and the second order modulus. Carpathian J. Math. **24**(3), 332–340 (2008)
103. H. Gonska, R. Păltănea, Quantitative convergence theorems for a class of Bernstein-Durrmeyer operators preserving linear functions. Ukr. Math. J. **62**, 913–922 (2010)
104. H. Gonska, R. Păltănea, Simultaneous approximation by a class of Bernstein-Durrmeyer operators preserving linear functions. Czechoslov. Math. J. **60**(135), 783–799 (2010)
105. H. Gonska, G. Tachev, Grüss-type inequality for positive linear operators with second order moduli. Mat. Vesnik **63**(4), 247–252 (2011)
106. H. Gonska, P. Piţul, I. Raşa, On differences of positive linear operators. Carpathian J. Math. **22**(1–2), 65–78 (2006)
107. H. Gonska, P. Piţul, I. Raşa, On Peano's form of the Taylor remainder, Voronovskaja's theorem and the commutator of positive linear operators, in *Numerical Analysis and Approximation Theory. Proceedings of the International Conference on Cluj-Napoca 2006*, ed. by O. Agratini, P. Blaga (Cluj-Napoca, Casa Cărţii de Ştiinţă, 2006), pp. 55–80
108. H. Gonska, D. Kacso, I. Raşa, On genuine Bernstein-Durrmeyer operators. Result Math. **50**, 213–225 (2007)
109. H. Gonska, P. Piţul, I. Raşa, General king-type operators. Results Math. **53**(3–4), 279–286 (2009)
110. H. Gonska, I. Raşa, M.-D. Rusu, Applications of an Ostrowski-type inequality. J. Comput. Anal. Appl. **14**(1), 19–31 (2012)
111. H. Gonska, M. Heilmann, A. Lupaş, I. Raşa, On the composition and decomposition of positive linear operators III: a non-trivial decomposition of the Bernstein operator, arXiv:1204.2723v1 (2012)
112. H. Gonska, I. Raşa, M.D. Rusu, Čebyšev-Grüss-type inequalities revisited. Math. Slovaca **63**(5), 1007–1024 (2013)
113. H. Gonska, I. Raşa, M.D. Rusu, Chebyshev-Grüss-type inequalities via discrete oscillations. Bul Acad. Stiinte Repub. Mold. Mat. **74**(1), 63–89 (2014)
114. T.N.T. Goodman, A. Sharma, A modified Bernstein-Schoenberg operator, in *Proceedings of the Conference on Constructive Theory of Functions, Varna 1987*, ed. by Bl. Sendov, et al. (Bulgarian Academy of Sciences Publishing House, Sofia, 1988), pp. 166–173
115. V. Gupta, A note on modified Baskakov type operators. Approx. Theory Appl. **10**(3), 74–78 (1994)
116. V. Gupta, An estimate on the convergence of Baskakov–Bézier operators. J. Math. Anal. Appl. **312**(1), 280–288 (2005)
117. V. Gupta, Some approximation properties on q-Durrmeyer operators. Appl. Math. Comput. **197**(1), 172–178 (2008)
118. V. Gupta, (p, q)-Baskakov-Kantorovich operators. Appl. Math. Inf. Sci. **10**(4), 1551–1556 (2016)
119. V. Gupta, (p, q) genuine Bernstein Durrmeyer operators. Boll. Un. Matt. Ital. **9**(3), 399–409 (2016)
120. V. Gupta, (p, q)-Szász-Mirakyan-Baskakov operators. Complex Anal. Oper. Theory **12**(1), 17–25 (2018)
121. V. Gupta, G.S. Srivastava, Simultaneous approximation by Baskakov-Szász type operators. Bull. Math. Soc. Sci. Math. Roum. Nouv. Ser. **37**(85), 73–85 (1993)

122. V. Gupta, M.A. Noor, Convergence of derivatives for certain mixed Szász beta operators. J. Math. Anal. Appl. **321**(1), 1–9 (2006)
123. V. Gupta, P.N. Agrawal, Approximation by modified Păltănea operators, communicated
124. V. Gupta, A. Aral, Convergence of the q analogue zász beta operators. Appl. Math. Comput. **216**, 374–380 (2010)
125. V. Gupta, R.P. Agarwal, *Convergence Estimates in Approximation Theory* (Springer, Cham, 2014)
126. V. Gupta, Th.M. Rassias, Hypergeometric representation of certain summation-integral operators, in *Topics in Mathematical Analysis and Applications* (Springer, New York, 2014), pp. 447–460
127. V. Gupta, T.M. Rassias, Lupas-Durrmeyer operators based on Polya distribution. Banach J. Math. Anal. **8**(2), 146–155 (2014)
128. V. Gupta, A. Aral, Bernstein Durrmeyer operators based on two parameters. Facta Univ. Ser. Math. Inform. **31**(1), 79–95 (2016)
129. V. Gupta, N. Malik, Approximation with certain Szász-Mirakyan operators. Khayyam J. Math. **3**(2), 90–97 (2017)
130. V. Gupta, G. Tachev, *Approximation with Positive Linear Operators and Linear Combinations* (Springer, Cham, 2017)
131. V. Gupta, G. Tachev, On approximation properties of Phillips operators preserving exponential functions. Mediterr. J. Math. **14**, Art. 177 (2017)
132. V. Gupta, A. Aral, A note on Szász-Mirakyan-Kantorovich type operators preserving e^{-x}. Positivity **22**(2), 415–423 (2018)
133. V. Gupta, P. Maheshwari, Approximation with certain post-Widder operators. Publ. Inst. Math. (Beograd) (to appear)
134. V. Gupta, Th.M. Rassias, E. Pandey, On genuine Lupas-beta operators and modulus of continuity. Int. J. Nonlinear Anal. Appl. **8**(1), 23–32 (2017)
135. V. Gupta, N. Malik, Th.M. Rassias, Moment generating functions and moments of linear positive operators, in *Modern Discrete Mathematics and Analysis*, ed. by N.J. Daras, Th.M. Rassias (Springer, Berlin, 2017)
136. V. Gupta, T.M. Rassias, D. Agrawal, Approximation by Lupaş-Kantorovich operators, in *Modern Discrete Mathematics and Analysis*, ed. by N.J. Daras, T.M. Rassias (Springer, Berlin, 2018)
137. M. Gurdek, L. Rempulska, M. Skorupka, The Baskakov operators for functions of two variables. Collect. Math. **50**(3), 289–302 (1999)
138. G. Grüss, Über das maximum des absoluten Betrages von $\frac{1}{b-a}\int_a^b f(x)g(x)dx - \frac{1}{(b-a)^2}\int_a^b f(x)dx \cdot \int_a^b g(x)dx$. Math. Z. **39**, 215–226 (1935)
139. A. Holhoş, The rate of approximation of functions in an infinite interval by positive linear operators. Stud. Univ. Babeş-Bolyai Math. **55**(2), 133–142 (2011)
140. M. Heilmann, G. Tachev, Commutativity, direct and strong converse results for Phillips operators. East J. Approx. **17**(3), 299–317 (2011)
141. T. Hermann, P. Vértesi, On an interpolatory operator and its saturation. Acta Math. Hungar. **37**(1–3), 1–9 (1981)
142. N. Ispir, C. Atakut, Approximation by modified Szasz-Mirakjan operators on weighted spaces. Proc. Indian Acad. Sci. Math. Sci. **112**(4), 571–578 (2002)
143. N. Ispir, I. Büyükyazıcı, Quantitative estimates for a certain bivariate Chlodowsky-Szasz-Kantorovich type operators. Math. Commun. **21**(1), 31–44 (2016)
144. F.H. Jackson, On a q-definite integrals. Q. J. Pure Appl. Math. **41**, 193–203 (1910)
145. D. Jackson, Über die Genauigkeit der Annäherung stetiger Funktionen durch ganze rationale Funktionen gegebenen Grades und trigonometrische Summen gegebener Ordnung, Ph.D. Thesis, Georg-August Univ. of Göttingen (1911)
146. A. Jakimovski, D. Leviatan, Generalized Szász operators for the approximation in the infinite interval. Math. (Cluj) **34**, 97–103 (1969)
147. V. Kac, P. Cheung, *Quantum Calculus* (Springer, New York, 2002)

148. L.V. Kantorovich, Sur certain développements suivant les polynômes de la forme de S. Bernstein, I, II. C.R. Acad. URSS **563–568**, 595–600 (1930)
149. J. Karamata, Inégalités relatives aux quotients et à la différence de $\int fg$ et $\int f \int g$. Bull. Acad. Serbe Sci. Math. Nat. A, 131–145 (1948)
150. J.P. King, Positive linear operators which preserve x^2. Acta Math. Hung. **99**(3), 203–208 (2003)
151. E.H. Kingsley, Bernstein polynomials for functions of two variables of class $C^{(k)}$. Proc. Am. Math. Soc. **2**(1), 64–71 (1951)
152. P.P. Korovkin, *Linear Operators and Approximation Theory, Moscow 1957* (Hindustan Publishing Corporation, Delhi, 1960, in Russian)
153. Y.C. Kwun, A.M. Acu, A. Rafiq, V.A. Radu, F. Ali, S.M. Kang, Bernstein-Stancu type operators which preserve polynomials. J. Comput. Anal. Appl. **23**(4), 758–770 (2017)
154. E. Landau, Über einige Ungleichungen von Herrn G. Grüss. Math. Z. **39**, 742–744 (1935)
155. H.G. Lehnhoff, A simple proof of A.F. Timan's theorem. J. Approx. Theory **38**, 172–176 (1983)
156. A. Lupaş, Die Folge der Betaoperatoren, Dissertation, Universität Stuttgart, 1972
157. A. Lupaş, A q-analogue of the Bernstein operator, in *Seminar on Numerical and Statistical Calculus* (Cluj-Napoca, 1987), pp. 85–92, Preprint 87–9 Univ. Babes-Bolyai, Cluj
158. A. Lupaş, The approximation by means of some linear positive operators, in *Approximation Theory. Proceedings of the International Dortmund Meeting IDoMAT 95, held in Witten, Germany, March 13–17, 1995*, ed. by M.W. Muller, M. Felten, D.H. Mache. Mathematical Research, vol. 86 (Akademie Verlag, Berlin, 1995), pp. 201–229
159. L. Lupas, A. Lupas, Polynomials of binomial type and approximation operators. Stud. Univ. Babes-Bolyai Math. **32**(4), 61–69 (1987)
160. D.H. Mache, Gewichtete Simultanapproximation in der Lp-Metrik durch das Verfahren der Kantorovic Operatoren, Dissertation, Univ. Dortmund (1991)
161. D.H. Mache, A link between Bernstein polynomials and Durrmeyer polynomials with Jacobi weights, in *Approximation Theory VIII*, ed. by Ch.K. Chui, L.L. Schmaker. Approximation and Interpolation, vol. 1 (World Scientific Publication, Singapore, 1995), pp. 403–410
162. H. Maier, M.Th. Rassias, The order of magnitude for moments for certain cotangent sums. J. Math. Anal. Appl. **429**(1), 576–590 (2015)
163. H. Maier, M.Th. Rassias, Generalizations of a cotangent sum associated to the Estermann zeta function. Commun. Contemp. Math. **18**(1), 89 (2016). https://doi.org/10.1142/S0219199715500789
164. H. Maier, M.Th. Rassias, The rate of growth of moments of certain cotangent sums. Aequationes Math. **90**(3), 581–595 (2016)
165. H. Maier, M.Th. Rassias, Asymptotics for moments of certain cotangent sums. Houst. J. Math. **43**(1), 207–222 (2017)
166. H. Maier, M.Th. Rassias, Asymptotics for moments of certain cotangent sums for arbitrary exponents. Houst. J. Math. **43**(4), 1235–1249 (2017)
167. N. Malik, V. Gupta, Approximation by (p, q)-Baskakov-Beta operators. Appl. Math. Comput. **293**, 49–53 (2017)
168. S. Marinković, P. Rajković, M. Stanković, The inequalities for some type of q-integrals. Comput. Math. Appl. **56**, 2490–2498 (2008)
169. F.L. Martinez, Some properties of two dimensional Bernstein polynomials. J. Approx. Theory **59**(3), 300–306 (1989)
170. Y. Matsuoka, On the degree of approximation of functions by some positive linear operators. Sci. Rep. Kagoshima Univ. **9**, 11–16 (1960)
171. D. Miclăuş, The revision of some results for Bernstein-Stancu type operators. Carpathian J. Math. **28**(2), 289–300 (2012)
172. D. Miclaus, On the GBS Bernstein-Stancu's type operators. Creat. Math. Inform. **22**, 73–80 (2013)
173. G.V. Milovanović, M.Th. Rassias (eds.), *Analytic Number Theory, Approximation Theory and Special Functions* (Springer, New York, 2014)

174. G.V. Milovanović, D.S. Mitrinović, Th.M. Rassias, *Topics in Polynomials: Extremal Problems, Inequalities, Zeros* (World Scientific Publishing, Singapore, 1994)
175. G.V. Milovanovic, V. Gupta, N. Malik, (p, q)-beta functions and applications in approximation. Bol. Soc. Mat. Mex. **24**(1), 219–237 (2018)
176. B.S. Mitjagin, E.M. Semenov, Lack of interpolation of linear operators in spaces of smooth functions. Math. USSR-Izv. **11**, 1229–1266 (1977)
177. D.S. Mitrinović, J.E. Pečarić, A.M. Fink, *Classical and New Inequalities in Analysis* (Kluwer, Dordrecht, 1993)
178. M. Mursaleen, K.J. Ansari, A. Khan, On (p, q)-analogue of Bernstein operators. Appl. Math. Comput. **266**, 874–882 (2015)
179. M. Mursaleen, S. Rahman, A. Alotaibi, Dunkl generalization of q- Szász-Mirakjan-Kantorovich operators which preserves some test functions. J. Inequal. Appl. (2016). https://doi.org/10.1186/s13660-016-1257-z
180. M. Mursaleen, K.J. Ansari, A. Khan, Erratum to on (p, q)-analogue of Bernstein operators. Appl. Math. Comput. **266**, 874–882 (2015), Appl. Math. Comput. **278**, 70–71 (2016)
181. M. Mursaleen, K.J. Ansari, A. Khan, Some approximation results for Bernstein-Kantorovich operators based on (p, q)-calculus (15 Jan 2016). arXiv:1504.05887v4[math.CA]
182. T. Neer, P.N. Agrawal, S. Araci, Stancu Durrmeyer type operators based on q-integers. Appl. Math. Inf. Sci. **11**(3), 767–775 (2017)
183. T. Neer, A.M. Acu, P.N. Agrawal, Approximation of functions by bivariate q-Stancu-Durrmeyer type operators. Math. Commun. **23**, 161–180 (2018)
184. G. Nowak, Approximation properties for generalized q-Bernstein polynomials. J. Math. Anal. Appl. **350**, 50–55 (2009)
185. S. Ostrovska, q-Bernstein polynomials and their iterates. J. Approx. Theory **123**, 232–255 (2003)
186. S. Ostrovska, The sharpness of convergence results for q Bernstein polynomials in the case $q > 1$. Czechoslov. Math. J. **58**(133), 1195–1206 (2008)
187. S. Ostrovska, On the image of the limit q Bernstein operator. Math. Methods Appl. Sci. **32**(15), 1964–1970 (2009)
188. A. Ostrowski, Über die Absolutabweichung einer differentiierbaren Funktion von ihrem Integralmittelwert. Comment. Math. Helv. **10**, 226–227 (1938)
189. A. Ostrowski, On an integral inequality. Aequationes Math. **4**, 358–373 (1970)
190. B.G. Pachpatte, A note on Grüss type inequalities via Cauchy's mean value theorem. Math. Inequal. Appl. **11**(1), 75–80 (2007)
191. R. Păltănea, A class of Durrmeyer type operators preserving linear functions. Ann. Tiberiu Popoviciu Sem. Funct. Equat. Approx. Convex. (Cluj-Napoca) **5**, 109–117 (2007)
192. R. Păltănea, Modified Szász-Mirakjan operators of integral form. Carpathian J. Math. **24**, 378–385 (2008)
193. R. Păltănea, Representation of the K-functional $K(f, C[a, b], C^1[a, b], \cdot)$- a new approach. Bull. Transilv. Univ. Braşov Ser. III Math. Inform. Phys. **3**(52), 93–100 (2010)
194. R. Păltănea, Simultaneous approximation by a class of Szász-Mirakjan operators. J. Appl. Funct. Anal. **9**(3-4), 356–368 (2014)
195. J. Peetre, *A Theory of Interpolation of Normed Spaces*. Notas de Matematica, vol. 39 (Instituto de Matematica Pura e Aplicada, Conselho Nacional de Pesquisas, Rio de Janeiro, 1968)
196. R.S. Phillips, An inversion formula for Laplace transforms and semi-group of operators. Ann. Math. **59**, 325–356 (1954)
197. G.M. Phillips, Bernstein polynomials based on the q-integers. The heritage of P.L. Chebyshev: a Festschrift in honor of the 70th-birthday of Professor T.J. Rivlin. Ann. Numer. Math. **4**, 511–518 (1997)
198. O.T. Pop, The generalization of Voronovskaja's theorem for a class of bivariate operators. Stud. Univ. Babes-Bolyai Math. **53**(2), 85–107 (2008)
199. I. Raşa, Discrete operators associated with certain integral operators. Stud. Univ. Babeş-Bolyai Math. **56**(2), 537–544 (2011)

200. I. Raşa, E. Stănilă, On some operators linking the Bernstein and the genuine Bernstein-Durrmeyer operators. J. Appl. Funct. Anal. **9**, 369–378 (2014)
201. Th.M. Rassias, *Approximation Theory and Applications* (Hadronic Press, Florida, 1998)
202. Th.M. Rassias, J. Šimša, *Finite Sums Decompositions in Mathematical Analysis*. Wiley-Interscience Series in Pure and Applied Mathematics (Wiley, Chichester, 1995)
203. Th.M. Rassias, V. Gupta (eds.), *Mathematical Analysis, Approximation Theory and Their Applications*. Series: Springer Optimization and Its Applications, vol. 111 (Springer, Cham, 2016)
204. Th.M. Rassias, H.M. Srivastava, A. Yanushauskas, Topics in polynomials of one and several variables and their applications (World Scientific Publication, Singapore, 1993), pp. 638
205. M.Y. Ren, X.M. Zeng, On statistical approximation properties of modified q-Bernstein-Schurer operators. Bull. Korean Math. Soc. **50**(4), 1145–1156 (2013)
206. R.T. Rockafellar, *Convex Analysis* (Princeton University Press, Princeton, 1970)
207. M.D. Rusu, On Grüss-type inequalities for positive linear operators. Stud. Univ. Babeş-Bolyai Math. **56**(2), 551–565 (2011)
208. M.D. Rusu, Chebyshev-Grüss- and Ostrowski-type inequalities, Ph.D. Thesis, Duisburg-Essen University (2014)
209. P.N. Sadjang, On the fundamental theorem of (p, q)-calculus and some (p, q)-Taylor formulas. arXiv:1309.3934 [math.QA]
210. P.N. Sadjang, On the (p, q)-Gamma and the (p, q)-beta functions (22 Jun 2015). arXiv 1506.07394v1
211. V. Sahai, S. Yadav, Representations of two parameter quantum algebras and p, q-special functions. J. Math. Anal. Appl. **335**, 268–279 (2007)
212. L.L. Schumaker, *Spline Functions: Basic Theory* (Wiley, New York, 1981)
213. H. Sharma, C. Gupta, On (p, q)-generalization of Szász-Mirakyan Kantorovich operators. Boll. Unione Mat. Ital. **8**(3), 213–222 (2015)
214. O. Shisha, B. Mond, The degree of convergence of sequence linear positive operators. Proc. Nat. Acad. Sci. U.S.A. **60**, 1196–1200 (1968)
215. M. Sidharth, N. Ispir, P.N. Agrawal, GBS operators of Bernstein-Schurer-Kantorovich type based on q-integers. Appl. Math. Comput. **269**, 558–568 (2015)
216. M. Sidharth, N. Ispir, P.N. Agrawal, Approximation of B-continuous and B-differentiable functions by GBS operators of q-Bernstein-Schurer-Stancu type. Turk. J. Math. **40**, 1298–1315 (2016)
217. M. Sidharth, A.M. Acu, P.N. Agrawal, Chlodowsky-Szasz-Appell type operators for functions of two variables. Ann. Funct. Anal. **8**(4), 446–459 (2017)
218. I.M. Sheffer, Some properties of polynomial sets of type zero. Duke Math. J. **5**, 590–622 (1939)
219. H.M. Srivastava, V. Gupta, Rate of convergence for the Bézier variant of the Bleimann–Butzer–Hahn operators. Appl. Math. Lett. **18**(8), 849–857 (2005)
220. D.D. Stancu, A method for obtaining polynomials of Bernstein type of two variables. Am. Math. Mon. **70**(3), 260–264 (1963)
221. D.D. Stancu, Approximation of functions by a new class of linear polynomial operators. Rev. Roum. Math. Pures Appl. **13**, 1173–1194 (1968)
222. D.D. Stancu, Asupra unei generalizari a polinoamelor lui Bernstein. Stud. Univ. Babes-Bolyai Math. **14**, 31–45 (1969)
223. E.D. Stănilă, On Bernstein-Euler-Jacobi operators. Ph.D. Thesis, Duisburg-Essen University (July 2014)
224. G. Tachev, V. Gupta, A. Aral, Voronovskaja's theorem for functions with exponential growth. Georgian Math. J. https://doi.org/10.1515/gmj-2018-0041
225. V.I. Volkov, On the convergence of sequences of linear positive operators in the space of continuous functions of two variables. Dokl. Akad. Nauk SSSR (N.S.) **115**, 17–19 (1957)
226. Z. Walczak, On approximation by modified Szász-Mirakyan operators. Glas. Mat. Ser. III **37**(2), 303–319 (2002)

227. D.V. Widder, *The Laplace Transform*. Princeton Mathematical Series (Princeton University Press, Princeton, 1941)
228. O.G. Yılmaz, V. Gupta, A. Aral, On Baskakov operators preserving exponential function. J. Numer. Anal. Approx. Theory **46**(2), 150–161 (2017)
229. X.M. Zeng, V. Gupta, Rate of convergence of Baskakov-Bézier type operators for locally bounded functions. Comput. Math. Appl. **44**(10), 1445–1453 (2002)

Index

V. Gupta et al., *Recent Advances in Constructive Approximation Theory*, Springer Optimization and Its Applications 138,
https://doi.org/10.1007/978-3-319-92165-5

Printed in the United States
By Bookmasters